The Good Life in a Technological Age

Routledge Studies in Science, Technology and Society

1 **Science and the Media**
Alternative Routes in Scientific Communication
Massimiano Bucchi

2 **Animals, Disease and Human Society**
Human-Animal Relations and the Rise of Veterinary Medicine
Joanna Swabe

3 **Transnational Environmental Policy**
The Ozone Layer
Reiner Grundmann

4 **Biology and Political Science**
Robert H Blank and Samuel M. Hines, Jr.

5 **Technoculture and Critical Theory**
In the Service of the Machine?
Simon Cooper

6 **Biomedicine as Culture**
Instrumental Practices, Technoscientific Knowledge, and New Modes of Life
Edited by Regula Valérie Burri and Joseph Dumit

7 **Journalism, Science and Society**
Science Communication between News and Public Relations
Edited by Martin W. Bauer and Massimiano Bucchi

8 **Science Images and Popular Images of Science**
Edited by Bernd Hüppauf and Peter Weingart

9 **Wind Power and Power Politics**
International Perspectives
Edited by Peter A. Strachan, David Lal and David Toke

10 **Global Public Health Vigilance**
Creating a World on Alert
Lorna Weir and Eric Mykhalovskiy

11 **Rethinking Disability**
Bodies, Senses, and Things
Michael Schillmeier

12 **Biometrics**
Bodies, Technologies, Biopolitics
Joseph Pugliese

13 **Wired and Mobilizing**
Social Movements, New Technology, and Electoral Politics
Victoria Carty

14 **The Politics of Bioethics**
Alan Petersen

15 **The Culture of Science**
How the Public Relates to Science Across the Globe
Edited by Martin W. Bauer, Rajesh Shukla and Nick Allum

16 **Internet and Surveillance**
The Challenges of Web 2.0 and Social Media
Edited by Christian Fuchs, Kees Boersma, Anders Albrechtslund and Marisol Sandoval

17 **The Good Life in a Technological Age**
Edited by Philip Brey, Adam Briggle and Edward Spence

The Good Life in a Technological Age

Edited by Philip Brey, Adam Briggle, and Edward Spence

Routledge
Taylor & Francis Group
NEW YORK LONDON

First published 2012
by Routledge
711 Third Avenue, New York, NY 10017

Simultaneously published in the UK
by Routledge
2 Park Square, Milton Park, Abingdon, Oxfordshire OX14 4RN

First issued in paperback 2014

Routledge is an imprint of the Taylor and Francis Group, an informa company

Library of Congress Cataloging-in-Publication Data

The good life in a technological age / edited by Philip Brey, Adam Briggle, and Edward Spence. — 1st ed.
p. cm. — (Routledge studies in science, technology and society ; 17)
Includes bibliographical references and index.
1. Information society. 2. Technological innovations—Social aspects. I. Brey, Philip. II. Briggle, Adam. III. Spence, Edward.
HM851.G663 2012
303.48'3—dc23
2011037807

ISBN13: 978-0-415-89126-4 (hbk)
ISBN13: 978-0-415-75452-1 (pbk)

Typeset in Sabon
by IBT Global.

For Elly. Fellow traveler, fellow wonderer. – Philip
For Amber and Gracie. My girls, my life. – Adam
For Kaye as always for her constant inspiration and loving support. – Edward

Contents

PART V
Medical and Agricultural Technology and Well-Being

PART VI
Technology Design and Policy

Figures

Tables

Acknowledgments

For us, writing and thinking together with friends is essential to the good life. We are grateful to all those who have contributed to making "GLiTA," as we affectionately came to call this project, a reality. We would like to thank our faculty and staff colleagues at our home institutions and networks around the world, especially the University of Twente and the 3TU Centre for Ethics and Technology in the Netherlands, the University of North Texas and the Center for the Study of Interdisciplinarity in the United States, and Charles Sturt University and the Centre for Applied Philosophy and Public Ethics in Australia.

We started this book when we were working together in the Philosophy Department at the University of Twente on a "Vici" research project funded by De Nederlandse Organisatie voor Wetenschappelijk Onderzoek (the Netherlands Organization for Scientific Research, NWO). We are thankful for the support of NWO and for the additional funds provided by the Centre for Telematics and Information Technology at the University of Twente. We could not have done this without our fellow Vici team members, Pak Wong, Johnny Søraker, and Omar Rosas. Thank you for your inspiring friendship—we are grateful that our team endures despite having been scattered now across four continents. Finally, we would also like to thank Gijs Houwen for helping to tie all the loose ends together and Faridun Sattarov for assisting with the index.

Introduction

Adam Briggle, Philip Brey, and Edward Spence

Technology changes the way we live, work, play, communicate, fight, love, and die. Yet, few works have systematically explored these changes in light of their implications for individual well-being and social welfare. Are these changes for better or worse? Are they good for some but not others? And how could we answer these questions—what evidence do we need and what standards can we use? In short, how can we conceptualize and evaluate the influence of technology on human well-being? And how can we combine disciplinary perspectives to do so? This volume tackles these questions.

Our ruling ideal of the good life seems to be inextricably bound up with technology and the means it offers to control the external world. It was Francis Bacon who first articulated this idea in the early 17th century: The road to happiness is far more a matter of controlling the outer world than controlling one's inner world of desires. In *The New Organon*, Bacon argues that knowledge and power should be combined and used for the "conquest of nature for the relief of man's estate" (Bacon 1620, LII). To use the term "technological age" only for our recent past and our future is not to deny that humans have always been tool-making and tool-using creatures. It is also not just to point out the more intensive and rapidly changing nature of our material surroundings. Yes, we are immersed like never before in a human-built world of machines, structures, pavement, and media. But what ultimately makes the modern world a technological age is this underlying vision of the good life first expressed by Bacon.

Earlier views were different. The Buddha teaches that to end suffering and achieve nirvana, one must eliminate attachment to the things of this world. Desires return again and again; trying to satisfy them is a futile quest. Similarly, Aristotle argues that eudaimonia, or flourishing, is a matter of instructing one's soul to desire rightly. For these schools of thought, achieving the good life consists far less in controlling nature than in purifying or perfecting the self. Consequently, technology is not essential to their vision of the good. This is not to deny the importance they place on certain technical skills or know-how, or what Michel Foucault calls "technologies of the self"—techniques whereby individuals perform operations on their bodies and souls to attain happiness. It is just to note that the whole point of a human life is not to be disburdened by technological machines and devices.

Thanks to the now-implicit faith in better living through technology, we find ourselves rapidly climbing the exponential curve of change. Our faith has become a new law of nature, what Raymond Kurzweil calls the law of accelerating returns. Technologies feed on one another to create exponential change. The kinds of things we take for granted now—reliable electricity, abundant consumer goods, aisles upon aisles of prepackaged processed foods in giant supermarkets, mobile personal electronics, global supply chains and transportation networks, ubiquitous media, vaccinations, pills, and organ transplants—were unthinkable a short time ago. We take lives for granted that humans just a few generations ago could scarcely imagine. If the exponential curve continues to bend upward, our children and grandchildren will lead lives that we cannot presently comprehend but that they will consider quite natural. They will struggle to imagine our primitive lives and laugh that we said we inhabited a "technological age."

We have devoted ourselves to the idea that the technological control of nature will liberate and enrich our lives. Whereas Buddha or Aristotle might reserve the term "development" for the formation of character, we now use the term to describe the process whereby someone is enrolled into a technological network that affords him a freezer to store his frozen dinner, a microwave to cook it, and a television to watch while eating it. This is the good life in a technological age, or so we largely assume.

But for many this assumption is wearing thin. A few perceptive minds spotted something rotten in the modern project from the start. In 1754, Jean-Jocques Rousseau argued that through habitual use, the conveniences afforded by technological progress had "lost almost all their pleasurableness, degenerating into genuine needs, and the deprivation of them became much more painful than the possession of them was pleasing; and people were unhappy in losing them without being happy in possessing them" (Rousseau 1754, II). We are unhappy sitting in our cars stuck in a traffic jam, but we would be even more miserable if someone took our car away because now we can't get to work, the grocery store, the movie theater, or any of the other far-flung places that require us to have cars.

Indeed, Rousseau's line of reasoning helps explain the well-documented "progress's paradox" of the past 60 years (e.g. Lane 2000). Subjective reports of happiness have flatlined since 1950. Despite having twice as many consumer goods and twice the standard of living and consuming twice the amount of energy, people are not any happier. But this is not really a paradox. One just has to note the tight connection between any given level of technological development and one's perception of a good life. Our horizon of desires is constantly moving along with the moving technological horizon. Contemporary social critics such as Daniel Sarewitz (1996) have pointed this out, but it is the same basic idea from Buddha and Aristotle about the futility of desire. No matter how wealthy they were, a family in 1950 could not microwave their dinner or download music onto their personal computer. But the absence of these technologies was not experienced as a deprivation.

The fact that teens today would be miserable in a 1950s home (what, no Facebook!?) does not mean teens of that age were miserable. It does not mean that teens today lead better lives by anything other than their own subjective, arbitrary, historically bounded standards. What is cheered by Kurzweil as exponential progress is reframed here as an endless run round the hedonic hamster wheel. Technologies create their own needs and desires. What seems like progress may be just the Red Queen's race: We must continually adapt to the changing material environment just to stay in place.

For all of our emphasis on new instruments and means of living, we may be losing sight of ends. What is the point of all this hyperactivity and all this ceaseless novelty? Are we leading good lives? Will our children be leading better lives than us? Is the grand experiment to pursue the good life via technology paying off? These are the questions that we hope this volume sheds some light on. We can hardly imagine more important or pressing questions for a world so committed to technology and so beset by problems arising from that commitment. Yet for being such significant questions, we hardly find our culture ruminating on them in any serious fashion. Yes, there are debates between technological optimists and pessimists, cornucopians and Luddites. But these are all either confined to a single discipline or restricted to certain kinds of technology, such as media or medicine, or limited to a single author's perspective.

AIM AND OVERVIEW

In this volume, we intend to change this situation. We bring together a variety of disciplines to study multiple technologies and their implications for well-being. In this interdisciplinary approach, we align ourselves with the science, technology, and society (STS) community broadly conceived to include those who study the social and human dimensions of technology but may not self-identify as STS scholars. We aim to present tools for studying the implications of technology for well-being, as well as ideas for generating better outcomes, through better design, use, and governance. In taking an interdisciplinary approach, we seek to recognize the study of well-being, quality of life, and happiness can no longer proceed along strict disciplinary lines but has become an interdisciplinary endeavor that requires collaboration between fields such as psychology, philosophy, economics, and sociology. All these fields are represented in the volume, and many of the chapters are interdisciplinary in nature.

The volume combines theory and application. It seeks strong theoretical bases for analyzing quality-of-life issues in relation to technology but also explores the practical consequences for the design of new technologies and for policies guiding the design and use of technology. This makes the volume of potential interest to those working in engineering design and in policy. The focus in the volume is both on technology in the broad sense and on specific technologies. We seek to develop general approaches that can be applied to a wide range of technologies. Yet at the same time, we recognize that specific technologies or technological artifacts present their own specific issues that need to be analyzed. We therefore explore a variety of representative contemporary technologies and their implications for quality of life. We specifically focus on information technologies, medical and agricultural technologies, and consumer technologies, and we present case studies and analyses that range from iPods to brain implants to genetically modified foods.

The volume is designed to move from a theoretical analysis of well-being, technology, and their relationships (Parts I and II); to more specific analyses and evaluations of some of the primary applied manifestations of technology and their practical implications for individuals and society (Parts III, IV, V); and finally to a theoretical and practical examination and evaluation of some key policy matters that should inform the design and regulation of emerging and future technologies (Part VI). With this structure, we aim to present more than a mere collection of chapters. We aim to present an organized, coherent series of chapters that build on each other by developing key themes, questions, and approaches.

Part I introduces central concepts and approaches in the study of well-being and technology. The first chapter, by Philip Brey, focuses on well-being. It introduces central concepts in, and major approaches to, the study of well-being from the three disciplines in which well-being has been studied most extensively: philosophy, psychology, and economics. It also considers the small amount of work that has been done on the topic of well-being and technology from within these fields. The second chapter, by Carl Mitcham and Adam Briggle, focuses on technology. It attempts to define what technology is and surveys major approaches to the study of technology and its social implications with an eye toward questions broached in the following section.

Part II introduces major approaches to the study of well-being from philosophy, economics, psychology, and sociology, and proposes how technology may be studied from within these approaches. Chapter 3, by Ruut Veenhoven, opens Part II with a macrosociological approach in which the quality of life in developed and developing societies is compared, both across nations and over time. Using surveys from 140 nations, and using findings from historical anthropology, Veenhoven argues that modern technology has made our lives significantly better. Yet, he also cautions that not every technological advance enhances well-being. The next chapter, by Justine Johnstone, takes a capability approach, which was originally proposed in economics by Amartya Sen and in philosophy by Martha

Nussbaum. Johnstone introduces the capability approach and shows how it can be applied to technology. She argues that technology can enhance human capability (and hence well-being) by serving as instrument, as a background condition that enhance capabilities, as well as by enabling qualitatively new capabilities. It can, however, also diminish capabilities.

Chapter 5 by Michael Steger and Joo Yeon Shin presents a psychological approach. The authors review psychological approaches to well-being, focusing on the debate between subjectivists and objectivists. They then present their own approach, the dual process model, which combines subjective and objective approaches ("pleasure" and "mattering"). This approach is subsequently used to evaluate the implications of various technologies for well-being. In the next chapter, Luigino Bruni presents an economic account of well-being and technology that focuses on the way in which markets and technology mediate social relationships in contemporary society. Bruni argues that although they can both play a civilizing role that has positive consequences for well-being, their dominance in contemporary life has led to a devaluation of relationships and the relational goods that they yield, which are an essential component of well-being. Anton Tupa, in the final chapter of Part II, presents a desire-satisfactionist approach, which is an approach from within philosophy according to which a good life is one in which one's desires are fulfilled. After introducing desire-satisfactionism and discussing various criticisms and refinements, he turns to technology. He argues that technology has three roles in desire-satisfaction. It can provide information that can help people to discover what is desirable to them, it can create new desires, and it may help to satisfy existing desires. He then gives examples of each role.

Part III focuses on consumer products and the fact that we live in a consumer culture. Much of the impact of technology on individual well-being is mediated by consumer products. Products such as automobiles, televisions, mobile phones, clothing, and light bulbs form a large part of the texture of people's everyday lives and can both be objects of desire and sources of pleasure. In a consumer society, well-being is defined in large part by consumer ideals that focus on the consumption of desired products and services. Consumer products are made possible by modern technology, which provides us with high-tech devices such as smartphones and motorcycles, but also enables low-cost, mass-produced artifacts such as t-shirts and scissors. Part III explores how consumer products and the consumer culture of which they are a part carry with them particular conceptions of well-being and evaluates whether these conceptions are indeed helpful to people in their pursuit of a good life.

Chapter 8 by Lindsey Patterson and Robert Biswas-Diener reviews psychological research on the role of material goods in the satisfaction of human needs. The authors argue that although material items may help satisfy certain human needs, consumers tend to adapt to their positive benefits and may start to desire more. They are running around on a "hedonic treadmill" in which the short-term satisfaction of desire turns to saturation and then dissatisfaction. However, some consumer goods, those that are experiential or that engage the consumer in enjoyable activities, can also have long-term positive impacts.

Pak Hang Wong then presents a view on consumer products that draws from social theory and philosophy. He adopts Zygmunt Bauman's theory of consumption, according to which contemporary society is characterized by the framework of consumerism, which induces particular understandings and valuations of the world and in which well-being is defined in terms of pleasure-seeking with an emphasis on immediacy and novelty. Wong analyzes how the consumerist value system is reflected in consumer products and discusses the extent to which alternative designs and uses of technological products can support alternative conceptions of well-being.

Edward Spence proposes a theory of the sustainable good life that relates consumption to sustainability. His theory is based on Epicurus' philosophy of happiness, according to which well-being consists in the satisfaction of simple and natural pleasures that are easy to fulfill and to abandon other desires, such as those for extreme wealth, power, and fame, which ultimately cause unhappiness. Spence argues that a consumer culture based on such a conception of well-being can be environmentally sustainable and still yield a high quality of life. Valerie Tiberius, in the final chapter of Part III, analyzes the implications for well-being of technological products such as cell phones and digital camcorders that mediate our experience. She presents a subjective theory of well-being, the value-based life-satisfaction account, which she then applies to mediating technologies. She uses this theory to show that various technologically mediated experiences are not good for us even if we tend to desire them, because they do not live up to our most cherished values.

Part IV examines the impact of information technologies on well-being. Most of us for most of the time now live in the "infosphere." Information and communication technologies (ICTs) and new media are fundamentally changing the ways in which information is conceived, created, produced, communicated, and used, creating profound and far-reaching effects on society. The exponential growth of computational power and capability is fast computerizing many different facets of our everyday lives that, in turn, generate an ever-increasing growth of digital information. ICTs are transforming key industries, such as food production and distribution, transport, travel, holidays, finance and banking, the national health system, education, telecommunications, science, safety and national security systems, social security, and welfare. Digital information, through the affordance of the Internet, can now potentially be created, accessed, disseminated, and used by anyone, anytime, and anywhere, worldwide. This has significant implications for individuals and society generally: epistemological (relating to knowledge), ethical (relating to moral principles, virtues, and values), and not least eudemonic (relating to notions of the good life, well-being, happiness, or self-fulfillment).

Shannon Vallor explores how new social media technologies such as Facebook and Twitter may impact the cultivation of certain communicative virtues essential to the good life. She observes that at present, the impact of emerging technologies on the virtues is not well understood. Her chapter attempts to address this lacuna by exploring the implications of new media practices for the communicative virtues of patience, honesty, and empathy. Vallor concludes that although these practices make communication easier, more flexible, and more rewarding, they may also be used as escape routes from the riskier and more difficult offline communicative practices that are essential for cultivating virtue.

Diane P. Michelfelder asks how everyday experience with the social Web helps to shape individual well-being and the quality of life. She examines that question through the lens of the distinction between community and commodity. She concludes that the ethical question as to whether the Web in its current stage of development, as the so-called Web 2.0 or "social" web, does or does not further the growth of community over commodity is a complex one. However, she does observe that some online informational activities such as those pursued on Facebook seem to align more with commodity than community, whereas by contrast online activities in virtual worlds such as the World of Warcraft seem to align more with community.

Omar Rosas explores the field of prudential psychology and examines some empirical research on the effects of different types of Internet use on the users' well-being and their consequences for the good life. He then provides suggestive interfaces between prudential psychology and corresponding philosophical approaches to the good life and concludes that prudential psychology and philosophical reflection can and ought to engage in interdisciplinary collaboration aimed at normatively assessing the effects of Internet technology on people's well-being and the good life.

In the final chapter of Part IV, Johnny Hartz Søraker clarifies the distinction between the "real" and the "virtual" by disclosing the many contentious and often tacit presuppositions that lie behind claims that we ought to choose what is "really good" in real life over what is merely "virtually good" in virtual life. He argues that what all those claims have in common is the belief that what is "really good" has a much higher value than what is merely "virtually good." For example, real friends are more valuable than virtual friends and hence contribute more to our well-being. He argues that such claims about the inferiority of the virtual can be reconstructed as a generic line of reasoning that requires many presuppositions to hold true. He concludes that most of them are inherently subjective and context-dependent and thus cannot be assumed to be true across all the particulars of our individual lives and conceptions of the good life.

Part V explores the ways in which technology is reshaping the medical and agricultural landscapes. Undoubtedly, the quality of health care and food are two of the most crucial factors affecting people's well-being. Consider just two examples of profound technology-driven change in these sectors taken from the U.S. Health care costs skyrocketed from $187 billion in 1965 to $1.9 trillion in 2005. In 1900, 41% of the workforce was employed in agriculture. By 2000, it was less than 2%. When the term "bioethics" was first conceived, it brought medicine and food together in the study of the human relationship to nature: our own human nature and the natural world around us. Indeed, the way we eat and how we treat the environment have direct bearing on human health. And both health and connectedness to nature are often thought to play essential roles in human flourishing. What is becoming of health and our connection to nature in a technological age? Can we and should we transcend all the limitations imposed on us by human nature and the natural world? What does it mean to be human and to live rightly on this planet?

David Kaplan begins by observing that what humans eat and drink has always been technologically transformed to some degree. Does this mean that there is no moral difference between natural and techno food? Echoing the "real" and "virtual" theme in the previous section, he argues that a case can be made for natural and minimally processed foods as superior to techno food, but the reasons are subtle and the differences are more social than scientific or metaphysical. We can decide whether or not to make and eat techno foods using only concepts that are in principle available to everyone, such as common sense and public debates about what is right and what is good. He explores both "hard" concepts of the right and "soft impacts" of the good to assess what, if anything, is wrong with techno food. His conclusion is that their production, distribution, and consumption can sometimes violate rights and fail to foster the good.

Bengt Brülde introduces the ethics of human enhancement by listing central problems and locating it on the larger map of the ethics of technology. He then investigates whether chemically induced mood enhancement is (if successful) likely to make humans happier over time. This entails consideration of (1) what effects mood-enhancing drugs are likely to have on the long-term happiness of the consumer, (2) how these drugs would affect the happiness of his or her immediate environment, and (3) what effects a widespread use of mood-enhancing drugs would have on society as a whole and how this would affect the happiness of its citizens.

Motivated by similar concerns about the technological enhancement of human nature, Peter-Paul Verbeek takes current discussions of technology and the good life back to their roots in ancient Greek thinking. This demonstrates how 'good life ethics' can articulate the so-called nonmodernist approaches to technology that have developed over the past decades. The central idea in these approaches is that humanity and technology need to be seen as fundamentally intertwined: They shape each other, and the one cannot be understood without the other. This makes the ethics of technology a complicated affair, because it makes it impossible to defend humanity against technology—even the criteria for assessing technology are technologically mediated in this approach. Yet he argues that good life ethics makes it possible to develop an alternative form of ethical reflection, which aims to accompany the development and use of technologies 'from within' rather than assessing them 'from outside.'

Tsjalling Swierstra is similarly concerned to work out a notion of the human as always evolving with technology. He uses Aldous Huxley's *Brave New World* to consider the fear of success: Perhaps humans might actually attain victory over nature only to find out too late that the result is something less than a humanly good life and something more like a shallow, blissful slavery. The novel does express something essential about the good life, but it neither does justice to technology nor to politics and modern culture. Huxley was mostly wrong when he prophesied a world where mass consumption would produce uniform and shallow mass consumers. The modern world allows for a large degree of individualism and diversity. He concludes that technology will not take the challenges, conflicts, dynamism, unpredictability, and resistance out of our lives, because technology itself is the source of new challenges, conflicts, dynamism, unpredictability, and resistance. The Romantics were right: too much control over the world is not good for humans. But, fortunately, we will not achieve control any time soon.

In the final chapter of Part V, Mark Coeckelbergh asks: Should we let robots care for the elderly? Can they contribute to their 'good life'? Although robots cannot live up to Aristotle's perfectionist view of friendship, they may provide companionship and careship and thus contribute to the *best possible* life (BPL). But do they deceive people by *appearing* to care? Proposing the concepts of 'virtual virtue' and 'virtual care,' he argues that appearance might be less morally problematic than this objection supposes and that professional caregivers do not necessarily meet the BPL criterion either. A comprehensive discussion about care robots should therefore involve a reassessment of the way society deals with the ill and elderly.

Part VI concludes the volume by considering the role of design and policy in matters of technology and well-being. Engineers and others who design artifacts and systems wield an enormous influence over the character of contemporary life. What obligations do they have for promoting human well-being? What conceptual and practical tools might they use for integrating a wider set of considerations into processes that are often construed narrowly in instrumentalist terms of maximizing profit? Policymakers in liberal democracies face similar challenges. Their decisions are supposed to be neutral with respect to the good life, yet they actively promote and regulate technologies in ways that are far from neutral. If technologies are not neutral tools, then can the state be in the business of managing technology and still claim neutrality? Policymakers must also consider which investments in science and technology best serve human welfare. How can progress in knowledge production and technology more reliably bring about social progress? And how is progress measured?

Ibo van de Poel discusses how an objective list account of well-being can be used to integrate considerations of well-being into engineering design. He focuses on two philosophical problems encountered in any such integrative attempt. One is the epistemological problem: How can designers know what constitutes well-being for users? The other is the aggregation problem: How can the well-being of different people be aggregated into an overall measure of well-being? He examines ways of dealing with both problems and argues that neither problem bars the possibility of designing for well-being.

Robert Frodeman, J. Britt Holbrook, and Kelli Barr begin with Aristotle's notion of contemplation or *theoria* as the good life. Contemplation has been traditionally understood as intrinsically valuable and pursued simply for its own sake. Viewed pragmatically, however, contemplation is useless. In today's university, increasing calls for accountability have led to a proliferation of what Heidegger called technological thinking—calculative approaches that are marked by the growth in the use of metrics to gauge academic performance. Some have suggested that academics offer an Aristotelian defense of academia as intrinsically valuable (and practically useless), a defense that is incorrect and unwise in neoliberal times. Instead, they argue that academics should demonstrate their accountability to society not only through calculative thinking and technology-dependent metrics, but via a more Platonic account of *theoria* that sees contemplation as essential to the healthy functioning of society.

Adam Briggle picks up on this question of accountability, or the relationship between knowledge production, technological change, and good social outcomes. His focus is on the U.S. National Institutes of Health (NIH), which seeks knowledge to improve health. But the connection between advances in scientific knowledge and improvements in health is not automatic or linear: as science and technology have progressed, health costs have skyrocketed and many health outcomes have stagnated. The NIH peer review process for assessing "good science" focuses far too much on advancing scientific research and far too little on assessing how such progress relates to the goal of improved health. The currently underdeveloped second-stage of peer review should be transformed into an open, adversarial science court for assuring that science serves the social good.

Jeroen van den Hoven notes that questions about ultimate goals are often marginalized in technology policy for two reasons. First, substantive debate about goods is displaced by instrumental technical discourse of risks, costs, and benefits. Second, the ideal of liberal neutrality restrains public debate about the good life. He develops an interpretation of neutrality that allows for systematic and careful political discussion and decision making about substantive goods. His argument draws from Jose Ortega Y Gasset's claims that designing or using technology amounts to implementing a conception of the good life. Ortega's work helps demonstrate how public policy concerning technology cannot be neutral regarding conceptions of the good life. It calls for institutions that allow for explicit and rational political discussions about the good life in a technological age. This requires a shift from preclusion to accommodation of disagreements.

In the final chapter of the volume, Govert Valkenburg argues that notions of the good life are generally considered part of our private comprehensive doctrines rather than a political matter. At the same time, ideas of the good life are intricately connected to technology: We cannot discuss technology without referring to the good life, and what we think of the good life is also dependent on the technologies surrounding us. He explains that technologies also bear on *how* we discuss the good life in public. Not only does technology pull considerations of the good life to the fore, it also suggests particular ways to suppress it.

REFERENCES

Bacon, Francis. (1620). *The New Organon.* Lisa Jardine and Michael Silverthorne, eds. 2000. Cambridge: Cambrideg University Press.

Lane, Robert. (2000). *The Loss of Happiness in Market Democracies.* New Haven, CT: Yale University Press.

Rousseau. Jean-Jacques. (1754). "A Discourse on the Origin of Inequality," in *The Social Contract and Discourses*, G.D.H. Cole, trans. 2001. London: Everyman.

Sarewitz, Daniel. (1996). *Frontiers of Illusion: Science, Technology, and the Politics of Progress.* Philadelphia, PA: Temple University Press.

Part I

Mapping the Landscape

1 Well-Being in Philosophy, Psychology, and Economics

Philip Brey

Well-being has traditionally been a topic of philosophical study, but in recent decades it has become an important concern in other fields as well, particularly psychology and economics. Empirical research on well-being and happiness has become very influential in the policy world, in which their measurement is increasingly used to guide social and economic policy. In this chapter, I will introduce and discuss the study of well-being and happiness in philosophy, psychology, and economics. These fields have been chosen because they are the ones in which well-being has been studied most extensively. Next to introducing the study of well-being in these fields, I will also discuss academic studies of the relation between well-being and technology.

In the next section, I will discuss how well-being has been taken up in philosophy, and I will introduce and critique major philosophical approaches to well-being, which include hedonism, desire-satisfactionism, and objective list theories (including perfectionism and capability theories). The section that follows focuses on the psychological study of well-being. It discusses its history, its central concepts and methods, such as the notion of subjective well-being, and the new field of positive psychology. This section is then followed by a review of the role of well-being in economics, which focuses on the emerging field of happiness economics and on happiness and quality of life indices. The chapter will be concluded with a discussion of the relation between well-being and technology, as it is studied (or has been failed to be studied) in philosophy, psychology, economics, and other fields.

PHILOSOPHICAL THEORIES OF WELL-BEING

In philosophy, well-being (the good life, happiness) has been studied since the ancient Greeks. It is an important theme in the works of Aristotle, as it is in ancient Greek philosophy throughout. In modern philosophy, well-being and happiness take center stage in the works of 18th- and 19th-century philosophers Jeremy Bentham and John Stuart Mill, who present a modern version of hedonism as part of their theory of utilitarianism. In contemporary philosophy, well-being remains an important topic that has received increased attention in the past 30 years. It is generally recognized that there are three major types of theories of well-being in philosophy: hedonist, desire-fulfillment, and objective theories (Parfit 1984; Griffin 1986; Sumner 1996). These will now be discussed in turn.

Hedonism (and Other Mental State Theories)

Hedonist theories hold that only pleasure is intrinsically good, and pain is the only intrinsic bad. A person's life therefore goes well to the extent that he or she is able to accumulate pleasure and avoid pain. To strive for well-being is to strive for the greatest balance of pleasure over pain. Modern hedonism finds its roots in the works of the utilitarian philosophers John Stuart Mill and Jeremy Bentham, who both argued for the centrality of pleasure in well-being. Earlier forms of hedonism date back to antiquity, including the teachings of Epicurus in the 3rd and 4th century B.C., who held that a good life is attained by maximizing pleasure and by avoiding protracted fear and bodily suffering.

Quantitative hedonism, or simple hedonism, is the view that the value of pleasure is only determined by its quantity (involving parameters such as duration and intensity) and not by its quality. It therefore does not matter what type of pleasure one has or what its source is. A better life is in all circumstances one in which the amount of pleasure is maximized and the amount of pain is minimized. Quantitative hedonism was originally proposed by Jeremy Bentham (1789). It has since then come under severe criticism. A powerful objection to it, brought forward by John Stuart Mill (1863), is that certain types of pleasure seem more desirable, or worth having, than others. As Mill argues, it is better to be an unsatisfied human than a satisfied pig. In other words, if one lives a deprived life, in which one only satisfies one's most primitive urges and pursues only bodily pleasures, then one's life is not necessarily good. One misses out on so-called higher pleasures that may involve experiences of friendship, knowledge, art, contemplation, and refinement in taste. These are pleasures that are more worthwhile than the so-called lower pleasures. *Qualitative hedonism*, proposed by Mill, holds that some pleasures are more valuable or pleasurable than others. A good life is therefore not merely a life in which the duration and intensity of pleasure is maximized, but one in which the highest pleasures are well-represented.

Hedonism, in either its qualitative or quantitative form, has been criticized for not answering sufficiently to our intuitions about what makes a life good. One of the most powerful objections is the *experience machine objection* of Robert Nozick (1974). Nozick hypothesizes the existence of an "experience machine" that uses neurostimulation to simulate a nonexistent world that provides the user with constant gratification. The user experiences any type of pleasure, without being able to distinguish the simulated world from a real one. Nozick argues that many people would choose not to plug into such a machine for the rest of their lives, because people value real events and experiences and want to live a real life, which they believe is not provided by being hooked up to an experience machine. This, he argues, demonstrates that hedonism does not provide us with an adequate account of well-being.

Hedonist theories are part of a larger class of *mental state theories* of well-being. Mental state theories hold that well-being depends on the presence of certain types of mental states in humans. These could be cognitive, affective, emotional, volitional, or mixed. Hedonism is by far the most prominent type of mental state theory. *Happiness theories of well-being* are a second prominent type. Proponents of such theories usually argue that happiness is different from mere pleasure, and that happiness is something more than or different from the mere experience of positive feelings. On a *cognitive view of happiness*, happiness involves a positive attitude to one's life as a whole, or the tendency to evaluate one's life positively. This is a cognitive state that need not be accompanied by positive feelings at all. *Hybrid views of happiness* hold that happiness is constituted by different types of cognitive states, for instance by a combination of pleasant feelings and positive evaluations of one's life. On *hedonistic views of happiness*, happiness is a purely hedonic state and therefore either equals pleasure or constitutes a particular form of it (Brülde 2007). On this view, a happiness theory of well-being is a particular type of hedonistic theory.

Desire-Fulfillment Theories

Desire-fulfillment theories, also called *preference-satisfaction theories*, hold that well-being lies in the fulfillment of one's desires. Desire-fulfillment theories emerged in the 19th century, in part as an outgrowth of welfare economics. Welfare economists wanted to have objective criteria for measuring well-being or utility in economic subjects. However, pleasure and pain are in people's heads and therefore cannot be easily measured. So economists instead came to a conception of well-being as the satisfaction of preferences or desires. People are normally able to state their preferences and rank them in relation to each other, which then allows economists to measure preferences and include them in 'utility functions' in which value is attached to the satisfactions of preferences.

Next to this practical advantage, a more fundamental advantage of desire-fulfillment theories over hedonist theories is that they are capable of avoiding the "experience machine" dilemma. This dilemma can be avoided because desire theories can account for the fact that people sometimes desire that one's experiences are real. For example, if one desires to be genuinely loved by people, and an "experience machine" simulates people that seem to love one, then one's desire is not really satisfied, because this love is only simulated and not real. Therefore, this love can be held to be of less value than genuine love.

Three major types of desire-fulfillment theories have been proposed (Crisp 2008). *Simple desire-fulfillment theories* (also called *present desire theories*) hold that one is better off to the extent that one's current desires are fulfilled, and the best life is one in which all one's actual desires are fulfilled. A major problem with this view is that many desires that people have seem to go against their long-term interests. An angry and impulsive adolescent, who chooses a life of crime and substance abuse and engages in reckless acts, may follow his short-term desires but is not likely to be better off by following them in the long run. He would have a better life by suppressing some of his impulsive desires and choosing different ones.

Reflective or *comprehensive desire-fulfillment theories* give priority to one's reflective preferences that concern one's life as a whole. If it is one's preference, on reflection, to have a long and happy life, then this preference would trump short-term preferences that seem to go against this goal. An objection to reflective theories is that they, also, may result in choices that are not optimal for a particular person. Take, for example, an orphan monk who has lived a very sheltered life, but then wins the lottery. He now has two choices: to either stay in the cloister and give away his prize, or to live a new life outside the cloister. Because he has no conception of what life may be like outside the cloister, he chooses, after much careful reflection, to remain a monk. It is conceivable that the monk did not make the choice that is best for him, because he was not properly informed about the alternatives. But if this is the case, then the reflective view does not give us the best account of well-being (cf. Crisp 2008).

Informed desire-fulfillment theories hold that the best life one could lead is the life in which all desires are fulfilled that one would have if one were fully informed of one's situation. Relevant information could include information about possible alternative courses of action and objects of desire, about possible and likely consequences of actions, and about one's own psychological makeup and behavioral tendencies. The idea is that if one is properly informed, one would be smart about the desires that one would have and strive to satisfy, both in terms of ensuring a good fit between one's desires and one's character and in terms of choosing (sets of) desires that one is sufficiently likely to be able to realize.

A general objection to desire-fulfillment theories is that they present a very abstract, formal theory of well-being that does not tell us anything about the sources of well-being. It tells us that a necessary and sufficient condition of well-being is that our desires are satisfied. But it does not tell us anything about the source of these desires. Yet, it would seem that it is not the case that things are good for us because we desire them, but rather that we desire things *because* they are good for us. Yet, desire theories tell us nothing about the reasons why things may be good for us. In addition, they have a difficult time accounting for so-called defective desires: desires that seem to be bad for one, such as base, poorly cultivated, and pointless desires. People may hold such desires even after being properly informed, but if this is the case, it is not clear that we should conclude that satisfying them improves the quality of their lives more so than alternative courses of action would.

Objective List Theories

Objective list theories hold that well-being is the result of a number of objective conditions of persons rather than the subjective experience of pleasure or the fulfillment of subjective desires. They propose that there are goods that contribute to our well-being even if we do not desire them or experience pleasure from them. Well-being is attained by living a life in which one is in possession of all or most of the goods on the list. The list is objective in the sense that items on the list increase the value of one's life independently of one's tastes, attitudes, traits, or interests. Acquiring goods or doing things on the list makes one's life better, even if one does not desire or want them.

Let us consider a few such lists. Derek Parfit has proposed a list that includes as items moral goodness, rational activity, the development of one's abilities, having children and being a good parent, knowledge, and the awareness of true beauty (Parfit 1984). James Griffin's list includes accomplishment, the components of human existence (autonomy, capability, and liberty), understanding, enjoyment, and deep personal relationships (Griffin 1986, p. 60). John Finnis, finally, has proposed a list that includes life, knowledge, play, aesthetic experience, friendship, practical reasonableness, and religion (Finnis 1980, p. 60).

Objective list theories can be distinguished not only by what items the list contains but also by the explanation, if any, of why they are on the list. Philip Kitcher (1999) has made a distinction between bare and explanatory objective theories of well-being. *Bare objective theories* merely present a list of items that have no single unifying element. Each item on the list contributes to well-being independently of the others. *Explanatory objective theories* identify a fundamental element that unifies all the items on the list and explains why they contribute to well-being. Most contemporary objective list theories are explanatory theories.

Perfectionism constitutes the most influential explanatory objective list theory. Perfectionist theories hold that what items on the list have in common is their contribution to the realization or perfecting of human nature. Whether it stems from biology or divinity, we have an intrinsic nature that provides us with a *telos,* or end, that defines what types of things are good and bad for us. For example, most perfectionist theorists would hold that it is part of our nature for us to flourish when we attain friendship or knowledge. The best possible life is one in which one's human nature is fully realized and one acts in such a way as to contribute to such a realization. Typically, perfectionist theories exhort people to put in effort to develop their faculties, so as to excel in art, science, culture, or sports, as well as to excel in everyday living.

One famous perfectionist theory is Aristotle's theory of *eudaimonia*. Aristotle held that the good for human beings is found in the cultivation of human virtues, that is, human capacities that are part of their *telos*, and particularly in the cultivation of rationality. This would result in the highest good for human beings, which he called *eudaimonia*, or personal flourishing (often translated as 'happiness'). Thomas Hurka (1993) has presented a well-known neo-Aristotelian perfectionist theory. He holds that human nature is geared toward three types of perfection. Physical perfection is the optimal functioning of our biological systems and finds its highest form in athleticism. Theoretical and practical perfection are the optimal formation of intentions and beliefs and our optimal way of acting on them. Well-being is realized through the physical, theoretical, and practical perfecting of ourselves.

Martha Nussbaum (2000) has presented a *capability theory* of well-being, a neo-Aristotelian theory according to which well-being is dependent on capabilities, which are real possibilities of persons to be or do something, such as capabilities to read, be healthy, or take care of others. Nussbaum has presented a list of 10 capabilities that she considers necessary for well-being and that derive from human nature. They include capabilities such as bodily health, bodily integrity, practical reason, play, and control over one's environment. Nussbaum does not claim that everyone must have all 10 capabilities to attain well-being. Rather, she claims that, normally, people would want to realize most of these capabilities but can also choose that some of them are less important for them.

A frequent objection to objective list theories is that they are paternalist by claiming that things may be good for people even if they do not want or value them. Critics hold that such a claim fosters elitism and may have undesirable political consequences if it is used to legitimize social policies. A related objection is that objective list theories deny that people are different, and that different things may be good for different people. Objective list theorists can accommodate this objection to some extent by pointing out that items on the list can often be realized in very different ways, and may themselves be subjective (e.g., enjoyment). They could also, as Nussbaum does, make certain items on the list dependent on individual characteristics or preferences. A final objection to objective list theories is that items on the list often seem incomparable. Is a life with little friendship but much knowledge better than a life with much friendship and little knowledge? This objection seems to be more difficult for bare objective theories than for explanatory ones.

Conclusion

Hedonism, desire-satisfactionism, and objective list theories have for long been the three main rival approaches to well-being in philosophy. A resolution between them does not appear to be in view. Some authors have attempted to resolve the issue by combining two or more of the three approaches. One way to create a hybrid version is to include pleasure or desire-satisfaction on the objective list. Another approach is to argue that some desires correspond with objective needs (Hamilton 2003). Desires are subjective and needs are objective, but some desires may be expressive of genuine needs, whereas others are not. Yet another approach is to argue that informed desire-fulfillment or informed qualitative hedonism would result in the selection of items on objective list approaches, or to suggest that objective list items are relative to persons and circumstances, thereby narrowing the gap between the approaches. In these ways and others, philosophers are trying to overcome the stalemate between the three approaches.

WELL-BEING IN PSYCHOLOGY

The psychology of happiness and well-being started to take off in the 1950s, when an increasing number of psychologists began studying positive emotions and feelings of well-being, next to negative emotional states that were their dominant object of study. Part of this research belonged to the new approach of humanistic psychology, founded by Abraham Maslow and Carl Rogers, which emphasized humans' uniqueness, human potential, and psychological growth. Additional roots are found in the 1960s, when American survey organizations started doing surveys of people's happiness and satisfaction. During the 1970s and 1980s, happiness emerged as a serious research topic in the social and behavioral sciences, and the number of journal publications on happiness and quality of life was on the rise. The psychology of happiness started taking shape as a field in the 1980s, due to the seminal work of Ed Diener and others. Diener coined the term "subjective well-being" (Diener 1984) and showed how to measure and study it, inspiring many psychologists to follow his lead. Diener has made major contributions to the psychological measurement of well-being and to the influence on well-being of temperament and personality, income, and culture. Further important developments in the 20th century were Mihaly Csikszentmihalyi's seminal publications on flow and optimal experience (Csikszentmihalyi 1990), the start of the *Journal of Happiness Studies* in 1999, and the publication of the book *Well-Being: Foundations of Hedonic Psychology* (Kahneman, Diener, and Schwarz 1999).

Another important occurrence in the 1990s was Martin Seligman's 1998 presidential address to the American Psychological Association (APA). In his address, Seligman argued that psychology had largely been preoccupied with studying and treating mental illness, and that it should balance this focus with an alternative approach with a focus on studying and improving people's positive functioning and well-being. He called this approach *positive psychology.* Its aim was defined as finding and nurturing talent and making normal life more fulfilling (Seligman and Csikszentmihalyi 2000). Positive psychology was quickly embraced by psychologists and has generated numerous studies and activities, among which was launch of the *Journal of Positive Psychology* in 2006.

Concepts and Methods

The psychological study of happiness and well-being is sometimes referred to as "happiness psychology" or "psychology of happiness." By far the most frequently used term nowadays, however, is "positive psychology." This will be the preferred term in the remainder of this chapter. Strictly speaking, the scope of positive psychology is broader than that of happiness psychology, as it does not only include well-being but also the study and nurturing of talent and genius. "Hedonic psychology" is a term that refers more specifically to the study of positive and negative feelings. It "is the study of what makes experiences and life pleasant or unpleasant, and concerns itself with feelings of pleasure and pain, of interest and boredom, of joy and sorrow, and of satisfaction and dissatisfaction (Kahneman, Diener, and Schwarz 1999, p. ix). Psychological research on happiness and well-being is sometimes also subsumed under more general scientific labels that include "happiness studies" and "the science of subjective well-being."

The central concept in the psychology of happiness is *subjective well-being,* often abbreviated as SWB. Subjective well-being is an individual's current evaluation of their happiness. SWB is sometimes contrasted with *objective well-being* or objective happiness. Daniel Kahneman has argued that psychology should study objective happiness, which he defines as people's real-time positive sensations and feelings of happiness (Kahneman 1999). Such sensations and feelings, however, are difficult to measure. Diener instead proposes to study people's own judgments of their feelings and degree of happiness. These are much easier to measure, for example, through surveys and interviews, but also in controlled experiments. He argues that people's judgments of their own happiness are generally reliable, because in general people are happy if they think they are happy. This assumption remains controversial, however, with critics such as Kahneman arguing that self-reports of happiness are not very reliable.

Diener has argued that there are three major components to SWB: positive affect (pleasurable feelings), negative affect (painful feelings), and *life satisfaction* (Diener and Lucas 1999). Life satisfaction is how people assess the balance between positive and negative affect in their lives as a whole, and how well their lives measure up to their aspirations and goals. It is often contrasted with happiness, which is considered to be a temporary state of positive affect, whereas life satisfaction is a cognitive appraisal that tends to change little over time. SWB is hence both affective and cognitive: It includes hedonic evaluations that are guided by emotions and feelings, as well as assessments of people's lives as a whole based on their aspirations and expectations.

Methodologically, the study of SWB involves obtaining people's reports on their experienced happiness, (life) satisfaction, and positive and negative feelings. Such reports are often collected by means of questionnaires that contain questions such as "How happy are you?" "How satisfied are you with your life as a whole?" and "Do you find beauty in most things?" Answers are often given on a sliding scale. Another important method is real-time "experience sampling," where the subjects are asked to report their current mood in real-time. Usually, the goal of studies in SWB is to find correlations between self-reported positive states and other variables, such as individual traits and social indicators like income and health. Many studies involve large-scale social surveys, but studies may also make use of smaller-scale surveys or controlled real-time experiments in which variables are modified. The study of objective well-being, on the other hand, requires the recording of physiological data in real time, such as brain activity or bodily responses, which are then interpreted as representing either positive or negative feelings and emotions.

Theories and Applications

Luigino Bruni and Pierluigi Porta (2005) argue that psychological research on happiness tends to use either a *hedonic* or an *eudaimonistic* notion of happiness. The hedonic tradition, exemplified by Kahneman, Diener, and Schwarz (1999), defines happiness in terms of pleasure and good feelings, and it attempts to measure such states, either through objective or subjective measuring procedures. The field of hedonic psychology is hence defined as:

> the study of what makes experiences and life pleasant or unpleasant. It is concerned with feelings of pleasure and pain, of interest and boredom, of joy and sorrow, and of satisfaction and dissatisfaction. It is also concerned with the whole range of circumstances, from the biological to the societal, that occasion suffering and enjoyment. (Kahneman, Diener, and Schwarz 1999, p. ix)

The hedonic tradition in psychology clearly corresponds with the hedonist tradition in philosophy, although its inclusion of satisfaction as a relevant parameter also makes it extend to desire-satisfaction approaches.

Eudaimonic approaches hold that happiness includes objective factors that go beyond subjective experiences and attitudes. Specifically, it includes the actualization of human potential. Eudaimonic approaches therefore correspond to objective list approaches in philosophy, and more specifically to perfectionist approaches. A very influential eudaimonic theory is Martin Seligman's theory of authentic happiness (Seligman 2002). Seligman holds that a good life is a life that integrates three types of lives: the pleasant life, the engaged life, and the meaningful life. The pleasant life is attained by having (and learning to have) positive feelings about the present, past, and future. The engaged life is a life in which engagement and involvement are pursued, in work, intimate relations, and leisure. In engaged activity, one's attention is completely focused and time loses meaning. One experiences a kind of "flow," as Csikszentmihalyi (1990) has called it. To pursue engaging activities, one must have character strengths and virtues ("signature strengths") that allow one to execute activities in an engaged manner. Finally, the meaningful life is a life in which one's signature strengths and talents are used in the service of things that one believes to be bigger than oneself. Seligman supplements his theory with an account of positive individual qualities, such as kindness, wisdom, and perseverance, that are held to contribute to a good life.

Research in positive psychology is applied with the aim of developing positive practices that enhance human well-being. These positive practices are aimed to support positive experiences, positive individual qualities, or positive social processes and institutions, such as nurturing families, healthy workplaces, and positive civic and political processes. Techniques that are developed include therapeutic techniques and programs, as well as social and organizational techniques and programs. Surveys in positive psychology are also increasingly being used to guide policy (Diener, Lucas, Schimmack, and Helliwell 2009). Concepts and data of positive psychology are also being used more and more in other fields of study in which an understanding of well-being is relevant. One of these fields is economics, to which we will now turn.

WELL-BEING IN ECONOMICS

Economics is a third field, next to philosophy and psychology, in which well-being has become an important research topic. Recent decades have seen the emergence of happiness economics, a new branch of economics that studies the economic conditions for happiness and well-being. Yet, notions of well-being have always played a role in economics. In what follows, I will first consider the role of well-being in classical and neoclassical economics, after which I will chart the emergence of the new approach of happiness economics and the recent interest in national and international quality of life indices as policy instruments for economic development.

Neoclassical and Welfare Economics

Well-being has for a long time played a role in economics, directly or indirectly, through the notion of utility. In neoclassical economics, the dominant approach to microeconomics since the late 19th century, economic behavior is typically explained in terms of attempts of individuals to increase their *utility*. Utility is a measure of relative satisfaction, and the utility of a good or service is the total satisfaction derived from consuming it. Neoclassical economics has been influenced by utilitarianism and 18th- and 19th-century doctrine championed by Jeremy Bentham and John Stuart Mill, in which happiness is conceptualized as utility, a quantitative measure of pleasure or satisfaction, and which prescribes that only those actions should be undertaken that maximize utility.

In neoclassical economics, the notion of utility mostly takes on an explanatory role, instead of the normative role it has in utilitarianism. Neoclassical economics aims to explain how prices, outputs, and income distributions are determined in markets as a function of supply and demand. Some of its central assumptions are that individuals aim to maximize their utility and that they have rational preferences among outcomes that can be associated with a value. In late classical and early neoclassical economics, it was assumed that utility could be cardinally measured, that is, that goods and services could be given numerical values that expressed the amount of satisfaction that individuals would receive from them and that would correspond with increases or decreases in their happiness.

However, since the 1930s, cardinal notions of utility were rejected in favor of ordinal notions. Ordinal utility only determines relative preferences between pairs of goods or services, but does not define any absolute measure of satisfaction for them. As a consequence of this development, some argued that economics no longer provided measures of happiness or satisfaction. As Bruno S. Frey and Alois Stutzer (2002, p. 43) put it: "Utility was deprived of all content and was reduced to a preference index reflecting revealed behavior." Happiness and (unquantified) individual preferences are, they claimed, distinct notions that may not always correlate strongly. For instance, knowing that people tend to prefer apples over pears may not tell you much about their significance for human well-being. Thus, although the new notion of utility still aims to say something about individual happiness or satisfaction, it does so in a more indirect and less reliable way.

Within neoclassical economics, the concept of utility is mostly used to explain economic behavior and is not, as it is in utilitarianism, a normative notion that prescribes how one should act. An important exception is *welfare economics*, a normative branch of microeconomics that aims to measure and improve *social welfare*. Welfare economics particularly aims to assist the public sector, which has always had promotion of the general welfare as a central aim. It does so by developing and evaluating economic solutions that aim to improve social welfare by generating outcomes that maximize overall utility. Social welfare is the overall welfare of society and is a function of the individual welfare or well-being of all individuals in society. Being a branch of neoclassical economics, welfare economics has also moved from cardinal to ordinal measures of utility, and it has been questioned whether these are adequate for determining social welfare (Cooter and Rappoport 1984).

Happiness Economics

Although well-being was not a central concern of early and mid-20th century economics, it reemerged as a topic in the 1970s (Bruni and Porta 2005). The 1970s were the stage for a debate about the *paradox of happiness* in economics. This debate was initiated by the American economist and demographer Richard Easterlin, who had found on the basis of surveys that people in high-income countries were not significantly happier than people in low-income countries (at least, ones in which incomes reached a certain threshold), and that rises of income above a certain threshold within countries did not seem to yield significant increases in happiness (Easterlin 1974). Easterlin's data showed that the correlation between happiness on the one hand and income and wealth on the other was not as strong as previously believed. Another important contribution was the publication of *The Joyless Economy* by Tibor Scitovsky (1976), in which he argued, against economic orthodoxy, that people's individual preferences in capitalist consumer societies would not necessarily generate happiness for them and that more wealth, income, and consumer products did not necessarily equal more happiness. The 1970s also saw other works critical of assumptions of neoclassical economics, by authors such as Fred Hirsch, George Akerlof, and Amartya Sen, that rethought rational choice theory, criticized preference rankings as a model for economic value, uncoupled the link between wealth and happiness, and infused other disciplines into economics, such as psychology and sociology.

In the wake of these developments in the 1970s, happiness emerged in economics as a topic of study in its own right. Under the assumption that an important, if not the only, value of economic activity lies in its contribution of individual happiness, economists engaged in studies of the correlation between economic processes and individual well-being. They considered how economic factors, such as income, wealth, unemployment, and social security, as well as social and institutional factors, such as freedom, relationships, and good governance, affect individual well-being. Some economists went even further to argue that happiness should become the new metric of economics, replacing monetary value or ordinal utility as the values that economics aims to optimize. On this conception of economics, economic and public policies should not aim to maximize GDP but should rather aim to maximize gross national happiness, as measured through some happiness index.

Happiness economics often incorporates psychological approaches and typically measures happiness in terms of self-reported SWB. It makes frequent use of large-scale surveys that enable it to study correlations between self-reported well-being and economic and social indicators. Its major application are in public policy, in which it is helpful in evaluating and improving policies that aim to improve social welfare or individual happiness.

Quality of Life Indices

In the past 20 years, happiness and quality of life indices have taken on a major role in public policy. Usually, these indices have a strong basis in economics, and they frequently include sociological and psychological perspectives as well. These indices are used to measure happiness or quality of life within nations and to make comparisons between nations. Some also include cities or regions.

Although the notions of happiness and quality of life are sometimes used interchangeably, quality of life may also refer to what Ruut Veenhoven (1995) has called "*livability*." Livability is not a quality of individuals but of environments or societies, and it refers to the extent that these allow the satisfaction of human needs and hence well-being. A country may have a high degree of livability but people may nevertheless fail to be happy because they do not make good use of the resources provided by their environment.

Indices that measure quality of life in terms of objective environmental conditions or resources may be called resource-based indices. External factors measured by them are, for instance, GDP per person, mean years of schooling, job security, and life expectancy. Other indices aim to measure subjective conditions in people that are believed to reflect their well-being. They include utility-based indices, which measure individual preferences, and indices that use psychological measures of subjective well-being. Yet another type of index makes use of Amartya Sen's (1980; 1993) and Martha Nussbaum's (2000) work in capability theory. In such indices, the development of major human capabilities is measured, such as the capability to read, to feed oneself, and to participate in political processes. These capabilities are believed by them to be a better indicator of human welfare than the availability of resources.

The *Human Development Index* (HDI) was established in 1990 by the United Nations Development Program and provides an annual ranking of countries by their level of human development. It was devised with the purpose of shifting the discussion of development from a focus on GDP to people-centered criteria. Sen's capability approach is the major inspiration behind the index. The current HDI indexes three variables, life expectancy at birth, mean years of schooling and expected years of schooling, and gross national income per capita, to arrive at a composite index for development. Human development, in turn, is believed to give an indication of the quality of life in indexed nations (United Nations Development Program 2011).

The Economist Intelligence Unit's *quality-of-life index* combines subjective life-satisfaction surveys with objective determinants of quality of life to arrive at a ranking of the quality of life in different countries. Objective determinants include health, family life, GDP per person, climate, and gender equality, among others (*The Economist* 2011). *Mercer's Quality of Living Survey* is an annual survey of the most livable cities. It uses 39 objective criteria, such as safety, education, hygiene, culture, environment, political-economic stability, and public transportation (Mercer 2010).

Gallup's *global wellbeing survey*, in contrast, only uses subjective data. It simply asks respondents from different nations to evaluate the quality of their lives on a scale from 0 to 10, where 0 indicates the worst possible life and 10 the best possible life. It then uses this survey to rank nations according to the percentage of people in it that are thriving (score of 7 or above for their current life and 8 or above for their expected life) (Gallup 2010). The recent Organisation for Economic Co-operation and Development (OECD) *Better Life Index* is interesting because it lets users decide for themselves how to weigh different indicators in ranking the quality of life in different countries. The index combines objective indicators such as housing, income, and work-life balance with the subjective indicator of life satisfaction (Organisation for Economic Co-operation and Development 2011).

Some countries have started to use happiness indices to guide national policies and are using them as a supplement or alternative to GDP as a measure of progress. Bhutan is the first country to measure its progress in terms of *gross national happiness* (GNH), using sophisticated surveys to measure the population's level of well-being. Other countries that are using, or are considering using, GNH indices include Thailand, China, Australia, Canada, France, and the United Kingdom. Critics of GNH argue that such indices can easily be slanted by governments who may define GNH in ways that suit their interest, and that the existence of different national indices will make international comparisons difficult.

WELL-BEING AND TECHNOLOGY

Despite the pervasive role of technology in modern society and contemporary life, very little research on well-being has focused on technology. Technology hardly registers as a topic in positive psychology or happiness economics, and in philosophy the situation is not much better. In what follows, I will review some of the work on well-being and technology that does exist, and I will look forward to the future study of this topic.

The benefits and disadvantages of modern technology for human life have been a topic of discussion since the Enlightenment in the 17th century. Enlightenment thinkers believed that the application of science had the potential of giving humanity unlimited control over nature; they believed that human reason was fully capable of comprehending reality and anticipating the future; and they believed that technology would mostly be used for good ends. As a result, they had a generally positive view of technology as a means for improving one's living conditions and for personal liberation. The 17th-century philosopher René Descartes, for example, held that the technological application of science would bring ever-more control of nature and would yield an unlimited number of devices that would allow people to enjoy without effort all that the earth could offer them (Descartes 1637).

The optimistic vision of technology of the Enlightenment is still well-represented in society. Technology is frequently seen as an instrument of social and economic progress that makes people's lives better. It is seen as a means that gives people additional powers to get what they want or need, thereby are better able to realize the good life. On a hedonist version of this perspective, technology is believed to reduce pain by reducing labor and hardship and improving health, and to increase pleasure by expanding the time and opportunities for pleasurable experiences. On a desire-satisfactionist view, technology is seen as a powerful means for the fulfillment of desires, such as desires for food, luxury products, travel, or social interaction. On an objective list or eudaimonic view, finally, technology is seen as a means to more successfully secure items on the objective list or to realize one's potential and perfect oneself.

Enlightenment optimism about technology and its implications for well-being was countered in the 20th century. The early 20th century saw the widespread expansion of the industrial sector, the establishment of rationalized production processes and labor patterns, rapid urban growth, and the emergence a consumer society based on mass production and marketing. It became clear during the 20th century that modern technology did bring achievements, but also had major drawbacks. Technology was used on a large scale for the purpose of war and persecution, and enabled the unprecedented destruction that resulted from World Wars I and II, including the atrocities of Auschwitz and Hiroshima. In addition, many technological developments have proven harmful to nature or to health and have created environmental problems that are a threat to humankind. The promised improvements in the quality of life often appeared to be ambiguous as well. Labor processes were rationalized and therefore became more monotonous and impersonal. And although the consumer society brought many benefits, it has also been claimed to yield exuberant materialism and a loss of spiritual values and feelings of community.

Social theorists and philosophers of the 20th century therefore often arrived at negative evaluations of modern technology and the industrial society that evolved with it. They criticized the Enlightenment's philosophy of control and the idea that technology was predominantly good. They emphasized the negative and destructive nature of technology and posited that, rather than being freed by it, humankind was being made subservient to technology. They also declared that humanity had lost control of technology, which had now developed according to its own logic, and that rather than being improved, the quality of life was often worsened by processes of rationalization, uniformity, alienation, and shallow consumption.

This negative attitude toward technology and the industrial society was voiced strongly in critical theory, an influential social-philosophical approach initiated by representatives of the so-called Frankfurt School, a group of

German thinkers that, from the 1930s onward, engaged in influential social criticism. Its members include Theodor Adorno, Max Horkheimer, Herbert Marcuse, and Jürgen Habermas. They built on the work of Karl Marx and Max Weber, who, as the founder of sociology, held that bureaucratic organizations built after the Industrial Revolution propagated a new form of suppression through their rationalization of labor practices, an 'iron cage' that limited human potential. One of the most influential works in critical theory was *One-Dimensional Man* by Herbert Marcuse (1964), a work that was one of the spearheads of the 1960s counterculture. In it, Marcuse argued that advanced industrial society has imprisoned humankind in a system of production and consumption in which people are held captive in monotonous jobs to buy ever-more new products, and which has rendered critical thinking and conduct impossible.

Martin Heidegger, one of the most influential philosophers of the 20th century, held a similarly pessimistic view, arguing that modern technology has infiltrated our entire way of thinking and feeling and had turned humankind and the world into standing reserves, commodities with a utility value. His vision has been taken up by neo-Heideggerians such as Albert Borgmann and Hubert Dreyfus. A similar vision is found in the work of philosopher and sociologist Jacques Ellul, who portrayed technology as an unstoppable autonomous force that constructed social and political institutions according to its own logic and undermines the self-determination of humanity. Modern technology has also been seen as negative by other 20th-century postmodernist philosophers, such Jean-François Lyotard and Jean Baudrillard.

In contemporary philosophy, Albert Borgmann is well-known for his critique of technology and of consumer culture. Borgmann (1984) argues that modern life is concentrated around devices whose sole function is to produce available commodities, such as water, transportation, and entertainment, and to do so rapidly, easily, safely, and ubiquitously. As a result, he argues, people lose engagement with their environment and with each other. To regain engagement and the good life, he argues, we must choose more often for focal practices, such as gardening, teaching, and social activities, and use low-tech, "focal" artifacts in them. Eric Higgs, Andrew Light, and David Strong (2000) present detailed critiques of Borgmann's views and develop alternative philosophical perspectives on the implications of modern technology for our pursuit of a good life.

In economics and psychology, the topic of consumer culture, with its emphasis on acquiring, using, and displaying consumer products, is perhaps also the main way in which technology is being related to well-being. Contemporary consumer products are, after all, the fruits of modern technology. Psychologists, economists, sociologists, historians, and cultural theorists have been developing mostly critical assessments of consumer culture and its implications for well-being.

In economics, the stage for such critical assessments was set in the 1970s by Easterlin's paradox of happiness and Scitovsky's *The Joyless Economy*, both of which questioned the existence of a correlation between well-being and affluence. Some of the recent work in happiness economics maintains their focus on income, affluence, and consumption. Not all of this research is critical of consumer society, however. American economist Stanley Lebergott (1993), for example, argues that people lead easier, happier lives in contemporary society than do people in low-tech societies and lauds the amount of choice and convenience offered in consumer societies.

Psychological studies of consumer culture and well-being focus on the manufacture and manipulation of desire and the implications of consumerism and materialism for self-esteem, emotional well-being, feelings of security and anxiety, mental stability, and social relations (Kasser and Kanner 2004). Kasser (2002) presents studies that indicate that people whose lives are focused on the accumulation of wealth and material possessions have a significantly greater risk of unhappiness, including anxiety, depression, low self-esteem, and problems with intimacy. Dittmar (2008) analyzes how much modern identities have become intertwined with material possessions and argues, like Kasser, that consumerism does not lead to happiness. She holds that consumer products are often acquired to repair or complete people's identities, but usually fail to do so.

The information revolution and the rise of the Internet have led to studies of the relation between information technology and well-being, particularly in psychology (Joinson 2003; Amichai-Hamburger 2009). These studies focus on the consequences of using information technology for social relations, intimacy, personal development, and emotional well-being. Psychologists have identified both positive and negative effects. Turkle (2011) is mostly critical, she argues that information technology (and technology generally) overwhelms and depletes us by the choices it offers and the lives it makes possible. In addition, it makes us more lonely, even if we become connected to more people. Joinson (2003) balances this perspective by emphasizing how online communities and relations can also enrich our lives and allow us to experiment with and develop our identity.

The relation between technology and well-being is also a theme in studies of the quality of working life, which consider job satisfaction, work stress, and work-related well-being. In this research, much attention has gone to the impact of automation and deskilling and the role of technology in work-related stress. Well-being and technology are also being studied in health care, where the concept of "health-related quality of life" is used. This is an individual's satisfaction or happiness with domains of life that fall under the responsibility of health care providers and health care systems. Some of these studies also consider the impact of new medical technologies. Well-being is sometimes also being considered in technology assessment, a field of study in which the effects of technology are evaluated and anticipated for industry, the environment, and society. In engineering design, finally, well-being is not often taken into account, but it has recently become a topic in value-sensitive design (Friedman, Kahn, and Borning

2006), an approach to design which seeks to integrate considerations of value into design processes.

Although there is a variety of studies of the relation between technology and well-being, much of this work is limited in scope or volume, and much progress is still to be made. In particular, many more studies are needed of the implications of particular technologies and technological products for different dimensions of well-being. Also, a better understanding is needed of the interplay between technology, culture, and social structure as they affect well-being. Currently, limitations also exist regarding how well-being can be made a factor in technology policy. Much progress is also to be made in engineering design, in which the relation between design and well-being is currently hardly being thematized. Much is to be gained, also, from more collaboration between the disciplines: philosophy, psychology, and economics, as well as policy studies and engineering.

REFERENCES

Amichai-Hamburger, Yair. (Ed.). (2009). *Technology and Psychological Well-being*. New York: Cambridge University Press.

Bentham, Jeremy. (1996). *An Introduction to the Principles of Morals and Legislation* (J. Burns and H. L. A. Hart, eds., Introduction by F. Rosen). Oxford: Clarendon Press. (Original work published 1789)

Borgmann, Albert. (1984). Technology and the Character of Contemporary Life: A Philosophical Inquiry. Chicago: University of Chicago Press.

Brülde, Bengt. (2007). "Happiness and the Good Life: Introduction and Conceptual Framework," *Journal of Happiness Studies*, vol. 8, no. 1, pp. 1–14.

Bruni, Luigino, and Pierluigi Porta. (2005). *Economics and Happiness: Framings of Analysis*. Oxford: Oxford University Press.

Cooter, Robert, and Peter Rappoport. (1984). "Were the Ordinalists Wrong about Welfare Economics?" *Journal of Economic Literature*, vol. 22, pp. 507–530.

Crisp, R. (2008) "Well-Being," in E. N. Zalta (ed.) Stanford Encyclopedia of Philosophy, <http://plato.stanford.edu/archives/win2008/entries/well-being/>

Csikszentmihalyi, Mihaly. (1990). *Flow: The Psychology of Optimal Experience*. New York: Harper and Row.

Descartes, Rene. (1994). *Discourse on Method*. New York: Everyman Paperbacks. (Original work published 1637)

Diener, Ed. (1984). "Subjective Well-Being," *Psychological Bulletin*, vol. 95, pp. 542–575.

Diener, Ed, and Richard E. Lucas. (1999). "Personality and Subjective Well-being," in *Well-Being: The Foundations of Hedonic Psychology*, edited by Kahneman, Daniel, Ed Diener, and Norbert Schwarz (New York: Russell Sage Foundation Press) pp. 213–229.

Diener, Ed, Richard Lucas, Uli Schimmack, and John F. Helliwell. (2009). *Well-Being for Public Policy*. New York: Oxford University Press.

Dittmar, Helga. (2008). Consumer Society, Identity, and Well-Being: The Search for the 'Good Life' and the 'Body Perfect.' European Monographs in Social Psychology Series, edited by Rupert Brown. London: Psychology Press.

Easterlin, Richard, A. (1974). "Does Economic Growth Improve the Human Lot? Some Empirical Evidence," in *Nations, Households, and Economic Growth:*

Essays in Honor of Mozes Abramowitz, edited by David, Paul A. and Melvin W. Reder (New York: Academic Press).
The Economist. (2011). The Economist Intelligence Unit's Quality-of-Life Index: The World in 2005. *The Economist*. Retrieved July 25, 2011, from http://www.economist.com/media/pdf/QUALITY_OF_LIFE.pdf
Finnis, John. (1980). *Natural Law and Natural Rights*. Oxford: Clarendon Press.
Frey, Bruno S., and Alois Stutzer. (2002). *Happiness and Economics*. Princeton, NJ: Princeton University Press.
Friedman, Batya, Peter Kahn, and Alan Borning. (2006). "Value Sensitive Design and Information Systems," in *Human-Computer Interaction in Management Information Systems: Foundations*, edited by Zhang, Ping and Dennis Galletta (Armonk, NY: M. E. Sharpe).
Gallup Inc. (2010). Gallup Global Wellbeing: The Behavioral Economics of GDP Growth. Retrieved from http://www.gallup.com/poll/File/126965/GlobalWellbeing_Rpt_POLL_0310_lowres.pdf
Griffin, James. (1986). *Well-Being: Its Meaning, Measurement, and Moral Importance*. Oxford: Clarendon Press.
Hamilton, Lawrence A. (2003). *The Political Philosophy of Needs*. Cambridge: Cambridge University Press.
Higgs, Eric, Andrew Light, and David Strong. (2000). *Technology and the Good Life?* Chicago: University of Chicago Press.
Hurka, Thomas. (1996). *Perfectionism*. Oxford: Oxford University Press.
Joinson, Adam. (2003). *Understanding the Psychology of Internet Behaviour: Virtual Worlds, Real Lives*. Basingstoke, UK: Palgrave Macmillan.
Kahneman, Daniel. (1999). "Objective Happiness," in *Well-being: Foundations of Hedonic Psychology*, edited by Kahneman, Daniel, Ed Diener, and Norbert Schwarz (New York: Russell Sage Foundation Press) pp. 3–25.
Kahneman, Daniel, Ed Diener, and Norbert Schwarz. (1999). *Well-Being: Foundations of Hedonic Psychology*. New York: Russell Sage Foundation Press.
Kasser, Tim. (2002). *The High Price of Materialism*. Cambridge, MA: MIT Press.
Kasser, Tim, and Allen Kanner. (Eds.). (2004). *Psychology and Consumer Culture: The Struggle for a Good Life in a Materialistic World*. Washington, DC: American Psychological Association.
Kitcher, Philip. (1999). "Essence and Perfection," *Ethics*, vol. 110, no. 1, pp. 59–83.
Lebergott, Stanley. (1993). *Pursuing Happiness: American Consumers in the Twentieth Century*. Princeton, NJ: Princeton University Press.
Marcuse, Herbert. (1964). *One-Dimensional Man*. Boston: Beacon Press.
Mercer. (2010). "Quality of Living Worldwide City Rankings 2010: Mercer Survey." Retrieved May 26, 2010, from http://www.mercer.com/press-releases/quality-of-living-report-2010
Mill, John S. (1998). *Utilitarianism* (R. Crisp, ed.). Oxford: Oxford University Press. (Original work published 1863)
Nozick, Robert. (1974). *Anarchy, State, and Utopia*. New York: Basic Books.
Nussbaum, Martha C. (2000). *Women and Human Development: The Capabilities Approach*. Cambridge: Cambridge University Press.
Nussbaum, Martha C., and Amartya K. Sen. (1993) *The Quality of Life*. Oxford: Clarendon Press.
Organisation for Economic Co-operation and Development. (2011). OECD Better Life Initiative. Retrieved from http://www.oecdbetterlifeindex.org/
Parfit, Derek. (1984). *Reasons and Persons*. New York: Oxford University Press
Scitovsky, Tibor. (1976). *The Joyless Economy: An Inquiry into Human Satisfaction and Consumer Dissatisfaction*. Oxford: Oxford University Press.
Seligman, Martin E. P. (2002). *Authentic Happiness: Using the New Positive Psychology to Realize Your Potential for Lasting Fulfillment*. New York: Free Press.

Seligman, Martin E. P., and Mihaly Csikszentmihalyi. (2002). "Positive Psychology: An Introduction," American Psychologist, vol. 55, pp. 5–14.
Sen, Amartya. (1980). "Equality of What?" in *The Tanner Lectures on Human Value*, edited by McMurrin, Sterling M. (Salt Lake City: University of Utah Press) pp. 195–220.
Sen, Amartya. (1993). "Capability and Well-Being," in *The Quality of Life*, edited by Nussbaum, Martha C. and Amartya Sen (Oxford: Oxford University Press) pp. 30–53.
Sumner, Leonard W. (1996). *Welfare, Happiness, and Ethics*. Oxford: Clarendon Press.
Turkle, Sherry. (2011). *Alone Together: Why We Expect More from Technology and Less from Each Other*. New York: Basic Books.
United Nations Development Program. (2011). "The Human Development Index." Retrieved from http://hdr.undp.org/en/statistics/hdi/
Veenhoven, Ruut. (1996). "The Cross-Cultural Pattern of Happiness: Test of Predictions Implied in Three Theories of Happiness," *Social Indicators Research*, vol. 34, pp. 33–68.

2 Theorizing Technology

Carl Mitcham and Adam Briggle

What is technology? The question is more complex than may initially appear. The term "technology" itself carries different meanings across science and engineering, the humanities, and the social sciences, giving rise to a variety of entanglements. Here, we review complexities with an eye toward advancing reflection on relations between technology and the good life.

THE NEED FOR THEORY

Technology has become our dominant way of taking up with reality: our eyes fixed on digital screens, our skin covered in synthetic fibers, our fingers engaged with keypads, our ears pressed against mobile phones, our stomachs filled with processed foods, and our blood laced with pharmaceuticals. We increasingly inhabit what historian of technology Thomas P. Hughes (2004) calls a "human-built world." Not only are ever-more humans residing in urban environments, but even the furthest reaches of the planet bear the traces of human making and using. Technology has become so influential that some geologists call our age the "Anthropocene" to signify that humanity is now the most powerful shaper of the Earth. Pessimists tend to side with Marshall McLuhan, who has written that humans are increasingly reduced to functioning as the "sex organs of the machine world" (McLuhan 1967, p. 46). Optimists may side with Ray Kurzweil (2005), who foresees a near future when humans will live forever with transformed bodies and cyber minds.

Most people, however, find themselves somewhere in between—or drawn first to one side then the other. Above all, we are bewildered by the complexity and speed of the technological age. Modern societies find themselves dependent on technology but unable to effectively manage its wily twists and turns. Practices of technology assessment (TA) and other efforts to guide technological development toward more desirable goals are on the rise. Yet for such efforts to succeed, we must have a way to see through the tangle and arrive at some understanding of what technology is and how it relates to good lives and good societies.

Numerous efforts have been made to present such a measured theoretical assessment of technology. The pursuit of theory in this sense is often denigrated as the impotent love of abstraction—yet such an assertion itself contains an implicit theory about what is and is not important. Theory aims to provide orientation in the face of a wealth of competing experiences that cannot be equally accepted or affirmed. Philosopher of technology Albert Borgmann translates the Greek *theoria* as a "calm and resourceful vision of the world" (Borgmann 1984, p. 6). In the midst of multiple countervailing interpretations of technologically mediated experience, the conscious articulation of theory is perhaps more important than ever before. We are in need of theory to navigate a jumble of automobiles, airplanes, and iPods: What does it all mean in terms of our ultimate concern, namely, living a good life?

Driven by this need for vision, many thinkers have proposed theories of technology. Highlighting examples since the late 19th century emergence of the philosophy of technology, the principle feature of technology has been argued to be:

- "organ projection" (Kapp 1877) as well as "organ relief," "organ substitution or replacement," and "organ strengthening or improvement" (Gehlen 1940)
- "the totality of means employed as opposed to the meaning or end to which the action is, in the last analysis, oriented" (Weber 1921), "how man does or makes" (Drucker 1959), "practice, the way we do things around here" (Franklin 1992), or "humanity at work" (Pitt 2000)
- the inventive realization of transcendent forms or the creation of "real being from ideas" (Dessauer 1927 and 1956)
- "tactics for living" (Spengler 1931), means for molding the environment (Jaspers 1949), or "the pursuit of life by other means" (Stiegler 1994)
- means for the realization of "the *Gestalt* of the worker" (Jünger 1932) or any supernatural self-conception (Ortega 1939)
- "self-initiated salvation" (Brinkmann 1945)
- a "provoking, setting-up disclosure of nature" (Heidegger, 1954)
- synonymous with "man" and "well-being" (Ortega 1961; see chapter 4 by Jeroen van den Hoven in this volume)
- the "*totality of methods rationally arrived at and having absolute efficiency* (for a given stage of development) in *every* field of human activity" (Ellul 1964, his italics)
- the pursuit of power (Mumford 1967, 1970)
- sensorimotor skills (Feibleman 1966)
- applied science (Bunge 1966)
- the pursuit of increased effectiveness (Skolimowski 1966)
- means for socially set purposes (Jarvie 1966)
- "systematic application of scientific or other organized knowledge to practical tasks" (Galbraith 1967) or "knowledge of techniques, methods, and designs that work" (Rosenberg 1982)

- "any systematized practical knowledge, based on experimentation and/or scientific theory, which enhances the capacity of society to produce goods and services, and which is embodied in productive skills, organization, or machinery" (Gendron 1977)
- "a problem-solving process" (DeGregori 1985) or the "invention, development, and cognitive deployment of tools and other artifacts, brought to bear on raw materials and intermediate stock parts, with a view to the resolution of perceived problems" (Hickman 2001)
- "practical implementations of intelligence" (Ferré 1988)
- "humankind's modification of its biological and physical surroundings" (or "the human modification of the natural world") (Tenner 1996) and of ourselves (Tenner 2003)
- constituted by "the creative activities, individual and collective, of craftsmen, mechanics, inventors, engineers, designers, and scientists" (Hughes 2004)

Given the resulting welter of sometimes overlapping theories, scholars have naturally attempted syntheses (see Mitcham and Briggle 2009). Some have rejected such efforts, as when Joseph Agassi argued that "technology in general is not definable in any narrow clear-cut definition" (Agassi 1985, pp. 23–24). Most philosophical reflections on technology nevertheless include some synthetic drafting of a core definition; on the grounds that failure to do so creates problems (see Winner 1977).

Robert McGinn, in one relevant effort to draft a synthetic description, treated "technology as a *form of human activity* [comparable to] science, art, religion, and sport" (McGinn 1978, p. 180). The key characteristics of this activity are that it (1) has material outcomes, (2) fabricates or is constitutive of those outcomes, (3) is purposive, (4) is resource based and resource expending, (5) utilizes or generates knowledge, (6) is methodological, (7) takes place in a sociocultural-environmental context, and (8) is influenced by individual practitioners' mental sets. Characteristics (1) and (2) can be interpreted as incorporating aspects of Dessauer's theory; (3) as able to incorporate Engelmeier, Drucker, Pitt, and others; (4) as referencing Heidegger; and so on. As McGinn summarizes in a later paper, "technology can be characterized as that form of cultural activity devoted to the production or transformation of material objects, or the creation of procedural systems, in order to expand the realm of practical human possibility" (Hannay and McGinn 1980, p. 27).

In like manner, Carl Mitcham argued that "Technology is pivotally engaged with the human." As such, it should be considered as part of a philosophical anthropology "with differences drawn between its manifestations in the mind, through bodily activities, and as independent objects that take their place in the physical and social world" (Mitcham 1994, p. 159). From this, Mitcham derived a fourfold taxonomy of technology as knowledge, volition, activity, and object. The first two modes refer to the human interior (cognition and willing). Knowledge can range across sensorimotor skills and rules of thumb to descriptive laws, technological rules, and theories. The volitions involved can include the will to survive, the will to control or power, the will to freedom, the pursuit of efficiency, and efforts to realize a *Gestalt*, idea, or ideal such as well-being or the good life. The third mode refers to the embodied actions of making (crafting, constructing, inventing, designing, manufacturing, working, operating, and maintaining), and the fourth to the artifacts that compose human material culture (clothes, utensils, structures, apparatus, utilities, tools, machines, and automata).

In still a third case in point, Don Ihde, on the ground that "definitions are not neutral," cautions against any narrow stipulative response to ambiguity and breadth. His own definition is very broad:

> First, we shall insist that a technology must have some concrete component, some material element. . . . [S]econd, a technology must enter into some set of praxes—"uses". . . . [T]hird, we shall take as part of the definition, a relation between the technologies and the humans who use, design, make, or modify the technologies in question. (Ihde 1993, p. 47)

Finally, after a survey of different approaches to a definition in general, Val Dusek adopts what he terms a "systems definition as a consensus definition." "The technological system is the complex of hardware (possibly plants and animals), knowledge, inventors, operators, repair people, consumers, marketers, advertisers, government administrators, and others involved in technology" (Dusek 2006, p. 35).

For Brian Arthur, technology should be understood variously as "*a means to fulfill a human purpose*," "an *assemblage of practices and components*," and "the entire *collection of devices and engineering practices available to a culture*" (Arthur 2009, p. 28, his italics). One paradoxical implication of any sorting of technology into different aspects is to affirm persistent multiplicity in technology itself. The one and the many of technology are dialectically inseparable. As Jacques Ellul argued, however, the many cannot be left as they are. Their manyness likewise points toward some kind of unity; they are a many of some kind. Although some scholars "declare that 'technology' does not exist, and that they know only plural technolog*ies*," this fails to recognize how "technology as a concept allows us to understand a set of phenomena that remain invisible even where technologies are perceptibly manifest." Nevertheless, Ellul admits, "while the concept may be indispensable for comprehension, it is by no means clear and simple in itself" (Ellul 1977, p. 23).

APPROACHING DEFINITION

Theories of technology involve definitions of technology. On the one hand, definitions are necessary to identify what is to be theorized; on the other, by selectively highlighting aspects of phenomena, definitions themselves fail to be theoretically neutral. Any definition implicates theory. Definitions are, as it were, theories writ small.

Recognizing this mutual implication between definition and theory, the construction of definitions has itself been theorized in terms of its own distinctions and methods. Definitions can be ostensive (pointing or exhibiting with examples of something) or connotative (identifying the basic features that something must have to allow it to be pointed out). Connotative definitions in turn can be etymological, essential, prescriptive, and linguistic. Etymological definitions appeal to the historical evolution of words as a way to elucidate and distinguish their connotations. Essential definitions are efforts to disclose phenomena by means of a genus and species structure. Prescriptive definitions are positive assertions about the denotative boundaries of words and their conceptual referents. Linguistic definitions explore and sort out current usages as a means to select reasonably acceptable semantic norms.

With regard to method, the scholastic tradition in European philosophy proposed a set of criteria for good definitions. They should fully cover the *definitum* (phenomenon being defined); the *definitum* should not reappear in the definition; and the definition should avoid ambiguous terms. A previous review of distinctions and method in relation to technology nevertheless concluded that

> even following these rules, definitions can sometimes be inadequate to their contexts. Pragmatic consideration of how a definition fits in or serves the context in and for which it is formulated thus constitutes another approach to definition—one that is able to utilize, as appropriate, etymological, essential, prescriptive, or linguistic approaches. (Mitcham and Schatzberg 2009, p. 32)

This argument suggests that in a context concerned with relations between technology and the good, it may useful to be aware as well about different theories of the good.

Theories of the good can themselves be sorted into teleological, deontological, and consequentialist approaches—depending on whether the focus is on actors, actions, or the results of action. In the first instance, the good is theorized in terms of virtue understood as the proper functioning of human nature; in the second in terms of the universalizability or rationality of the maxims guiding action; in the third as aggregate social happiness understood in terms either of subjective well-being (the extent to which people feel happy) or objective social indicators (GDP, life span, etc.).

In an analogous manner, technology can be approached in terms of physical artifacts, processes or techniques, and social implications. In the initial case, technology is thought as a distinctive kind of physical object (something other than nature); in the second as a distinctive type of human action (often emphasizing functional efficiency); in the third as a distinctive social order understood either objectively (mass production) or subjectively (consumer society).*Techne*, Technology, and Engineering

The term "technology" combines two Greek words, *techne* (productive skill or art) and *logos* (rational speech). For the Greeks, however, there was never any serious effort to develop a *logos* or logic of *techne* or making in the sense so characteristic of modernity. Instead, *techne* or technics was understood as an intellectual virtue (*arête*). A virtue is a *hexis* (Latin *habitus*) or stable disposition to act in a way that realizes a perfection that otherwise remains only potential. The human potencies at issue can be appetitive or rational. Exemplifying the appetitive virtues, courage is constituted by the *habitus* of possessing the right mixture of fear and daring. Exemplifying intellectual virtues, *techne* is "a *habitus* possessing true *logos* concerned with making (*poesis*)" (Aristotle *Nicomachean Ethics* VI, 4; 1140a11). Although *techne* possesses *logos*, there is no *logos* of *techne* any more than there could be a fear or daring of courage. For the Greeks, making may have exhibited its own inherent rationality but there was no rational or systematic way to produce new forms of making.

Human well-being or flourishing—a better translation of *eudaimonia* than the common "happiness"—takes form through character, which is constituted in turn by whatever mixture or proportion of the virtues may be present in a particular person. The good of a social order is likewise manifest through whatever balance of character types predominates within it. The point at issue is initially descriptive. Yet insofar as it is possible to argue that some forms of flourishing and character are superior to others, the issue becomes normative.

The Greek understanding of *techne* or skillful artisanship situates it as one aspect of flourishing. According to Aristotle, the three general forms of flourishing are those dedicated to physical pleasure, social honor, and intellectual knowledge. Augustine and subsequent philosophers imbued with revelation, added a fourth, dedication to faith. Insofar as *techne* is an intellectual virtue ordered toward making, its artisan possessor partakes of the goods of both knowledge and pleasure. When Socrates questions the poets, politicians, and artisans about their wisdom, he even acknowledges that of the three, artisans are the most truly knowledgeable (*Apology* 22d). But in the Greek context—and for two thousand years afterward in the European tradition—*techne* was so embedded in cultures of social honor (as among the Romans) or faith (among Jews, Christians, and Muslims) that it remained a strictly limited virtue. *Techne* was never a virtue that predominated in more than a restricted number of subordinate individuals, nor was it able to characterize any social order as a whole.

The emergence of technology in the modern sense of a systematically pursed making and using took place in conjunction with a shift in emphasis from human flourishing in terms of honor or faith toward a view that privileged physical pleasure in terms of use and convenience. This shift is particularly apparent in the displacement of the traditional technics-using artisan by the technological profession of the engineer. The former could be described as a maker and user of tools, the latter as the designer and operator of machines.

Preparing the way for the rise of engineering as a new form of technology was an argument for the transformation of knowledge or science that would integrate it with productive activity. Francis Bacon, for instance, argued that knowledge should be transformed into an indefinite, self-validating project to conquer nature for "the relief of man's estate" (*Novum Organum* 1620, LII). Rejecting the more contemplative science of the ancients, Bacon argued that "human knowledge and human power meet in one" and that "the nature of things betrays itself more readily under the vexations of art than in its natural freedom" (Bacon, "Plan of the Work," paragraph 21). René Descartes similarly articulated a promise of the new sciences to "render ourselves the masters and possessors of nature," which will provide "an infinity of arts and crafts which enable us to enjoy without any trouble the fruits of the earth and all the good things which are to be found there" (*Discourse on Method* 1637, VI).

The prophecy of a formerly limited technics expanding indefinitely set the stage for the rise of technology as engineering. The first people explicitly denominated "engineers" (from the Latin *ingeniatorum*, one who is ingenious) were members of the military who designed and operated fortifications and various "engines of war," such as battering rams and catapults. The first institutions of engineering education were created by national governments and closely linked with the military (in contrast to universities that had been closely linked to the church). Engineers began to detach themselves from the military context during the Industrial Revolution (1750–1850 in Great Britain).

In the 1828 charter formally establishing the Institution of Civil Engineers (including any engineers who did not work for the military), engineering was defined as "the art [read *techne*] of directing the great sources of power in nature [as conceived and discovered by the new sciences of Bacon and Descartes] for the use and convenience of man" (quoted from Mitcham and Muñoz 2010, p. 3). The term "use and convenience" was an almost technical term associated with utilitarian ethics during the same period. David Hume, for example, observed that "a machine, a piece of furniture, a vestment, a house well contrived for use and conveniency" is for that very reason considered beautiful; he further argued for use and convenience under the general heading of "utility" as the foundation of morals (Hume 1751, II, part 2). According to Hume, "the ultimate reason for every [moral] rule [is] the convenience and necessities of mankind" (Hume 1751, III, part 2). For the Institution of Civil Engineers, use and convenience indicate a nonproblematic conception of human flourishing that is to be met by engineering embedded in industrial and commercial interests.

TECHNICS, INDUSTRIAL ARTS, TECHNOLOGY, AND MORE

Enmeshed historically in the English-language discourse about engineering as technology are a host of related terms, such as "machines," "mechanical arts," "mechanization," "industry," and "industrial arts"—and, more recently, "advanced-" and "postindustrial society." Even more significant are terminological differences from other major European languages that have distinct philosophical implications. Although the English "engineering" can be translated into the German and French cognates, *Ingenieurwesen* and *ingénierie*, respectively, the German *Technik* and French *génie* were, until the mid-20th century, more common equivalents. The German and French cognates of "technology," that is, *Technikologie* and *technologie*, were both reserved to reference the general study of *Technik* or *technique*. German, French, and other European languages thus allowed an easy distinction between technics and technology or the logical analysis of technics that is at best awkward in English (see Sebestik 1983, and Salmon 1984). As historians Ruth Oldenziel (2006), Eric Schatzberg (2006), and others have pointed out, the English term "technology" only acquired its current English valence in the early 1900s.

The replacement in English of "industrial arts" and related terms with "technology" was the work of a single thinker. As Schatzberg summarizes the issue, when U.S. economist Thorstein Veblen transplanted early 20th century German discussions about the social impact of *Technik* into the North American context, he subtly replaced the more common rendering of *Technik* as "industrial arts" with "technology," in part to emphasize the historical power and scientific associations of mechanized forms of production. As Oldenziel further observes, "the ascendency of *technology* as a keyword in the United States neatly parallels the emergence of America as a superpower committed to *technology* as the key tool for development in the rest of the world" (Oldenziel 2006, p. 485). Lewis Mumford's effort to hold onto "technics" as a transliteration of *Technik* (Mumford 1934), and then to distinguish between premodern "biotechnics" and modern "monotechnics" (Mumford 1967, 1970) instead of between technics and technology or premodern and modern technology was ultimately unsuccessful.

It was Veblen, too, who called American intellectual attention to and helped stimulate American development of the European cultural criticism of the industrial arts or *Technik* or technology. In England, John Ruskin and William Morris, criticizing the quality and cultural influence of mass production goods, gave rise to efforts to revive and adapt artisan traditions in the arts and crafts movement. In Germany, related criticisms, sometimes with a conservative Catholic flavor, fed into economic criticisms of *Technik*, stimulated engineering apologetics, and contributed to efforts to formulate a distinctive machine or technological aesthetic as exemplified in Bauhaus design. In France, promotion of *technologie* and rationalization of culture took a more nationalistic form. *Liberté, égalité*, and *fraternité* were complemented with *rationalité*. These multiple embeddings of technology or its cognates in distinct cultural discourses are easy to overlook when thinking about relations between technology and the good.

Under such circumstances, it is difficult to semantically disembed "technology" in ways that appropriately identify phenomenological dimensions selected as pragmatically relevant to particular arguments. In response, a number of authors have simply coined or adopted alternative terms, such as "technocracy" (beginning with Smyth 1919), *Technizismus* (or "technicism," Scheler 1926; Stanley 1978; Janicaud 1985), "technicification" and "technicization" (Huxley 1927), "technological rationality" (Marcuse 1964), "technostructure" (Galbraith 1967), "technotronic society" (Brzezinski 1970), *système techniciene* (or "technological system," Ellul 1977), "technopolis" (Calder 1979), "high-technology" (numerous socioeconomic studies from 1970s on), "technopoly" (Postman 1992), *technoscience* (or "technoscience," Hottois 1978/1996; Latour 1987), *technicité* (or "technicity," Combarnous 1984; Bradley and Armand 2006), and "technological consciousness" (Barry 1991). More recently, "innovation" often replaces "technology" in political and economic discourse. However, no alternative term has taken hold in any significant respect.

THEORETICAL VISIONS: DEPENDENCE, INDEPENDENCE, INTERDEPENDENCE

Technology remains framed in terms of distinct theoretical visions. We can identify three kinds: (1) technology is thought dependent on or embedded in other phenomena; (2) technology is thought independent of or disembedded from other phenomena; (3) the technological and nontechnological are thought interdependent.

Dependence

Illustrative of the first cluster of theories is the common saying, "Guns don't kill people; people kill people." Technologies are dependent means used by people to accomplish some predetermined end: a microwave heats food; birth control pills prevent pregnancy; automobiles transport people. Technologies manifest values of effectiveness with respect to a given function. They are otherwise neutral with regard to the wider practices and contexts in which they are deployed. They can be put to good or bad uses by good or bad people: a hammer can be used to construct a bookshelf or to bash in someone's head. The emphasis in such a view is on individual or social institutional users, on the ground that the technology is dependent and value-neutral.

Dependence theories of technology further emphasize the social determination of technology. There are at least two important options here. First, critical social theorists, such as Andrew Feenberg (1991), draw inspiration from Karl Marx to argue that what matters is not technology itself, which he admits has a quasi-independent character, but the social or economic system in which it is embedded. Modern technology has the potential to make wealth and freedom possible for everyone, but its historical appearance realized this dream only for the few capitalists. If the means of ownership could be reformed, then the same technologies would bring about radically different—and better—results. Second, the social construction of technology (SCOT) view argues that technologies are underdetermined by their technical criteria. There are multiple pathways open for any given problem, and social actors make the final choice from a batch of options. Technologies exhibit "interpretive flexibility": Facts about them are not objectively given but rather dependent on the interpretations of actors (Bijker, Hughes, and Pinch 1987).

Independence

The second cluster of theories sees technology as a phenomenon manifesting imperatives largely independent of context. Langdon Winner (1986), for instance, has defended a theory of semiautonomous technology by arguing that certain technologies *in themselves* have political properties. Railroads and nuclear power plants, for example, call forth hierarchical, centralized, and tightly coordinated command structures. Winner acknowledges with dependence theory that some technologies have an "original flexibility" and may not wholly determine a given social outcome. But he contends that the latitude of choice is often narrower than dependence theorists suggest. In the abstract, human relations may be unaffected by the introduction of hand guns; with or without hand guns, people pursue personal interests such as physical welfare, social recognition, power, or money with more or less effectiveness. But in reality, the introduction of guns has its own influence on human behavior and gives the emotionally excited otherwise absent means for inflicting serious harm in the course of pursuing personal interests, thus increasing the number of homicides in any given context. Indeed, in something of a paradox, the same people who think guns are neutral also regularly argue they are good because their widespread availability makes individuals able to defend themselves against criminals and protects against governmental tyranny.

Prior to Winner, Ellul (1954) advanced an analytic "characterology" of modern *Technique* (as opposed to traditional *techniques*) as exhibiting, across all social contexts, the features of rationality, artificiality, automatism, self-augmentation, unicity, universalism, and autonomy. On Ellul's character identification, all seven key characteristics can be found operative in the economy, the state, and even in the practice of "human techniques" associated with education, entertainment, sports, and medicine. The same year, Martin Heidegger (1954) advanced an ontological argument that

Technik is no simple means but a mode of revealing reality. Modern *Technik* is a "challenging revealing" that reduces the world to *Bestand,* or a standing reserve of materials available for exploitation. The hydroelectric plant sets on the Rhine River, transforming it into a power-supplier; the forest becomes board feet of lumber for homes and cellulose for paper. The greatest threat is that human beings are also challenged and ordered in this way to become "human resources."

In a lecture delivered shortly after World War II, however, Heidegger went so far as to equate modern factory agriculture with "the manufacturing of corpses in gas chambers" and the "production of hydrogen bombs" (quoted in Rockmore 1992, p. 241). These are all different expressions of the same *Gestell* or enframing that compels humans to inhabit a world of *Bestand,* or resources. But for all the problems with mechanized agriculture and nuclear weapons, such an easy equation of modern technology with mass murder seems to obscure more than it elucidates, while making it theoretically impossible to take action to moderate a sovereign influence. Dependence theories respond by calling attention to the real existence of important distinctions between, for instance, coal-fired power plants, solar panels, and wind turbines as sources of electricity, each of which has quite distinctive impacts on nature—and are likewise differentially promoted by distinctive governmental policies in diverse countries. Technology presents a bewildering complexity that is properly treated in a piecemeal, analytic, and context-sensitive fashion attuned to the empirical manifestations of different technologies (see Achterhuis 2001; Verbeek 2005).

INTERDEPENDENCE

Taking this argument farther, and in an effort to escape the dependent/independent duality, Bruno Latour (1993) proposed what may be termed a theory of interdependence. For Latour theories emphasizing both dependence and independence rely on impossible purifications of the categories of objects (nature, things) and subjects (culture, humans): nonhumans versus humans. Dependence theories claim that the latter dominate the former, whereas independence theories postulate the opposite. By contrast, Latour argues that just as nature and culture are interdependent categories, so are technological artifacts properly conceived as nodes in a network of interactions. Technological reality is composed of interdependent hybrids that themselves manifest interdependencies. For example, making an airplane fly requires a complex ensemble of humans and nonhumans. Indeed, even nonhuman things are "actants" that make a difference in the world: a traffic light makes cars stop just as a police officer does. So technologies are not the value-neutral and passive objects supposed by dependence theories. Yet whereas technologies "act," there is no unifying logic throughout the complex human-thing hybrids that would justify independence theory. To return to the archetypical gun, for Latour (1999) it is the "citizen-gun" or "gun-citizen" that kills. This composite actant takes on its own unique program of action or script and goal, both of which are different from the scripts and goals of any citizen or gun considered apart from each other.

A final way to consider interdependence theory is through technological metaphors. In the 1950s and 1960s, technological possibility tended toward the metaphorical form of a mushroom cloud of destruction and a DDT-silenced spring, leading to searches for limits to human dependence on technology. By contrast, since the 1990s, alternative guiding metaphors such as the networked computer and an ever-rising stock market have shifted interest toward surfing and catching the wave of the next big thing—and even stimulating waves, if necessary. To affirm the dependence of technology on our own social constructions becomes a paradoxical affirmation of our own dependence on technology. It now must expand for us to succeed; limits should be limited. Ethics should be used to help liberate and advance the process, because the technical and the human are hopelessly intertwined.

THEORETICAL VISIONS RELATED TO THE GOOD LIFE

Dependence and interdependence theories tend to argue a continuous historical thread leading from the current ensemble of technologies back to simple tools—even into nature. John Dewey, for instance, argued for the presence of proto-instrumentality (technology) in the most rudimentary forms of life. Any organism in the very process of living "struggles to use surrounding energies in its own behalf. It uses light, air, moisture, and the material of soil. To say that it uses them is to say that it turns them into means of its own conservation" (Dewey 1916, p. 1). Human technologies are simply extensions of this fundamental process. Dependence and interdependence theories further assume technological pluralism, criticizing any notion of a unified phenomenon called technology (in the singular and sometimes with a capital T). Human beings are tool makers and users who extend their manifold capacities through artifacts.

By contrast, independence theories often argue the existence of some radical break between premodern and modern technology—and to adopt a measure of holism, at least with regard to modern technology. Agreeing with Dewey, Hans Jonas (1966) granted organic roots to traditional technics, but then questioned whether this is equally true with regard to modern technology. At some point in history, the multiplicity of technics gave way to modern technology as a more unified phenomenon that now develops according to its own internal logic and exercises some degree of independent influence over individuals and on society. As there is "a distinction between technology as enslaving our nature versus technology as extending our nature" (Arthur 2009, p. 216; see also Fromm 1968). It is precisely on this basis that Jonas develops the imperative of responsibility: "Act so that the effect of our action are compatible with the permanence of genuine human life" (Jonas 1984, p. 11).

The most important point here, however, is that dependence and interdependence theories tend to be compatible with if not to reinforce liberal, individualist pluralism with regard to pursuit of the genuinely human or good life. The liberal state supposedly offers a neutral framework that allows a plurality of conceptions of the good to flourish—conceptions that each can choose from both a basic and an ever-expanding array of technological means to achieve their particular goods. For these theories, technology and liberalism partake in a straightforward relationship of providing and protecting individual freedom to pursue a plurality of goods.

In response, Albert Borgmann argues that "technology, in fact, does not take shape in a prohibitively complex way" (1984, p. 11). Dependence and interdependence theorists (which he terms "pluralists"), with their "minute and roving scrutiny," miss the forest for the trees. Borgmann argues that inherent to modern technology is a pattern (which he terms the "device paradigm") that is neither obvious nor exclusively dominant. Modern technology promises to liberate and enrich human lives. It realizes this promise by making commodities available in a characteristic manner that is instantaneous, ubiquitous, safe, and easy. In contrast to the work of chopping wood and building a fire, the central heating system makes heat available with the mere press of a button.

Borgmann's theory hinges on a distinction between "things" and "devices." A thing, such as the wood-burning stove, is dependent on its context and calls forth social and bodily engagement. A device, such as the furnace, independently produces heat in a way that disburdens users from the demands embedded in the context of chopping, hauling, and stoking. Living with devices leads to what Borgmann calls a "commodious existence," which constitutes an overarching pattern to modern life that becomes a default and predominant answer to the question of the good life (see Higgs, Light, and Strong 2000). (Note how the commodious life is a modern conception of the good life, as it emphasizes disburdenment from menial labor or the relief of hardship, whereas the ancients and Borgmann emphasize the active cultivation of virtues as central to the good life.) Insofar as liberal governments promote technological advance, they are far from neutral with respect to the good. In fact, they promote a commodious existence as the default answer to what it means to flourish and, in so doing, they marginalize other forms of life.

Borgmann's identification of a general pattern and critical assessment of it align with independence theory. But Borgmann does not argue that individuals should abandon modern technology or resign themselves to hopelessness. The experience of wilderness as a focal thing or practice, for example, actually calls for a technological context. Being attuned to such heightened experiences requires an intelligent and selective use of technology in the form of high-tech mountaineering gear—and the subsequent linking of artifacts into a new interdependent network. If chosen wisely, technology is able to be pursued via and to promote a more engaged, skilled, and intimate contact with the world. But this is a choice that can only be practiced with mindfulness. We shape technologies that also shape us. Additionally, we need to recognize how this co-construction takes place, not only as a collaboration between humans and nonhumans, but also among humans who, in order to practice their humanity, also think about what they are doing and making.

> We are always and already engaged in drawing the outlines of a common way of life, and we have to take responsibility for this fact and ask whether it is a good life, a decent life, or a lamentable life that we have outlined for ourselves. (Borgmann 2006, p. 6)

Kevin Kelly similarly notes how technologies such as the television distract people from higher pursuits and how others such as the automobile seem to get people to serve them and prod them into needing still more technologies, such as freeways, strip malls, and fast food joints. He felt that his well-being and authenticity required a minimalist approach to technology. Yet at the same time, "Once you gain your voice around technology and become more sure of what you want, it becomes obvious that some technologies are simply superior to others. . . . I became fascinated by the challenge of picking the few tools that might elevate my spirit" (Kelly 2010, pp. 2–3). It is the ultimate hope for a theory of technology that it provides a vision resourceful enough to allow individuals to, as it were, pick the right tools.

REFERENCES

Achterhuis, Hans. (Ed.). (2001). *American Philosophy of Technology: The Empirical Turn*. Bloomington: Indiana University Press.

Agassi, Joseph. (1985). *Technology: Philosophical and Social Aspects*. Dordrecht, Netherlands: D. Reidel.

Aristotle. (1999). *Nicomaclean Ethics*. Martin Ostwald, trans. Upper Saddle River, NJ: Prentice Hall.

Arthur, W. Brian. (2009). *The Nature of Technology: What It Is and How It Evolves*. New York: Free Press.

Bacon, Francis. (1996). "Plan of the Work" and "The New Organon" in *Collected Works of Francis Bacon*, James Spedding, ed. London: Routledge/Thoemmes Press.

Bijker, Wiebe, Thomas P. Hughes, and Trevor Pinch. (Eds.). (1987). *The Social Construction of Technological Systems: New Directions in the Sociology and History of Technology*. Cambridge, MA: MIT Press.

Borgmann, Albert. (1984). *Technology and the Character of Contemporary Life*. Chicago: University of Chicago Press.

Borgmann, Albert. (2006). *Real American Ethics: Taking Responsibility for Our Country*. Chicago: University of Chicago Press.

Bradley, Arthur, and Louis Armand. (2006). *Technicity*. Prague: Litteriaria Pragensia.
Brinkmann, Donald. (1945). Mensch *und Technik: Grundzüge einer Philosophie der Technik*. Berne, Switzerland: Franke.
Brzezinski, Zbigniew. (1970). *Between Two Ages: America's Role in the Technotronic Era*. New York: Viking Press.
Bunge, Mario. (1966). "Technology as Applied Science," *Technology and Culture*, vol. 7, no. 3, pp. 329–347.
Calder, Nigel. (1979). *Technopolis: Social Control of the Uses of Science*. London: MacGibbon and Kee.
Combarnous, Maurice. (1984). *Les techniques et la technicité*. Paris: Editions sociales.
Cooper, Barry. (1991). *Action into Nature: An Essay on the Meaning of Technology*. Notre Dame, IN: University of Notre Dame Press.
DeGregori, Thomas R. (1985). *A Theory of Technology: Continuity and Change in Human Development*. Ames: Iowa State University Press.
Descartes, René. (1968). *Discourse on Method and the Meditations*. F. E. Sutcliffe, trans. new York: Penguin.
Dessauer, Friedrich. (1927). *Philosophie der Technik: Das Problem der Realizierung*. Bonn: F. Cohen.
Dessauer, Friedrich. (1956). *Streit um die Technik*. Frankfurt am Main: Knecht.
Dewey, John. (1916). *Democracy and Education: An Introduction to the Philosophy of Education*. New York: Macmillan.
Drucker, Peter F. (1959). "Work and Tools," *Technology and Culture*, vol. 1, no. 1, pp. 28–37.
Dusek, Val. (2006). *Philosophy of Technology: An Introduction*. Malden, MA: Blackwell.
Ellul, Jacques. (1954). *La Technique ou l'enjeu du siècle*. Paris: Armand Colin.
Ellul, Jacques. (1964). *The Technological Society* (John Wilkinson, trans.). New York: Random House.
Ellul, Jacques. (1977). *Le Système technicien*. Paris: Calmann-Lévy.
Feenberg, Andrew. (1991). *Critical Theory of Technology*. New York: Oxford University Press.
Feibleman, James K. (1966). "Technology as Skills," *Technology and Culture*, vol. 7, no. 3, pp. 318–328.
Ferré, Frederick. (1988). *Philosophy of Technology*. Englewood Cliffs, NJ: Prentice Hall.
Franklin, Ursula M. (1992). *The Real World of Technology*. Toronto: Anansi.
Fromm, Erich. (1968). *The Revolution of Hope: Toward a Humanized Technology*. New York: Bantam Books.
Galbraith, John Kenneth. (1967). *The New Industrial State*. New York: Houghton Mifflin.
Gehlen, Arnold. (1940). *Der Mensch: Seine Natur und seine Stellung in der Welt*. Berlin: Junker and Dunnkampt.
Gendron, Bernard. (1977). *Technology and the Human Condition*. New York: St. Martin's Press.
Hannay, Bruce, and Robert McGinn. (1980). "The Anatomy of Modern Technology: Prolegomenon to an Improved Public Policy for the Social Management of Technology," *Daedalus*, vol. 109, pp. 25–53.
Heidegger, Martin. (1954). "Die Frage nach der Technik," in *Vorträge und Aufsätze* (W. Lovitt, trans.). Pfullingen, Germany: Neske.
Heidegger, Martin. (1977). *The Question Concerning Technology and Other Essays* (W. Lovitt, trans.). New York: Harper.
Hickman, Larry. (2001). *Philosophical Tools for Technological Culture: Putting Pragmatism to Work*. Bloomington: Indiana University Press.

Higgs, Eric, Andrew Light, and David Strong. (Eds.). (2000). *Technology and the Good Life?* Chicago: University of Chicago Press.

Hottois, Gilbert. (1996). "Ethique et techno-science," *Entre symboles et technosciences*. Paris: Presses Universitaires de France. (Original work published 1978).

Hughes, Thomas P. (2004). *The Human-Built World: How to Think about Technology and Culture*. Chicago: University of Chicago Press.

Hume, David. (1983). *An Enquiry Concerning the Principles of Morals*. J. B. Schneewind, ed. New York: Hackett.

Huxley, Aldous. (1927). *Proper Studies*. London: Chatto and Windus.

Ihde, Don. (1993). *Philosophy of Technology: An Introduction*. New York: Paragon House.

Janicaud, Dominique. (1985). *La puissance du rationnel*. Paris: Gallimard.

Jarvie, Ian C. (1966). "The Social Character of Technological Problems: Comments on Skolimowski's Paper," *Technology and Culture*, vol. 7, pp. 384–390.

Jaspers, Karl. (1949). *Von Ursprung und Ziel der Geschichte*. Zurich: Artemis-Verlag.

Jonas, Hans. (1966). *The Phenomenon of Life: Toward a Philosophical Biology*. New York: Harper and Row.

Jonas, Hans. (1984). *The Imperative of Responsibility: In Search of an Ethics for the Technological Age* (Hans Jonas and David Herr, trans.). Chicago: University of Chicago Press.

Jünger, Ernst. (1932). *Der Arbeiter: Herrschaft und Gestalt*. Hamburg: Hanseatische Verlagsanstalt.

Kapp, Friedrich. (1877). *Grundlinien einer Philosophie der Technik: Zur entstehungsgeschicte der Cultur aus neuen gesichtspunkten*. Braunschweig, Germany: Westermann.

Kelly, Kevin. (2010). *What Technology Wants*. New York: Viking.

Kurzweil, Ray. (2005). *The Singularity Is Near: When Humans Transcend Biology*. New York: Viking.

Latour, Bruno. (1987). *Science in Action: How to Follow Scientists and Engineers through Society*. Cambridge: Harvard University Press.

Latour, Bruno. (1993). *We Have Never Been Modern*. Cambridge, MA: Harvard University Press.

Latour, Bruno. (1999). *Pandora's Hope: Essays on the Reality of Science Studies*. Cambridge: Harvard University Press.

Marcuse, Herbert. (1964). *One-Dimensional Man: Studies in the Ideology of Advanced Industrial Society*. Boston: Beacon Press.

McGinn, Robert. (1978). "What Is Technology?" in *Research in Philosophy and Technology*, vol. 1, edited by Paul Durbin (Greenwich, CT: JAI Press) pp. 179–197.

McLuhan, Marshall. (1967). *Understanding Media: The Extensions of Man*. London: Sphere Books. Mitcham, Carl. (1994). *Thinking Through Technology: The Path between Engineering and Philosophy*. Chicago: University of Chicago Press.

Mitcham, Carl, and Adam Briggle. (2009). "The Interaction of Ethics and Technology in Historical Perspective," in Handbook of the Philosophy of Science, vol. 9: *Philosophy of Technology and Engineering Sciences*, edited by Anthonie Meijers (Amsterdam: Elsevier) pp. 1147–1191.

Mitcham, Carl, and David Muñoz. (2010). *Humanitarian Engineering*. San Rafael, CA: Morgan and Claypool.

Mitcham, Carl, and Eric Schatzberg. (2009). "Defining Technology and the Engineering Sciences," in Handbook of the Philosophy of Science, vol. 9: *Philosophy of Technology and Engineering Sciences*, edited by Anthonie Meijers (Amsterdam: Elsevier) pp. 27–63.

Mumford, Lewis. (1934). *Technics and Civilization*. New York: Harcourt Brace.

Mumford, Lewis. (1967, 1970). *The Myth of the Machine*. 2 vols. New York: Harcourt Brace and World.

Oldenziel, Ruth. (2006). "Signifying Semantics for a History of Technology," *Technology and Culture*, vol. 47, no. 3, pp. 477–485.

Ortega Y Gasset, José. (1939). "*Meditación de la técnica*." In *Obras completas*, vol. 5. Madrid: Alianza.

Ortega Y Gasset, J. (1961). "Man the Technician," in *History as a System and Other Essays Toward a Philosophy of History* (H. Weyl, trans.) (New York: W. W. Norton) pp. 87–164.

Pinch, Trevor, and Wiebe Bijker. (1987). "The Social Construction of Facts and Artifacts: Or How the Sociology of Science and the Sociology of Technology Might Benefit Each Other," in *The Social Construction of Technological Systems: New Directions in the Sociology and History of Technology,* edited by Bijker, W. E., T. P. Hughes, and T. J. Pinch (Cambridge, MA: MIT Press) pp. 17–50.

Pitt, Joseph C. (2000). *Thinking About Technology: Foundations of the Philosophy of Technology*. New York: Seven Bridges.

Plato. (1997). "Apology," in *Plato: Complete Works*. John M. Cooper, ed. Indianapolis, IN: Hackett.

Postman, Neil. (1992). *Technopoly: The Surrender of Culture to Technology*. New York: Knopf.

Rockmore, Tom. (1992). *On Heidegger's Nazism and Philosophy*. Berkeley: University of California Press.

Rosenberg, Nathan. (1982). *Inside the Black Box: Technology and Economics*. Cambridge: Cambridge University Press.

Salmon, Jean-Jacques. (1984). "What Is Technology? The Issue of Its Origins and Definitions," *History and Technology*, vol. 1, no. 2, pp. 113–156.

Schatzberg, Eric. (2006). "*Techik* Comes to America: Changing meanings of *Technology* before 1930," *Technology and Culture*, vol. 47, no. 3, pp. 486–512.

Scheler, Max. (1926). *Die Wissensformen und die Gesellschaft*. Leipzig: Der neue Geist.

Sebestik, Jan. (1983). "The Rise of the Technological Sciences," *History and Technology*, vol. 1, no. 1, pp. 25–43.

Skolimowaski, Henryk. (1966). "The Structure of Thinking in Technology," *Technology and Culture*, vol. 7, no. 3, pp. 371–383.

Smyth, William Henry. (1919). "'Technocracy'—Ways and Means to Gain Industrial Democracy," *Industrial Management*, vol. 57.

Spengler, Oswald. (1931). *Mensch und die Technik: Beitrage zu einer Philosophie des Lebens*. Munich: Beck.

Stanley, Manfred. (1978). *The Technological Conscience: Survival and Dignity in an Age of Expertise*. New York: Free Press.

Stiegler, Bernard. (1994). *La technique et le temps, I: La faute d'Epiméthée*. Paris: Galilee.

Tenner, Edward. (1996). *Why Things Bite Back: Technology and the Revenge of Unintended Consequences*. New York: Knopf.

Tenner, Edward. (2003). *Our Own Devices: The Past and Future of Body Technology*. New York: Knopf.

Verbeek, Peter-Paul. (2005). *What Things Do: Philosophical Reflections on Technology, Agency, and Design* (Robert P. Crease, trans.). University Park: Pennsylvania University Press.

Von Foerster, Heinz. (1984). *Observing Systems*. Seaside, CA: Intersystems.

Weber, Max. (1921). *Wirtschaft und Gesellschaft*. Tübingen, Germany: J. C. B. Mohr.

Winner, Langdon. (1977). *Autonomous Technology: Technics-out-of-Control as a Theme in Political Thought*. Cambridge, MA: MIT Press.

Winner, Langdon. (1986). *The Whale and the Reactor: A Search for Limits in an Age of High Technology*. Chicago: University of Chicago Press.

Part II

Theoretical Approaches

3 Quality of Life in a Technological Society

A Macrosociological Approach

Ruut Veenhoven

The human species has lived for most of its time in simple hunter-gatherer societies. Agrarian societies developed less than 5,000 years ago and it is only in the last 200 years that a 'modern' industrial society has come into being. Today this industrial society is rapidly transforming into a global information society. Technology is a driving force in societal evolution and is highly intertwined with wider modernization (Lenski, Nolan, and Lenski 1995).

THE ISSUE

Is this development a change for the better? There has always been much controversy over this question, and currently the dispute seems more intense than ever, possibly for the reason that we are more aware today that society is of our making and because social change is taking place at an ever-increasing rate. Progress optimists believe that we live better now than earlier generations, whereas pessimists argue that life is getting worse. Technology is an important issue in this discussion.

The Positive View

The idea that life is getting better draws on several achievements of modern technological society. One is the unprecedented rise in the material standard of living; the average citizen lives more comfortably now than kings did a few centuries ago. Mechanical technology is seen as a main source of these benefits. Another major improvement is that the chance of an untimely death is greatly reduced; fewer people die in epidemics. Medical technology is credited for much of that advance. Moreover, a number of social evils have been abated, such as poverty, inequality, ignorance, and oppression. This social progress is seen to result both from unintended consequences of technological development and from deliberate social engineering, using newly developed 'social' technology.

A recent statement of this positive view can be found in *It's Getting Better All The Time* by Moore and Simon (2000). This notion of improvement is typically part of an evolutionary view in which society is seen as a human tool that is gradually perfected. This idea developed during the Enlightenment period in the 18th century and lives today. The idea that life is getting better breaks with the traditional religious view of earthly life as a phase of penance awaiting paradise in the afterlife. It is deemed possible to reduce suffering by creating a better world, and societal development is seen to head in that way, be it with some ups and downs.

The Negative View

The view that life is getting worse is typically fueled by concern about new developments. Mechanical technology is seen to drive out traditional craftsmanship and to have 'degraded' work (e.g., Braverman 1974), and specialization and bureaucratization are believed to cause widespread 'alienation' (e.g., Marx 1871). Technological development is also seen to involve ever-greater risks, which reflects in a widespread sense of insecurity (e.g., Beck 1992).

Related social developments are seen to undermine the quality of life as well. Many sociologists see increasing loneliness and feelings of meaninglessness in modern society, for example, Emile Durkheim (1897), David Riesman (1950), George Ritzer (1993), and Robert D. Putnam (2000). In psychology, Sigmund Freud (1930) provides an outspoken example of the theory that life is getting worse. In his 1923 book *Unbehagen in der Kultur* (*Society and Its Discontents*), Freud asserts that any social organization requires the repression of instinctual urges, and that the development of modern society necessitates ever-more repression of natural impulses. Hence, he believed that societal civilization is antithetical to human happiness and that we are typically less happy than our primitive forefathers. A recent statement of the negative view can be found in Gregg Easterbrook (2003).

This notion of decay is often part of the idea of society drifting away from human nature, because society has changed a lot, whereas human nature has not (e.g., Nesse 2004, p. 1343). In this view, society is not a piece of equipment, but rather, an uncontrollable force that presses humans into a way of life that does not really fit them. The notion of decay also fits with the supposition that there is great wisdom and morality in tradition, which modernization destroys. The idea that life is getting worse fits a long tradition of social criticism and apocalyptic prophecies. In this view, paradise is lost and is unlikely to be restored.

Dominance of the Negative View

The negative view prevails in most discussions and, in particular, in social scientific discourse (e.g., Lane 2000). Sociologists have become even more negative about modern society over the last decade; an analysis of the Sociological Abstracts showed a doubling in the use of gloomy words between 1970 and 2000 (Elchardus 2004, p. 507).

The negative view has also gained the upper hand in public opinion. Survey studies in modern nations show wide support for the idea that life was better in the 'good old days.' In the U.S., the majority agrees with the statement: "In spite of what some people say, the lot of the average man is getting worse, not better." Support for this idea is growing, the percent affirmative answers rose from about 55% in the 1970s to 70% in the early 1990s. Curiously, this idea is not reflected in self-reports of an individual's own life, because most people feel that the quality of their personal life has improved (Hagerty 2003).

Approach

There is some plausibility in both these views, and possibly all the effects mentioned exist to some extent. Yet it is often impossible to assess the relative strength of these effects and especially difficult to grasp their interactions and variations over time and place. For that reason, we will never be able to settle this debate on the details. What we can do is assess the balance of effects of technological development on the final quality of human life. When people live better in low-tech conditions than in high-tech conditions, the negative effects apparently outweigh the positive ones; other things being equal. In that line, this chapter compares quality of life in more and less developed societies, both across nations and over time. An advantage of this approach is that it includes all the consequences of all technologies. A disadvantage is that it may also capture effects of societal modernization that are independent of technology.

This approach can be characterized as 'macro-sociological,' because it focuses on society as a whole and considers long-term developments (Lenski, Nolan, and Lenski 1995). In that context, quality of life is approached in terms of 'livability theory,' which assumes that happiness and health depend on the degree to which external conditions allow innate needs to be met (Veenhoven 1995).

MEASUREMENT OF QUALITY OF LIFE IN NATIONS

How can we assess quality of life in society? This first requires a clear definition of quality of life, and next a feasible way of measurement.

Concepts of 'Quality of Life'

The term 'quality of life' serves as a catchword for different notions of the good life. It is used, in fact, to denote a bunch of qualities of life, which can

Table 3.1 Four Qualities of Life

	Outer qualities	*Inner qualities*
Life chances	**Livability of environment**	**Life-ability of the person**
Life results	**Utility of life**	**Satisfaction**

be ordered on the basis of the following two distinctions. A first distinction is between opportunities for a good life and the outcomes of life. This distinction is quite common in the field of public health research. Preconditions for good health, such as adequate nutrition and professional care, are seldom mixed up with the concept of health. A second difference is between external and inner qualities. In the first case, the quality is in the environment; in the latter, it is in the individual. This distinction is also quite common in public health. External pathogens are distinguished from inner afflictions. Combining of these two dichotomies yields a fourfold matrix, presented in table 3.1.

In the upper half of the table, we see, next to the outer opportunities in one's environment, the inner capacities required to exploit these. The environmental conditions can be denoted by the term livability, the personal capacities with the word life-ability. This difference is not new. In sociology, the distinction between 'social capital' and 'psychological capital' is sometimes used in this context; and in the psychology of stress, the difference is labeled negatively in terms of 'burden' and 'bearing power.'

The lower half of the table is about the quality of life with respect to its outcomes. These outcomes can be judged by their value for one's environment and by their value for oneself. The external worth of a life is denoted by the term utility of life. The inner valuation of a life is called appreciation of life. These matters are of course related. Knowing that one's life is useful will typically add to one's appreciation of life. Yet useful lives are not always happy lives, and not every 'good-for-nothing' is unhappy.

Livability of the Environment

The left top quadrant denotes the meaning of good living conditions, which I call 'livability.' One could also speak of the 'habitability' of an environment, although that term is also used for the quality of housing in particular. Elsewhere I have explored the concept of livability in more detail (Veenhoven 1996, pp. 7–9).

Ecologists see livability in the natural environment and describe it in terms of pollution, global warming, and degradation of nature. Currently, they associate livability typically with environmental preservation. City planners see livability in the built environment and associate it with sewer systems, traffic jams, and ghettos. Here the good life is seen to be the fruit of human intervention. In public health, this all is referred to as a 'sane' environment. Society is central in the sociological view. First, livability is associated with the quality of society as a whole. Classic concepts of the 'good society' stress material welfare and social equality, sometimes equating the concept more or less with the welfare state. Current communitarian notions emphasize close networks, strong norms, and active voluntary associations. The reverse of this livability concept is 'social fragmentation.'

Life-Ability of the Person

The right top quadrant denotes inner life chances. That is, how well we are equipped to cope with the problems of life. I call this 'life-ability,' which contrasts elegantly with 'livability.' This quality is central in the 'capability approach,' in particular in Martha C. Nussbaum's (2000) variant. Elsewhere I have discussed that approach in more detail (Veenhoven 2010).

The most common depiction of this quality of life is an absence of functional defects. This is 'health' in the limited sense, sometimes referred to as 'negative health.' In this context, doctors focus on unimpaired functioning of the body, whereas psychologists stress the absence of mental defects. This use of words presupposes a 'normal' level of functioning. A good quality of life is seen to be the body and mind working as designed. This is the common meaning used in curative care.

Next to absence of disease, one can consider excellence of function. This is referred to as 'positive health' and associated with energy and resilience. Psychological concepts of positive mental health also involve autonomy, reality control, creativity, and inner synergy of traits and strivings. This broader definition is the favorite of the training professions and is central to the 'positive psychology' movement.

Utility of Life

The left bottom quadrant represents the notion that a good life must be good for something more than itself. I refer to these external turnouts as the 'utility' of life. When evaluating the external effects of a life, one can consider its functionality for the environment. In this context, doctors stress how essential a patient's life is to their intimates. At a higher level, quality of life is seen as a contribution to society. Historians see quality in the additions an individual can make to human culture and rate, for example, the lives of great inventors higher than those of anonymous peasants. Moralists see quality in the preservation of the moral order and would deem the life of a saint to be better than that of a sinner. In this vein, the quality of a life is also linked to effects on the ecosystem. Ecologists see more quality in a life lived in a 'sustainable' manner than in the life of a polluter.

Enjoyment of Life

Finally, the bottom right quadrant represents the inner outcomes of life. That is, the quality of life in the eye of the beholder. As we deal with conscious humans, this quality boils down to subjective appreciation of life. This is commonly referred to by terms such as 'subjective well-being,' 'life-satisfaction,' and 'happiness' in a limited sense of the word.

Humans are capable of evaluating their life in different ways. We have, in common with all higher animals, an ability to appraise our situation affectively. We feel good or bad about particular things and our mood level signals overall adaptation. As in animals, these affective appraisals are automatic, but unlike other animals, humans can reflect on this experience. We have an idea of how we have felt over the last year, whereas a cat does not. Humans can also judge life cognitively by comparing life as it is with notions of how it should be.

Happiness can be defined as the degree to which a person evaluates the overall quality of his or her present life-as-a-whole positively. In other words, how much the person likes the life he or she leads. This evaluation appears to draw on affective information in the first place, if people appraise how happy there are, they estimate how well they feel most of the time (Veenhoven 2009b).

Analogous Concepts in Biology

Similar distinctions are made in evolutionary biology (see Table 3.2). External living conditions are referred to as the 'biotope' or 'habitat.' A biotope can be a more or less suitable for a species, depending on for example, availability of food, shelter, and competition. This is analogous to what I call 'livability.' An organism's capability to survive in the environment is called 'fitness' by biologists. This latter term acknowledges the fact that the capabilities must meet (fit) environmental demand. This is equivalent to what I call 'life-ability.' With respect to outcomes of life, biologists also distinguish between external and internal effects. External effects are various ecological functions, such as being prey for other creatures and the continuation of the species. This is analogous to what I call the 'utility' of life. The outcome of life for the organism itself is depicted as 'survival,' which is seen to result from the fit between capabilities and

Table 3.2 Analogues Concepts in Biology

	Outer qualities	*Inner qualities*
Life chances	**Biotope**	**Fitness**
Life results	**Continuation**	**Survival**

environment. This notion corresponds to what I call 'enjoyment of life.' As we will see, this is more than mere correspondence, because subjective enjoyment is also a result of fit between individual abilities and environmental demands.

MEASURING QUALITY OF LIFE

Quality of life in nations is usually measured using indexes that involve indicators from each of the quadrants in table 3.1, for instance, the Human Development Index (1990) includes income per head (top left), education (top right), and life expectancy (bottom right). Yet this makes no sense and the schemes help us to see why not.

Comprehensive Measurement Not Possible

Quality of life cannot be measured by totaling quadrants. There is no point in combining the qualities in the upper and the lower half of the scheme, because this involves the adding of chances and outcomes. Combining the qualities at the left and the right makes little sense either and, in particular, not in the case of life chances, where it is not the sum that matters, but rather the fit between external conditions and inner capacities.

Still another problem is that three of these four qualities cannot be measured very well. We can only make guesses about the features that constitute the livability of an environment, and it is also quite difficult to establish what abilities are most required. Although it is clear that some necessities must be met, it is not so clear what is required on top of these, and in what quantities and in what mix. Measuring the utility of life is not really feasible either, because external effects are quite diverse and often difficult to assess. Due to this lack of sound scientific criteria, any measurements strongly depend on assumption and ideology, and hence there is little agreement how to measure these qualities of life.

Measuring happiness is less problematic however. Because happiness is an overall judgment of life, we cease to have the problem of trying to add and compare apples and oranges: as happiness is a state of mind for an individual, we can assess it rather easily by asking that person how happy he or she feels.

Most Inclusive Measure Is How Long and Happy People Live

In biology, 'survival' is assumed to result from the 'fit' between the abilities of the organism and environmental demands. This fit cannot be observed as such, but is typically inferred from survival rates. If an organism perishes before its programmed lifetime, there is apparently something wrong with this chance constellation.

In this line, we can also infer the life chances in a human society from the outcomes in happiness. If people live happily, their environment is apparently sufficiently livable and their abilities appropriate. This may not appeal to supporters of the theory that happiness is a culturally constructed illusion, but it fits well with the view that happiness is a biological signal of how well we thrive.

In simple animals, good adaptation reflects only in survival, in higher animals, good adaptation also reflects in hedonic experience. Negative affect is indicative of poor adaptation and tends to inhibit the organism, whereas positive affect is indicative of good adaptation and works as a 'go' signal (Frederickson 1998; Nesse 2004). So, an animal that does not feel good is probably not doing well. This inner experience is no great issue in biology, because we cannot assess how animals feel. Still, there is ground to see hedonic experience as an additional manifestation of good adaptation, and in this vein one could argue that an animal that feels well most of its lifetime seems to be better adapted than an animal that lives equally long but feels less well.

Humans are capable of reflecting on their experiences and can condense positive and negative affects into an overall appraisal of happiness. They are also capable of communicating that appraisal to investigators. Hence, in the case of humans, we can use the additional sign of good adaptation and assess how long and happy they live.

The degree to which people live long and happy is denoted in the right bottom quadrant in table 3.1 and is the most inclusive measure of outcomes of life for the individual. It is also indicative for the qualities denoted by the two top quadrants. If people live long and happy, their environment is apparently sufficiently livable and their life abilities adequate enough. So, this measure covers in fact three of the four quadrants and is therefore the most comprehensive measure of quality of life available.

Fairy tales often end with the phrase "and they lived long and happily ever after," which is to say that the characters of the story had a good life. The concept of quality of life is then operationalized as a 'long and happy life.' The previous reasoning shows that this makes sense. I have underpinned this position in more detail elsewhere and distinguished this measure of 'apparent' quality of life from currents counts of 'presumed' blessing (Veenhoven 1996; 2000e).

Also the Most Relevant in this Context

This chapter is about the question of whether life is getting better in technological society. As such, it concerns the actual outcomes of life and not assumed chances for a good life. What we want to know is how modern chance constellations work out on the final results of life. Likewise, the question is not about the outcomes for society, but about the outcomes for individuals. If we do not keep these matters apart, societal development is easily equated with a better life. So this is another reason to focus on the right bottom quadrant, which is best measured by how long and happy people live.

MEASURE OF HAPPY LIFE YEARS

The degree to which people live long and happy in a society can be measured by combining two sources of information: average longevity in the country and average happiness.

Measurement of Longevity in Nations

How long people live in a country can be assessed using civil registration and by assessing the average number of years between birth and death. This will give an adequate measure for past generations, but not those still alive. Hence, a next step is to estimate how long the living remain alive, and these estimates can be generalized to give a general population average. This estimate is called 'life expectancy' and is commonly used in world health statistics. Data is available for almost all countries in the world, and yearly updates are published in the Human Development Reports (United Nations Development Program 2009).

1	2	3	4	5	6	7	8	9	10
Dissatisfied									Satisfied

Measurement of Happiness in Nations

Happiness has been defined as something that we have in mind. Consequently, happiness can be measured using single direct questions. A common survey question is:

Taking all together, how satisfied or dissatisfied are you currently with your life as a whole?

Since the 1970s, such questions have been included in many surveys worldwide, and there is now a growing body of data on happiness in nations. Presently, there are comparable surveys in 90 nations. The data have been brought together in the 'World Database of Happiness' (Veenhoven 2009e).

There are many qualms about such simple self-reports of happiness, in particular about their validity and about comparability across nations. Elsewhere I have considered the objections and inspected the empirical evidence for claims about bias (Veenhoven 1993; 1997). I found no proof for any of the objections, so I assume that happiness can be measured in this way. Others have come to the same conclusion (Diener 1994; Saris, Scherpenzeel, and Van-Wijk 1998). Suffice to note that cross-national differences in happiness correspond in the predicted way with national rates of depression (Van Hemert, VandeVijver, and Poortinga 2002), and suicide (r = -.46; own analysis).

Combination with Longevity

How long and happy people live in a country can be measured by combining information about length of life, drawn from civil registrations of births and deaths, with data on average appreciation of life as assessed in surveys. The following simple formula can be applied:

Happy-Life-Years = Life-expectancy at birth x 0–1 happiness

Suppose that life expectancy in a country is 60 years, and that the average score on a 0 to 10-step happiness scale is 5. Converted to a 0–1 scale, the happiness score is then 0.5. The product of 60 and 0.5 is 30. So the number of happy life years is 30 in that country. If life expectancy is also 60 years but average happiness 8, the number of happy life years is 48 (60 x 0.8).

Theoretically, this indicator has a broad variation. The number of Happy Life Years is zero if nobody can live in the country, and will be endless if society is ideal and/or its inhabitants immortal. The practical range is between about 10 and 70 years. Presently at least, life expectancy at birth in nations varies between 40 and 80 years, whereas average happiness varies between 0.3 and 0.8. The number of Happy Life Years (HLY) will always be lower than standard life expectancy. It can equal real length of life only if everybody is perfectly happy in a country (score 1 on scale 0 to 1).

A high HLY means that citizens live both long and happily; a low HLY implies that the life of the average citizen is short and miserable. Medium HLY values can mean three things: (1) both moderate length-of-life and moderate appreciation-of-life, (2) long but unhappy life, and (3) short but happy life. I treat these intermediate outcomes as equal, but one can of course prefer one to the other.

I have described this indicator in more detail elsewhere (Veenhoven 1996; 2009a; 2009e). It scored highest in a scholarly review of social indicators (Hagerty et. al., 2001). In a similar way, the World Health Organization measures 'Disability Adjusted Life Years' (DALYs) in nations (World Health Organization 2004). In this case, life expectancy is weighted with a disability score, which is typically also derived from survey data. Because disabilities increase with age, age composition in nations must be taken into account in cross-national comparisons of DALYs. In the case of HLY, this is not required, because happiness is not related to age in the same way. There is a slight U-shaped pattern in the relation between happiness and age, happiness tending to be somewhat lower around middle age, yet this effect is not so robust that it requires correction.

QUALITY OF LIFE IN MODERN SOCIETY

Using this indicator, we can now answer the question whether life is getting better or worse in modern technological society. For this purpose, I will first compare the quality of life across contemporary nations and next consider the available data on trends over the last decades.

Difference between More and Less Developed Nations

The present world counts about 180 nations and for 144 of these we know how long and how happy its citizens lived in the 1990s. See the list on the World Database of Happiness (Veenhoven 2009a). These cases represent about 90% of the world's population, because all the big countries are included.

The level of technological development of these nations can be measured in different ways. One way is to assess the number of Internet users per 1,000 inhabitants. Another way is assessing the degree of industrialization, which is commonly measured by its reverse, which is the share of agriculture in the national income. The level of education is also indicative of technological development, as is the degree of urbanization. All

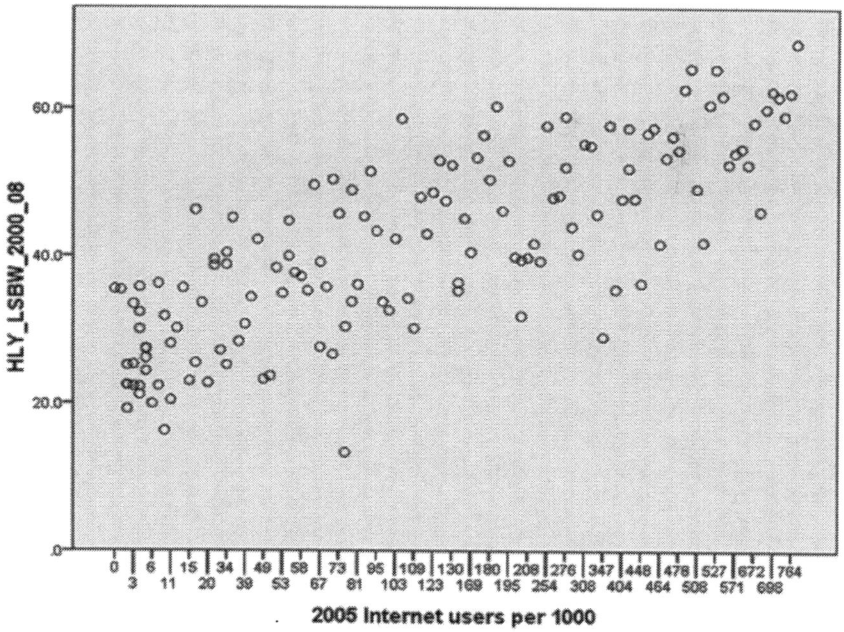

Figure 3.1 Happy Life Years and internet penetration in 138 nations around 2005.
Source: World Database of Happiness: States of Nations (Veenhoven 2009d)

measures of technological development are highly correlated with economic development.

In the scattergram of figure 3.1, the number of happy life years is plotted vertically and Internet connections horizontally. One can easily see that there is a strong positive correlation, HLY being systematically higher in nations with the highest Internet penetration. The correlation is +.73. Reversely, there is an equally strong negative correlation with the share of agriculture in the national income (r = -.74), which indicates technological underdevelopment. Likewise, there is a strong positive correlation with education (r = +.70) and urbanization (r = +.69). The correlation with economic development is again positive (r = +70).

The pattern is similar if we consider happiness and longevity separately. The correlation of technological development as measured with Internet penetration and happiness is +.63 and the correlation with life-expectancy +.66. These effects are largely independent: controlling for life expectancy, the partial correlation with happiness is still +.32; and controlling for happiness, the partial correlation of Internet penetration with life expectancy is +.45! This independence of the effects is another justification for the use of this combined measure of happy life years.[1*]

TREND OF MODERN NATIONS OVER THE PAST DECADES

The differences observed in the previous cross-sectional analysis could be due to intervening variables, for example, they may be due to the fact that modern nations are mainly found in moderate climate zones or to variation in genetic endowment. Such distortions can be controlled for if we compare over time within separate nations. If life is getting better, this must also manifest in a positive trend.

Assessment of the development in HLYs in nations requires trend data on happiness and life expectancy. In the case of life expectancy this is no problem, because considerable time series are available for many nations. Time series on happiness are less abundant however. Series of 30 years or longer and based on identical survey questions are available for only 10 nations and these are all highly developed ones.

Eight of these nations are the early EU members, which have participated in the Eurobarometer survey since 1973. These nations are Denmark, Belgium, France, West Germany, Italy, Luxembourg, the Netherlands, and the U.K. Scheme 8.4 presents the trend in HLYs in these countries from 1973 to 2008. A clear positive trend emerges; Europeans gained 4.6 more happy life years in this 35-year period. See figure 3.2.

If this trend continues, West-Europeans will live 62.2 happy years in 2073, which would mean a gain of 15.7 years in less than a century. Similar trends are observed in the other developed nations for which data are available. HLY rose 5 years in Japan and 6.4 years in the U.S. (Veenhoven 2009d).

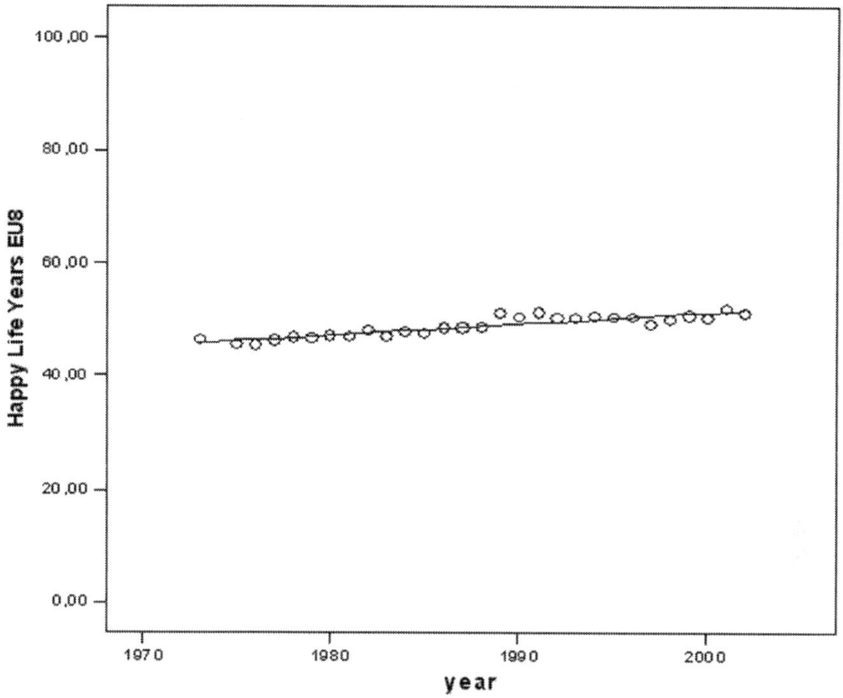

Figure 3.2 Trend Happy Life Years in 8 EU-nations 1973–2008.
Source: World Database of Happiness, collection Happiness in Nations (Veenhoven 2009c).

This upward trend is not the result of the happy getting happier, but rather the result of a reduction in the number of very unhappy people in the population. This manifests in a lowering of the standard deviations in all nations, also in the ones where the average remained at the same level (Veenhoven 2005). Likewise, the gains in life expectancy are greater at the bottom of the distribution than at the top. So the rise in HLY went together with a reduction of inequality in quality of life.

This is not to say that life has got better in all countries during the last decades, HLY has not risen in Belgium, due to a decline in happiness in the 1980s. Happiness and longevity dramatically plunged in the former communist countries in the 1990s, in particular in Russia (Veenhoven 2001), probably in response to the sudden transformations in these societies. The latest data suggest that the postcommunist nadir is passed. One must also realize that the effects of economic development are less smooth than suggested in figure 3.1. In some cases at least, early industrialization was accompanied by a shortening of life and even by a reduction in average body size (Komlos 1998). The wide variation in the left half of figure 3.1 can be interpreted as an indication that similar things are happening in today's developing nations. Further, it is beyond doubt that progress causes the 'pain of incomprehension' during periods of transition (Hays 1994).

QUALITY OF LIFE OVER HUMAN HISTORY

The previous evidence concerns contemporary societies and does not rule out that quality of life has been better in earlier times. All the cases considered are modern to some extent. Hence, these data cannot settle the question of whether we would have lived better in an ancient pretechnological society.

One way to check this would be to look at the quality of life in present day 'primitive' societies. Yet there are few such societies today, and the few that remain live typically live in poor ecological conditions. Moreover, the last century's anthropological research does not give us a clear picture of the quality of life in the primitive societies that were surveyed. There are accounts of positive features in such societies, such as Mead's (1953) description of relaxed sexual practices in Samoa, but there is also evidence of rampant evils, such as violence and superstition (Edgerton 1992). Anthropologists have never attempted to assess happiness, possibly because of their belief in cultural relativism.

Still, we find important clues in historical anthropology and, in particular, in work by Alexandra Maryanski and Jonathan H. Turner (1992) and by Stephen K. Sanderson (1995). This literature departs from the insight that the human species developed in the context of hunter-gatherer bands and that this type of social organization prevailed for most of the 100,000 years that Homo sapiens has existed. More complex kinds of societies seem to have developed only fairly recently in human history, first horticultural societies, then agrarian societies, and finally our present day industrial society, which is rapidly becoming postindustrial in the West (Lenski, Nolan, and Lenski 1995, chapter 1). There are good indications that these types of societies were not equally livable, and, in particular, that the agrarian phase marked a historic dip in human quality of life.

Less Free in Agrarian Society

One indication is that the development of freedom seems to have followed a U-curve over human history. Hunter-gatherer societies can impose few constraints on their members, because dissenters can support themselves for a while and join other bands. Accumulation of wealth and power is difficult, if not impossible, in these conditions, and hence this kind of society tends to be free and egalitarian.

This changes profoundly in an agrarian society, where survival requires control of the land and people became more dependent on their family and vulnerable to exploitation by a warrior caste. According to Alexandra Maryanski and Jonathan H. Turner (1992), this drove mankind into the 'social cage' of collectivistic society. In their view, such strong social bonds are less required in the context of industrial existence, not only because individuals get access to more relational alternatives, but also because the ongoing division of labor involves a shift of dependencies to anonymous institutions, such as the state. Emile Durkheim (1897) described that latter phenomenon as the change from 'mechanic solidarity' to 'organic solidarity.'

This theory fits the previous observation that people live happier in the most modern societies of this time and, in particular, it fits the observed relation between happiness and freedom in nations (Veenhoven 1999). It also provides an explanation for the ongoing migration from the land to cities.

Less Healthy in Agrarian Society

Another sign of the low quality of life in agrarian society can be found in anthropometric indicators of health. The average health of past generations can to some extent be reconstructed from human remains. On the basis of excavated bones and teeth we can estimate how long people have lived and to some degree how healthy they were when they lived.

Research along these lines suggests that people lived about equally long in early hunter-gatherer societies and later agrarian societies, but that they lived more healthily in the former than in the latter. Hunter-gatherers appear to have been better-nourished and less disease ridden than historical agrarian populations, and they seem to have been less burdened with work. Much of the literature on this subject is reviewed in Sanderson (1995, pp. 340–343). Apparently, the obvious advantages of a sedentary pastoral life are counter-balanced in some way, among other things probably by increased exposure to disease and to social stress. Present day industrial societies score better on anthropometric indicators than both of the other types do. We live longer and healthier than ever before. We also grow taller than our forefathers ever did.

Such historical data cover health and longevity and can be combined in an index of healthy life years that concurs with the measure of 'Disability Adjusted Life Years' that is currently used by the World Health Organization (2004). This resembles the 'Happy Life Years' I used in the previous analysis of contemporary nations, but it is not quite the same. Because survey research is a recent invention, we will probably never know how happy people were in the past. Hence, we must make do with the available data on health and longevity. Still, these matters appear to be strongly correlated with happiness and both can be seen as manifestations of human thriving.

If we consider the data on longevity, it is also easy to see that our forefathers cannot have lived as many happy years as we do now. The average length of life was about 45, both in a hunter-gatherer society and in an agrarian society. This means that the HLY could not be higher than 45, even if everybody was perfectly happy. This is clearly below the level in present day modern nations, where HLY varies between 50 and 60.

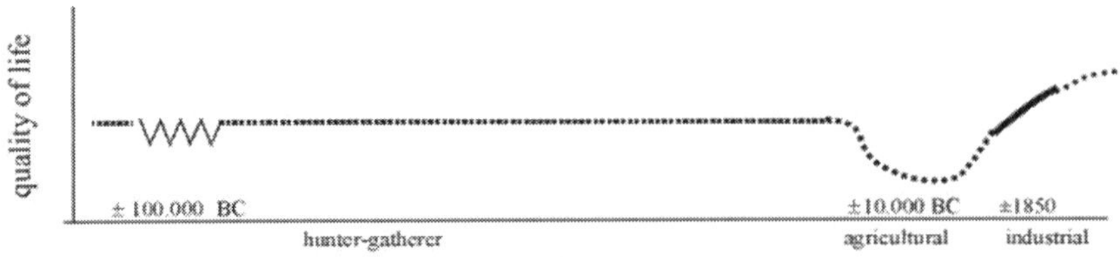

Figure 3.3 Quality-of-life over human history.
Source: Veenhoven 2010b.

Long-Term Pattern

Together these data suggest that societal evolution has worked out differently on the quality of human life, first negatively in the change from hunter-gatherer existence to agriculture and next positively in the recent transformation from an agrarian to an industrial society. This pattern is depicted graphically in figure 3.3.

Meanwhile, industrial society develops into an 'information society' (e.g., Castells 2000). This does not seem to change the pattern of rising quality of life. At least for the time being, HLY is still on the rise in the most developed nations of this time, as scheme 3.2 shows. It is not unlikely that the upward trend will continue in the future, both happiness and longevity are likely to rise. Currently average happiness ranges between 3.2 (Tanzania) and 8.4 (Denmark). This means that there is still much to win. There is no reason to assume that Tanzania can never reach the level of Switzerland. Possibly, average happiness can even go above an average of 8.5. The maximally possible average may be close to 9. An average of 10 in a nation is clearly not possible, not only because human life involves inevitable suffering but also because no society can serve everybody's needs equally well.

It is also quite likely that longevity will continue to rise. The underdeveloped nations are catching up at an unprecedented rate, and there is reason to expect that longevity will further extend in the developed world, among other things as the result of medical technology. There is speculation that the average life might be extended to 100 years or more (Manton, Stallard, and Tolley 1991; Vaupel and Lundstöm 1993). For the time being at least, gains in length of life have not come at the cost of quality of life. Elderly people are slightly happier than middle-age people. The dotted trend line at the right side of figure 3.3 depicts these projections.

DISCUSSION

These findings raise several questions: one is about compatibility with claims to the contrary and, in particular, with reports of alarming rises in rates of depression in modern society. A second question is about the mechanisms behind the recent rise in the quality of human life and the role of technology in that context.

Claims to the Contrary

As noted in the introduction, several social scientists believe that life is getting worse in modern society, and much of these claims are summarized in Gregg Easterbrook's (2003) 'Progress paradox.' It would lead too far to discuss all misery counts. Below I will consider two claims.

Easterlin Paradox

Income per capita has risen spectacularly in Western nations over the last 50 years, but according to Richard A. Easterlin (1974; 1995; 2005), this has not made us any happier; average happiness remaining stagnant. This claim may apply to some countries, including the U.S. Yet happiness is rising in most countries of the world (Veenhoven and Hagerty 2006). Possibly, an otherwise positive effect of economic and technical development was counterbalanced by different things, such as dwindling incomes in the middle class (Fischer 2007). Even if average happiness has remained the same, longevity has increased. Given the high level of happiness, this has resulted in a substantial increase in happy life years.

Depression Epidemic

Mental illness is no exception in modern societies, about 16% of the U.S. citizens have experienced episodes of serious depression, and every year depression interferes with effective functioning of some 6% of them for two weeks or more (Kessler and Wang 2002). There are signs of increasing rates of depression, especially among youth, and there is talk about a depression epidemic (Seligman 1990, p. 10). How does this fit with a rising number of happy life years?

The first thing to note is that depression is strongly correlated with happiness. Depressed individuals are clearly less happy, and there is a strong correlation between rates of depression and average happiness in nations (VanHemert, VandeVijver, and Poortinga 2002). A second point to keep in mind is that depression is temporary in most cases, and good times can balance the bad times. Third, we are not sure that the rate of depression has risen. There is an increase in the numbers for people diagnosed as being depressed, but this may due to better recognition by professionals of the signs of depression and by better treatments being available to treat depression. It is possible that depression is less recognized in underdeveloped nations, not only because the people are less aware of the syndrome, but also because bad feelings can be more easily attributed to bad conditions.

However, it is also possible that in modern societies people are more aware of how they feel, because they have more choice and use how they feel more to help them to assess what they want. If there is a real rise in rates of depression in modern society, that can still co-exist with a rise in average happiness. Modernization can be to the advantage of a majority, but can come at the expense of a minority who are pushed into depression; no society can suit the needs of all equally well. If this has happened at all, it has not resulted in a split between happy and unhappy, because the dispersion of happiness is lessening in modern societies (Veenhoven 2005).

WHY IS LIFE GETTING BETTER?

The observed growth of years lived happily can be attributed to several factors. One is obviously that several common evils of the past have been overcome in modern societies or at least much abated. For instance, few in the West die of hunger anymore, and the chance of untimely death is greatly reduced. Technology has evidently contributed to that development. A second factor is in the increased freedom in modern individualized society. The social system allows us more opportunity to choose, and we have also become more capable of making choices which, taken together, has increased our chance that we will live a life that fits our individual needs (Veenhoven 1999). Technological development has fostered freedom indirectly, because it pressed to an ever-more complex division of labor, which has resulted in greater personal autonomy in a context of greater mutual dependence (e.g., Lenski, Nolan, and Lenski 1995).

This links up with a third explanation, which is that modern society provides a challenging environment that fits an innate human need for self-actualization. In this view, the human species evolved in rather tough conditions and therefore typically thrives in modern society with its complexities, competition, and choices. One of the reasons for the low quality of life in agrarian society is probably its boring technology. Probably, increased self-understanding has also contributed to the quality of life of modern man, and information technology is likely to have facilitated that development. Last, a more 'critical' explanation could be that the modern nations successfully exploit the rest of the world, using their superior military technology. There may be some truth in this contention, but life is also getting better in most non-Western nations. Most of the poor countries became less poor over the last decade, and life expectancy is also rising in most countries (United Nations Development Program 2002; 2004).

CONCLUSION

Technological development has not always improved quality of life. The agrarian revolution seems to have brought humankind from the fire into the frying pan. But the later Industrial Revolution has brought a change for the better. We now live longer and are healthier than ever before in human history, and we are probably also happier. This upward trend in quality of life seems to continue in present day information society.

NOTES

1. The data used in this analysis are described in the appendix table.

Table 3.3 Variables Used in Comparison across Contemporary Nations (section 3.1)

Variable	*Measurement*	*Source*
Happiness	Average response to question about life satisfaction on Scale 0-10	World Database of Happiness, States of nations, variable HappinessLSBW_2000.08
Life expectancy	Estimated longevity of present generation from birth on	Human Development Report 2007 table 1
Happy Life Years	Estimated longevity x average happiness scale 0-1	World Database of Happiness, States of nations, variable HLY_LSBW_2000.08
Internet penetration	Internet users per 1000 inhabitants	Human Development Report 2007 table 13
Education	Index of Adult literacy (counted 2/3) and school enrolment (counted 1/3)	Human Development Report 2007 table 1
Urbanization	% population living in urban areas	Human Development Report 2007 table 5
Industrialization	Agrarian share of gross nation income in %	World Development Indicators 2007, table 6.14
Economic Affluence	Income per capita in purchasing power parities	Human Development Report 2007 table 1

Source: All these variables are in the data file 'States of Nations 2009' (Veenhoven 2009d)

REFERENCES

Beck, Ulrich. (1992). *Risk Society: Towards a New Modernity.* New Delhi: Sage.

Braverman, Harry. (1974). *Labor and Monopoly Capital: The Degradation of Work in the 20th Century.* New York: Monthly Review Press.

Castells, Manuel. (2000). *The Rise of the Network Society. The Information Age: Economy, Society and Culture,* vol. 1. Malden, MA: Blackwell.

Diener, Ed. (1994). "Assessing Subjective Well-Being: Progress and Opportunities." *Social Indicators Research*, vol. 31, pp. 103–157.
Durkheim, Emile. (1897). *Le suicide: étude de sociologie*. Paris: Alcan.
Easterbrook, Gregg. (2003). *The Progress Paradox: How Life Gets Better While People Feel Worse*. New York: Random House.
Easterlin, Richard A. (1974). "Does Economic Growth Improve the Human Lot?" In *Nations Households AND Economic Growth: Essays in Honor of Mozes Abramowitz*. New York: Academic Press.
Easterlin, Richard A. (1995). "Will Raising the Income of All Increase the Happiness of All?" *Journal of Economic Behavior and Organization*, vol. 27, pp. 35–48.
Easterlin, Richard A. (2005). "Feeding the Illusion of Growth and Happiness: A Reply to Hagerty and Veenhoven." *Social Indicators Research*, vol. 74, pp. 429–443.
Edgerton, Robert B. (1992). *Sick Societies: Challenging the Myth of Primitive Harmony*. New York: Free Press.
Elchardus, Mark. (2004). "Sociologie in de onbehaaglijke samenleving" [Sociology in uncomfortable society]. *Sociologische Gids*, vol. 51, pp. 507–515.
Fische, Claude (2008). "What Wealth-Happiness Paradox? A Short Note on the American Case." *Journal of Happiness Studies* 9: 219–226.
Frederickson, Barbara L. (1998). "What Good Are Positive Emotions?" *Review of General Psychology*, vol. 2, pp. 300–319.
Freud, Sigmund. (1961). *Das Unbehagen in der Kultur* [Civilization and Its Discontents]. Abriss der Psychoanalyse series. New York: Norton. (Originally published 1930)
Hagerty, Michael R. (2003). "Was Life Better in the 'Good Old Days'? Inter Temporal Judgment of Life-Satisfaction." *Journal of Happiness Studies*, vol. 4, pp. 115–139.
Hagerty, Michael R., Robert Cummins, Abbott L. Ferriss, Kenneth Land, Alex C. Michalos, Mark Peterson, Andrew Sharpe, Joseph Sirgy, and Joachim Vogel. (2001). "Quality of Life Indexes for National Policy: Review and Agenda for Research." *Social Indicators Research Series*, vol. 55, pp. 1–96.
Hays, Danica G. (1994). "On the Painfulness of Progress." *Cross-Cultural Research*, vol. 28, pp. 322–326.
Kessler, Ronald C., and Philip S. Wang. (2002). "The Epidemiology of Depression." In *Handbook of Depression*, edited by Gotlib, I. H. and C. L. Hammen. New York: Guilford Press.
Komlos, John. (1998). "Shrinking in a Growing Economy? The Mystery of Physical Stature During the Industrial Revolution." *The Journal of Economic History*, vol. 58, pp. 779–802.
Lane, Robert. (2000). *The Loss of Happiness in Market Democracies*. New Haven, CT: Yale University Press.
Lenski, Gerhard, Patrick Nolan, and Jean Lenski. (1995). *Human Societies: An Introduction into Macro Sociology*. New York: McGraw-Hill.
Manton, Kenneth E., Eric Stallard, and Dennis Tolley. (1991). "Limits to Human Life Expectancy: Evidence Prospects, and Implications." *Population and Development Review*, vol. 17, pp. 603–637.
Marx, Karl. (1983–1984). *Capital: A Critique of Political Economy*. London: Lawrence and Wishart. (Originally published 1871)
Maryanski, Alexandra, and Jonathan H. Turner. (1992). *The Social Cage, Human Nature and the Evolution of Society*. Stanford: Stanford University Press.
Mead, Margaret. (1953). *Coming to Age in Samoa: A Psychological Study of Primitive Youth for Western Civilization*. New York: Modern Library.
Moore, Stephen, and Julian L. Simon. (2000). *It's Getting Better All the Time: Greatest Trends of the Last 100 Years*. Washington, DC: Cato Institute.

Nesse, Randolph M. (2004). "Natural Selection and the Elusiveness of Happiness." *Philosophical Transactions of the Royal Society, Biological Sciences,* vol. 359, pp. 1333–1347.

Nussbaum, Martha C. (2000). *Women and Human Development: The Capabilities Approach.* Cambridge: Cambridge University Press.

Putnam, Robert D. (2000). *Bowling Alone: The Collapse and Revival of American Community.* New York: Simon and Schuster.

Riesman, David. (1950). *Lonely Crowd: A Study of the Changing American Character.* New Haven, CT: Yale University Press.

Ritzer, George. (1993). *The McDonaldization of Society.* Thousand Oaks, CA: Pine Forge Press.

Sanderson, Stephen K. (1995). *Social Transformations.* Oxford: Blackwell.

Saris, Willem, Annette Scherpenzeel, and Theresia VanWijk. (1998). "Validity and Reliability of Subjective Social Indicators." *Social Indicators Research,* vol. 45, pp. 173–199.

Seligman, Michael E. (1990). *Learned Optimism.* New York: Pocket Books.

United Nations Development Program. (1990, 2000, 2004, 2009). *Human Development Report.* New York: Oxford University Press.

VanHemert, Dianne, Fons VandeVijver, and Ype Poortinga. (2002). "The Beck Depression Inventory as a Measure of Subjective Well-Being in a Cross-National Study." *Journal of Happiness Studies,* vol. 3, pp. 257–286.

Vaupel, James W., and Hans Lundström. (1993). "The Future of Mortality at Older Ages in Developed Countries." In *The Future Population of the World: What Can We Assume Today?* edited by W. Lutz. International Institute for Applied Systems Analysis series. London: Earth Scan.

Veenhoven, Ruut. (1991). "Is Happiness Relative?" *Social Indicators Research,* vol. 24, pp. 1–34.

Veenhoven, Ruut. (1993). *Happiness in Nations.* Rotterdam: Risbo. Retrieved (with 2002 update) from http://www2.eur.nl/fsw/research/veenhoven/Pub1990s/93b-con.html

Veenhoven, Ruut. (1995). "The Cross-Cultural Pattern of Happiness Test of Predictions Implied in Three Theories of Happiness." *Social Indicators Research,* vol. 34, pp. 33–68.

Veenhoven, Ruut. (1996). "Happy Life Expectancy: A Comprehensive Measure of Quality-of-Life in Nations." *Social Indicators Research*, vol. 39, pp. 1–58.

Veenhoven, Ruut. (1997). "Progrès dans la compréhension du bonheur." *Revue Québécoise de Psychologie,* vol. 18, pp. 29–74. English version retrieved from http://www2.eur.nl/fsw/research/veenhoven/Pub1990s/97c-full.pdf

Veenhoven, Ruut. (1999). "Quality-of-Life in Individualistic Society: A Comparison of 43 Nations in the Early 1990s." *Social Indicators Research*, vol. 48, pp. 157–186.

Veenhoven, Ruut. (2000). "The Four Qualities of Life: Ordering Concepts and Measures of the Good Life." *Journal of Happiness Studies*, vol. 1, pp. 1–39.

Veenhoven, Ruut. (2001). "Are the Russians as Unhappy as They Say They Are?" *Journal of Happiness Studies*, vol. 2, pp. 111–136.

Veenhoven, Ruut. (2005). "Return of Inequality in Modern Society? Test by Dispersion of Life-Satisfaction Across Time and Nations." *Journal of Happiness Studies, Special Issue on Inequality of Happiness in Nations*, vol. 6, pp. 457–487.

Veenhoven, Ruut. (2009a). *Happy Life Years in 144 Nations 2000–2008: How Long and Happy People Live.* Rotterdam: Erasmus University. Retrieved from http://worlddatabaseofhappiness.eur.nl/hap_nat/findingreports/RankReport_HappyLifeYears.php

Veenhoven, Ruut. (2009b). "*How Do We Assess How Happy We Are?*" In *Happiness, Economics, and Politics,* edited by Dutt, A. and B. Radcliff (Cheltenham, UK: Edward Elgar) pp. 45–69.

Veenhoven, Ruut. (2009c). *States of Nations*. Rotterdam: Erasmus University. Retrieved from the World Database of happiness http://worlddatabaseofhappiness.eur.nl/statnat/statnat_fp.htm

Veenhoven, Ruut. (2009d). *Trend Happy Life Years in Nations 1946–2008*. Retrieved from the World Database of happiness http://worlddatabaseofhappiness.eur.nl/hap_nat/findingreports/TrendReport_HappyLifeYears.pdf

Veenhoven, Ruut. (2009e). *World Database of Happiness: Continuous Register of Research on Subjective Enjoyment of Life*. Rotterdam: Erasmus University. Retrieved from http://worlddatabaseofhappiness.eur.ul

Veenhoven, Ruut. (2010). "Capability and Happiness: Conceptual Differences and Reality Link." *Journal of Socio-Economics*, vol. 39, pp. 344–350.

Veenhoven, Ruut, and Michael R. Hagerty. (2006). "Rising Happiness in Nations 1946–2004: A Reply to Easterlin." *Social Indicators Research*, vol. 79, pp. 421–436.

World Health Organization. (2004). *World Health Report 2004*. Author: Geneva, Switzerland.

4 Capabilities and Technology

Justine Johnstone

A poor man is murdered in a sectarian riot, forced by his poverty to work in an area where he was an outsider and an easy target; a widow battles to survive in a society that barely recognizes her right to exist without a husband—cases like these that exemplify deprivation and insecurity, bad rather than good lives, are the starting point for the capability approach (CA). Pioneered by the development economist Amartya Sen and feminist philosopher Martha Nussbaum, both dissatisfied with conventional approaches to public policy and development, CA is an ambitious project to elaborate a broad liberal consensus on the fundamental requirements of a decent human life as a basis for international comparison and policy formation.

Over the last 10 to 20 years, CA has become influential in fields such as health economics, international development, social policy, and human rights, and recently it has begun to be used as a way of thinking critically about technology by judging it in terms of its contribution to human well-being rather than to economic growth or according to its own internal trajectory. This work is very new and much of it is still experimental and exploratory, but CA's potential in this area is already starting to attract attention in a variety of quarters, including policy, development, and research communities. Information and communication technology (ICT) has tended to be the major focus so far but rising concerns about sustainability, agriculture, climate change, water, and the environment generally suggest that we can in the future expect to see CA tackling a much broader range of contexts in which technology plays a major role.

Although relatively new, CA has an extensive literature and is based on a sophisticated set of ideas that need to be understood before its application to technology can be appreciated. The first two sections of this chapter set out the ideas as developed by the two leading CA theorists, Amartya Sen and Martha Nussbaum. The third section discusses the place of technology within CA and how CA can be used to develop new thinking about technology and its place in society. The final section briefly discusses some problems and future prospects.

AMARTYA SEN: 'DEVELOPMENT AS FREEDOM'

Amartya Sen's is probably the best known and most commonly used version of CA in practical contexts such as measuring inequality or assessing policy, perhaps because Sen was already well-known when he started developing his ideas about capabilities. Sen won the Nobel Prize for economics in 1998 and made his name with work showing that famines result from the failure of social and distributive systems rather than simple food shortages. His concern for the plight of the poorest in society and for ensuring that mechanisms of social distribution take account of their needs has been a fundamental motivation and can be seen as a driving force behind CA.

Sen's account of capabilities starts from dissatisfaction with traditional welfare economics and its emphasis on the distribution of income and material resources. For Sen, this approach is too simplistic because, whether one is trying to measure social deprivation and disadvantage or to do something about it, money and material goods are neither themselves what we are ultimately concerned about, nor are they necessarily very reliable indicators of what we *are* concerned about.

Consider first the detection and measurement of poverty and inequality. Everyone is familiar with measures such as gross domestic product (GDP) and per capita average income to rank countries as rich or poor, with notions such as 'breadline' or 'minimum wage' and 'a dollar a day,' which calibrate advantage and disadvantage in monetary terms. Such measures are highly attractive because data are relatively easy to collect, and relatively objective: income is income; you do not have to interpret it or allow for subjective distortions as you might if you asked people about their happiness or sense of well-being. Income measures are numerical and can be directly compared: if one person earns £10 per hour and another £20 per hour, we are in no doubt as to who is doing better income-wise. But for Sen, this very simplicity is a major weakness, as income alone does not necessarily reveal enough for us to judge how well people are doing in *all* the ways that matter, either in absolute terms or relative to one another. It is certainly not inconceivable that earning £10 per hour working at something you love might be preferable to earning £20 per hour in a dangerous, demeaning, or stressful job.

This problem transmits beyond the level of individual comparisons, to those of groups and countries. It is well-known, for example, that rich countries do not always score more highly on measures such as child health and infant mortality, and nor are they always democratic, free, or productive. Although a certain amount of money and resources are essential for both individuals and societies to prosper, it is not the case that having more necessarily means that a person, group, or country is doing better than those with less. Above a certain minimum, then, we must be careful about using wealth as a general measure of individual or social well-being, however convenient it may be. Sen argues that the public policy and development communities need to refocus their thinking and to consider what it is we actually value in social, political, and economic arrangements and not just what is easy to measure. Thinking about this leads us to realize that there will be multiple outcomes that matter and that to assess how well people are doing, we will need to take into account a much more diverse and richer set of data than simple measures such as income. One of the first applications of this view and still one of the most influential is the *Human Development Index* (HDI), which instead of using per capita GDP to rank countries, also incorporates measures of education and of life expectancy. A quick glance at the index shows that these other measures do not vary with income in any reliable way.

Now consider trying not just to measure but to *do* something about poverty or disadvantage. Here again it does not take much to realize that simply giving people money or goods is not necessarily going to solve their problems. This is because disadvantage tends to be complex and to have multiple dimensions. If you are poor, homeless, addicted to alcohol or heroin, illiterate, or physically disabled, your life is affected in all sorts of ways, many of which may not be improved by money or goods of any simple sort. Sometimes additional money even creates new problems. A homeless person will not have a bank account or a place to keep cash safely; getting regular funds may make her a target for criminals. Of course, resources often do help, but unless they are accompanied by other measures, they may not provide a permanent solution, and it may take a great deal of resources to make a real or lasting difference. The following true story illustrates the limited power of money and goods to change a person's life.[1]

David (not his real name) grew up on the streets of a large city in a middle-ranking developing country. He never knew his parents and seems to have been either orphaned or abandoned as a young child. At some stage, he was taken in by a poor but kind family, and in return, he helped out and contributed to the household when he was able to earn small amounts of money. Most of David's income came from working for a boat owner who took tourists for trips on the river. David's earnings were meager because all the fares went to the boat owner, while David kept merely the tips. One day, some tourists took a trip with David and started asking about his life. They were moved by his story and decided to help by getting together enough money to buy him his own small boat, something he could otherwise never have dreamed of.

But it soon became apparent that being a boat owner meant David needed even more money. He had to buy registration and insurance, rent a mooring, and pay for paperwork and bribes. David's tourist friends helped but still life was difficult. There were far more boats offering river trips than the tourist business could support. Most of the tourists were on organized tours, and the tour operators used the extreme competition among boat owners to drive down prices. Independent travelers were few and far between, and doing business with them was difficult. The police tried to stop boatmen like David from approaching tourists in the street. Once he was caught touting for business and spent the night in prison. But even so, David did manage to earn a little more money now that he had his own boat. Most of the money went straight to the family who had taken him in, to support their many children.

Things got worse when the father of the family became ill and died, leaving David the only breadwinner. His tiny income, although greater than before, was stretched more than ever, especially as the girls of the family were reaching marriageable age and needed dowries if they were to find husbands. As their brother and the eldest man in the family, it was David's duty to provide. He tried many things to improve his lot, from helping a friend to set up a camp for independent travelers, to working in a shop in a seaside resort hundreds of miles away, to traveling to markets far afield to buy and sell clothes. Some of these efforts made a little money, some lost him money, none brought a secure income or enough to support all the others, let alone to look forward to one day having a home and family of his own.

David's tourist friends still call him and occasionally visit or send small amounts of cash. His boat is still his pride and joy and because of the help he received he is a little better off than he would have been. But he is far from secure, and it is hard to see how the problems that make his life so difficult can be fixed. For it is much more than money that constrains David's situation and those of millions like him; it is a whole array of difficulties, ranging from chronic underemployment, illiteracy, large family size, and an extreme shortage of land and housing to corrupt officialdom, callous police, and exploitative tour companies.

David's story is a true one but it could be a textbook illustration of the limited power of resource redistribution to combat poverty. Because problems of poverty and underdevelopment tend to be systemic, Sen points out, the very same reasons that make it difficult for people to acquire resources in the first place tend also to undermine their ability to benefit from them once acquired. He proposes that we turn our attention instead to the ultimate reasons for which we want resources, the good life itself. Influenced by neo-Aristotelian notions of the excellent or flourishing life as one characterized by the tendency to certain types of activity, Sen argues that instead of judging how well off people are just on the basis of wealth or resources—which would count David's acquisition of a boat as a major improvement in his position—we should instead look at the *functionings* that resources enable, what it is that people can actually *do* or *be* as a result. In David's case, a boat made relatively little difference in these terms: it did not enable him to earn a secure income or provide well for his family, and in some ways, it imposed additional burdens on him and made life harder. Such examples litter the history of development and public action and are a salutary lesson for all those who wish to do good.

In turning toward functionings rather than resources, Sen is part of a general trend in welfare economics toward a well-being or outcome rather than an income or input focus. Of course, this move does complicate matters, because we need not only to establish what the outcomes should be but also how to attain them. Attainment will nearly always involve some kind of (re)distribution of resources and income but in a more informed way. Furthermore, attainment also depends on the personal qualities of individuals, and on their preferences and choices. Perhaps David would have benefited more from his boat had he deserted his adoptive family and struck out on his own, especially after the father died, or if he had refused to finance dowries for the daughters. Perhaps these thoughts occurred to him as choices that he rejected, or perhaps he never even considered them as options; perhaps the reasons were to do with his character or with his culture, or perhaps with his sense of duty, or perhaps with convention. We shall never know, but clearly his ability to derive functional benefits from his boat was constrained by personal as well as social factors.

This creates a problem because not all aspects of someone's personal life are appropriate arenas for intervention. It is true that the tourists who paid for David's boat were somewhat dismayed to see him spend most of his income for several years on marrying off his sisters, but should they have intervened and if so what would the result have been? Would David have been able to continue as part of the family? Would he still have been liked and respected in his community? And what about his own autonomy to determine how to spend his income? Such difficulties can be seen writ large in the problems that confront welfare states, and they motivate a distinctive turn in Sen's approach, away from directly aiming at functional outputs and toward *capabilities* instead, by which Sen means a person's *ability* to achieve a high level of functioning (irrespective of whether it is actually achieved). The idea here is to respect personal choice and to limit the sphere of legitimate intervention only to ensuring that people have the *possibility* of attaining a good life, not requiring that they actually do so.

For Sen, capability is best understood in terms of what he calls 'real freedoms'—the fundamental requirements, both positive (freedoms to—to receive an education, earn a living, express oneself, form relationships, etc.) and negative (freedoms from—from oppression, violence, censorship, arbitrary arrest, etc.), for determining and realizing a good life. Sen describes capabilities as

> . . . the various combinations of functionings (beings and doings) that a person can achieve . . . a set of vectors of functionings, reflecting the person's freedom to lead one type of life or another . . . to choose from possible livings . . . (Sen 1992, p. 40)

Not all possible functionings count as capabilities, only those that a person 'has reason to value,' setting CA apart from approaches that focus on subjective measures, such as satisfaction or happiness. Capabilities are clearly related to choices, but only those among rationally valued alternatives. Sen holds off from presenting a specific conceptualization of what it is rational to value. Instead, he advocates a process of public reasoning and social choice within a liberal commitment to value pluralism. However, he does accept that freedoms must be specified and ranked if CA is to have practical application, and in his book *Development as Freedom,* he provides some indications, such as:

> . . . being able to avoid such deprivations as starvation, undernourishment, escapable morbidity and premature mortality, as well as the freedoms that are associated with being literate and numerate, enjoying political participation and uncensored speech . . . (Sen 1999, p. 36)

Capabilities like these are complex phenomena. They depend on resources, personal characteristics, and also a person's social and material environment. They have both what Sen terms process and opportunity aspects, the former relating to broad social conditions and the functioning of institutions, and the latter to things closer to individuals, such as the resources they can access, and their personal skills, knowledge, and abilities. Either or both can compromise a person's capability—political participation, for example, requires not only having the opportunity to vote but also a broader political process that provides a meaningful choice between candidates, unbiased media and a fair system of vote counting. Capabilities are thus built up out of numerous lesser or instrumental freedoms and achievements relating to resources, personal attributes, and social processes, and their mutual interactions. Trying to bring about real improvements is very likely to require intervention on several fronts: as we saw David's problems were largely the result of systemic social process failures rather than his difficult personal situation or lack of resources or opportunities.

So Sen, although he leaves us with a deliberately underspecified and open-ended picture, does provide us with a sophisticated normative framework within which to develop conceptions of the good life in particular circumstances and for particular purposes.

MARTHA NUSSBAUM: UNIVERSAL VALUES AND CENTRAL FUNCTIONAL CAPABILITIES

The American philosopher Martha Nussbaum offers a version of CA similar to Sen's in many ways and compatible with it (Sen 1993) but arising from ethical and political traditions of thought rather than economics. Nussbaum's approach initially emerged out of a feminist concern for the plight of poor women in developing countries and a desire to ground a normative analysis of their situation in universal values that at the same time could be flexibly applied to particular contexts. Developed in a number of works (Nussbaum 2006, 1993, 1992), her capability thesis is perhaps most clearly set out in her books *Women and Human Development* (*WHD,* 2000) and *Frontiers of Justice* (2006). Like Sen, Nussbaum grounds her approach in the Aristotelian idea of human flourishing and the good life as a life of virtuous practice and action. Her work is perhaps more philosophical and less analytical than Sen's in tone, and where Sen's concerns are largely to do with the assessment and comparison of advantage, Nussbaum directs her attention more at political goals of securing rights and guarantees of minimum standards for all, advocating a human rights-like interpretation of capabilities.

Nussbaum's methods and arguments are also rather different from Sen's. Instead of employing economic logic, she offers an account of capabilities that is narrative and developmental, with biological, interpersonal, and environmental factors all playing a part in enabling a person to reach her full potential and to lead the best life that she can. Nussbaum distinguishes three types of capabilities or functionings that are constitutive of a well-lived life: basic capabilities that (all being well) are the innate endowment of all human beings, such as sight, hearing, and sociability; internal capabilities that develop with learning and maturation, such as the ability to form opinions, to express oneself, and to empathize with others; and combined capabilities that are internal capabilities coupled with an environment in which they can be expressed—a society that permits freedom of expression, association, and so on.

As with Sen, Nussbaum does not see all possible functionings as capabilities but only those that are considered necessary for the determination and realization of a good life. Capabilities thus have special ethical significance and impose ethical constraints on social and political arrangements, which must ensure at least a minimum level for all. This view is based on ideas of human dignity and the existence of universal human values at the deepest level, while accepting that they can be realized in a variety of different ways in actual contexts. Nussbaum goes further than Sen in specifying capabilities, providing a list of ten 'central human functional capabilities,' but despite her metaphysical commitment to universal values, she offers a methodological defense for the list, on the grounds that it was derived from cross-cultural discussions that, over many years, generated consensus on the core categories. Adapted and condensed somewhat from *WHD,* these are as follows:

Life: Being able to live a normal life span, not dying prematurely.

Health: Being able to secure bodily and reproductive health, nourishment, shelter.

Bodily integrity: Being able to move about; being able to avoid violence, abuse, and assault; having opportunities for sexual satisfaction and reproductive choice.

Senses, imagination, and thought: Being able to use one's faculties in a fully human way, informed by culture and education; being able to participate in art, music, religion, and culture; having freedom of expression; being able to search for life's meaning in one's own way; being able to have pleasurable experiences and avoid unnecessary pain.

Emotions: Being able to have attachments to things and people; to love, grieve, and experience normal human emotions; being able to develop emotionally and not have one's life blighted by neglect, fear, or abuse.

Practical reason: Being able to form a conception of the good and to plan one's own life.

Affiliation: Being able to live with others, engage in social interaction, being able to feel compassion and to act with justice and friendship; having the social basis of self-respect and being treated with dignity and equal worth (e.g., in having protection from discrimination).

Other species: Being able to live with concern for and in relation to animals, plants, and nature.

Play: Being able to laugh, play, and enjoy recreation.

Control over one's environment: Being able to participate in political choices, having freedom of speech and association, having the real opportunity to hold land and other property on an equal basis with others, having equal rights to employment, having freedom from unwarranted search and seizure.

Although the categories are related to one another in many ways, Nussbaum is clear that they are to be seen as independent and that there must be limits to trade-offs between the different items. In particular, she upholds a moral imperative to ensure a minimum level of each capability for everyone: no amount of economic growth can justify wholesale destruction of the local environment, for example, nor can security be used as an excuse for arbitrary arrest or violent suppression of free speech. There will also be a need to compensate for differences in natural endowments, especially in cases where social mechanisms cannot provide the capability itself but merely the social and material basis for it (the typical situation with positive goods, such as health).

Sen and Nussbaum's accounts are perhaps more different in style than in substance. Both are politically liberal and both locate their notion of the good life in the realm of individual functioning. The good society, for both of them, is the society that enables individuals to flourish; and it is on this basis that they would evaluate and direct intervention in social arrangements. They both recognize the need to consider a range of factors, including the provision and distribution of resources and the development of just institutions. Significantly for our purposes, they both emphasize the role of the external environment, material as well as social, in the development and expression of capabilities, and thus at least implicitly they pave the way to a new normative framework for the application and assessment of technology.

THE PLACE OF TECHNOLOGY IN THE CAPABILITY APPROACH

CA prizes apart any simplistic association between human well-being and technological progress. Arising in part from a humanist critique of economic approaches to development, with their emphasis on technology-driven modernization and growth, it is perhaps not surprising that technology features relatively little in the founding texts of CA and has taken rather longer than domains such as health to receive attention. There is, however, a growing interest in applying CA to technology policy and projects. Much of this work takes the form of assessments of technology post-deployment, but there are also attempts to use CA prospectively in design or policy formation stages.

The place of technology in CA at first sight appears relatively straightforward and entirely instrumental. In Sen's version, technology falls naturally within the category of commodities—goods and services that coupled with the right personal, social and material circumstances may form the basis for enhanced capabilities. Thus, a mobile phone can enhance my ability to maintain important relationships, earn a living, stay safe, and so on—provided of course that relevant 'conversion factors' apply, such as that I have the skill to use the phone, can afford the tariffs, and live in an area covered by the network. Sen himself and subsequent commentators frequently use the bicycle as an example of how commodities expand human capability—in this case by giving users enhanced freedom to move around.

Unlike Sen, Nussbaum does not distinguish commodities from the rest of the external environment, but instead emphasizes the role of 'material and social circumstances' in general, both in the development of internal capabilities and in the ability to express them as combined capabilities:

> . . . developing an internal capacity usually requires favourable external conditions . . . A child raised in an environment without freedom of speech or religion does not develop the same political or religious capabilities as a child who is raised in a nation that protects these liberties. . . . I insist on the twofold importance of material and social circumstances, both in training internal capacities and in letting them express themselves once trained . . . (Nussbaum 2000, p. 85–86)

So in both Sen and Nussbaum's versions it is clear that there is space for technology—along with other human and social creations, such as buildings, landscapes, institutions, organizations, practices, laws, works of art, and objects of all kinds—to play an important role in a person's capabilities. Of course, the picture is not a simple one because, whether one prefers Sen's account, with its threefold analysis of commodities, personal factors, and broader social and material context, or Nussbaum's account, with its levels of capability development, and generic material and social conditions, the contributions that any particular technology makes to capability will always be intertwined with a host of other factors.

In addition, because capabilities are multiple, complex, and incommensurable, it is quite possible that one and the same technology may have both capability enhancing and capability diminishing effects, even for the same person (think about the side effects of treatments such as chemotherapy or the repetitive strain injuries suffered by athletes and musicians). Or it may have positive effects for some and negative effects for others (consider military technologies or the relocation of jobs brought about by information technology), or positive effects in the short term but negative effects in the longer term (such as transport or heating systems that burn fossil fuels). Some of the negative effects may be foreseeable and preventable, either by adapting the technology or by intervening in other parts of the picture (training, regulation, infrastructure, providing incentives, etc.), but others may not even be detected until damage has already been done.

Furthermore, technology is also deeply entangled in the broader social and material context. Technology is not only bicycles and iPods, websites and home blood pressure monitors—artifacts that people encounter as users and know that they are using. It is also a vast background infrastructure, a multiplicity of systems that affect life for millions of people, regardless of whether they are themselves users of it or even aware that it exists. Technology in this broader sense affects us through interactions with others who are users, perhaps most notably professionals whose services we rely on—not only can I monitor my own blood pressure but my doctor can feed my measurements into a database and compare my results with other cases, discuss my condition with experts, look up rare complications, and in general help me to be more healthy through *her* use of technology.

The use of technology in society as a whole can also affect our capabilities independently of our own use—we can eat a varied diet all year round because of high-tech agriculture and refrigerated transport; we can avoid diseases such as cholera and typhoid thanks to modern sanitation; a family in a rural African village can receive remittances from the other side of the world through international banking systems based on modern telecommunications. In each case, the capability benefits (and of course harms) may be felt by people who are in no sense users of technology but merely in some way interacting with a system that is in part technological. There is certainly scope for analyzing these broader, more systemic dimensions of technology within CA but to do so means looking critically at technology's interactions with the broader process dimensions of capability as well as the narrower opportunity aspects.

We also have to recognize that technology offers not only new ways to do old things, but whole new forms of functioning—flying, emailing, surfing the Internet, making videos. This raises the question whether such functionings could sometimes be important enough to be considered capabilities, freedoms we have reason to value, essential components of a well-lived life. Some, perhaps most, are what Sen would call agency freedoms, things that people may wish to do but that are not essential to their well-being. Others, however, seem to transcend this position and to have a claim to being considered substantive well-being goods in their own right. Use of the Internet now seems to be attaining this status in many places, and Sen identifies India as a case in point.

An obvious objection to this argument is that even the fanciest new technologies are really only instrumental goods, and that behind them lie other more important goods—it is not the mobile phone that people really value but maintaining social contacts, for example. Such issues connect with longstanding debates about what if anything is intrinsically valuable, because one can ask of any proposed ultimate good 'why do you want it?' CA, however, imposes a less demanding requirement, accepting as substantive goods all those things that we have reason to value or on which we can reach widespread consensus as essential for well-being. It also accepts that these substantive goods can be mutually instrumental and that they can change over time. Literacy may be more or less essential in modern society, but there have been times when it was rare and, of course, even when it did not exist. Being able to use the Internet may now be starting to take on a similar kind of importance, and many would no doubt think of access to at least basic health care or sanitation systems as components of rather than means to good health. Disentangling constituting and causally contributing aspects of capabilities is a difficult matter whether technology is involved or not, and we may do better to think of a continuum rather than a sharp distinction, perhaps depending on such factors as:

- Whether something is the *only* way to achieve some aspect of well-being.
- Whether something is a *very much cheaper, more efficient, or more reliable* way of accomplishing an aspect of well-being.
- Whether something enables people to achieve *multiple* aspects of well-being that would otherwise all have to be done through a variety of other means and impose much greater costs.
- Whether something enables one to achieve a type of well-being that is *instrumental* to other types of well-being, so-called 'fertile functionings' (Wolff and de-Shalit 2007).

Seen in this light, some technologies seem tantamount to capabilities, not just resources that may enhance the ability to do something important but resources without which the ability would not exist or that promote more general or robust increases in capabilities. Think of what antiretroviral drugs mean to an AIDS patient or a prosthetic limb to an amputee or remote sensing and earthmoving equipment to a trapped earthquake survivor.

So technology can be related to capability in a variety of ways, ranging from instrumental use (opportunity) to being a part of the broader context (process) to, at times, amounting to the very existence of capability (constitutive). The way is thus open for a potentially very wide range of applications of CA, from high-level policy in areas such as technology promotion, regulation, infrastructure, and public service provision; to the detailed planning, implementation, and assessment of specific projects; to broad analyses of sociotechnological change. It is still in its early days and relatively little has been published, but already there are indications that CA is making an impact.

We have, for example, recently seen it used for such diverse purposes as reframing technology-based development projects in a way that promotes freedom and empowerment rather than predetermined outcomes (Kleine 2011), analyzing and evaluating ICT interventions in developing countries (Grunfeld, Pin, and Seng 2010; Ratan and Bailur 2007; Gigler 2004; Madon 2004), assessing the use of artificial intelligence assistive technologies in health care (Coeckelbergh 2010), articulating a theory of 'appropriate technology' (Oosterlaken under review), identifying social exclusion in the information society (Zheng and Walsham 2008), and for dealing with systemic issues in computer ethics (Johnstone 2007). At the same time, CA is starting to inform technology policy, for example, in the European Commission (Condagnone, Cilli, Biagi, Foley, and De Luca 2009) and United Nations (Hamel 2010) and to influence grassroots development work and activism (see for example Pearson 2005). Perhaps most significant is the appearance of anthologized collections of work applying CA to the study of technology, notably a special issue of the journal *Ethics and Information Technology* (volume 13, number 2, 2011) and a forthcoming volume entitled *The Capability Approach, Technology and Design* edited by Jeroen van den Hoven and Ilse Oosterlaken.

PROBLEMS AND PROSPECTS

CA is of course by no means free of problems and has attracted critical comment for a variety of reasons. A full discussion is beyond the scope of this chapter, but it is worth drawing attention to four issues that pose particular challenges. First, and perhaps most immediately is the very general and abstract nature of the underlying work. Although the lack of specification is intentional, it leaves researchers and practitioners with much to do to develop operational tools.

Second, there is the very complexity of the approach itself with its attempt to encompass the full range of capability determinants, both internal and external, opportunity and process related, and their multiple interactions. Capturing all those that are relevant in a particular case presents a major challenge and opens the door to a potentially vast range of data.

Third, there is the question of whether to focus on achieved or possible functionings (assuming that they can even be separated) and if the latter how to detect when a real but unachieved possibility exists (conjuring up the specter of enquiries into the nature of choice and free will). Skeptical concerns can also be raised about inferring the existence of capability from actual functioning because functioning can be accidental or unchosen. Possibly the process component of the analysis can deal with this by introducing some notion of reliability or reproducibility, but it is so far unclear how this is to be done.

Finally, there are normative questions: can we really hope to get agreement on a general list of goods and on their contextual specifications, even among a small group of people? What is the relationship between the metaphysical commitment to universal values and the advocated methodologies of consensus, participation, and dialogue? How do we know that these or other methods will in fact yield goods that instantiate underlying universal values? How are we to decide on threshold levels of functioning? And how are we to make socially just choices between different outcomes above the threshold level?

CA is still a relatively new and underdeveloped approach to social evaluation and intervention, so it is not surprising that so many questions remain. Its philosophical sophistication coupled with radical empirical underspecification is a challenging albeit exciting and potentially rewarding combination, and we are seeing a steadily increasing degree of influence within social policy and research. At the same time, technology is emerging within the CA community as an area of rapidly expanding interest. In 2009, a Technology and Design thematic group was formed within the Human Development and Capabilities Association (2009), and innovation was taken as the central theme of the association's 2011 annual conference. These trends look set to continue, but CA has yet to prove itself in practice. Its future will depend on showing the added value of its potentially enormous expansion of the informational basis of technology assessment, and in terms of intervention, how it complements or improves on existing sociotechnical and human-centered approaches.

NOTES

1. The story of David is based on the personal experience of the author, who witnessed the events described over a number of years.

REFERENCES

Coeckelbergh, Mark. (2010). "Health Care, Capabilities and AI Assistive Technologies," *Ethical Theory and Moral Practice*, vol. 13, pp. 181–190. doi: 10.1007/s10677-009-9186-2

Condagnone, Cristiano, Valentina Cilli, Federico Biagi, Paul Foley, and Annalisa De Luca. (2009). *Vienna Study on Inclusive Innovation for Growth and Cohesion: Modelling and Demonstrating the Impact of eInclusion*. Report prepared for the European Commission, DG Information Society and Media, ICT for Inclusion Unit. Retrieved from http://ec.europa.eu/information_society/activities/einclusion/library/studies/docs/eco_report.pdf

Gigler, Björn-Sören. (2004). "Including the Excluded: Can ICTs Empower Poor Communities? Towards an Alternative Evaluation Framework Based on the Capability Approach." Paper presented at the 4th International Conference on the Capability Approach, University of Pavia, Italy, September 5–7, 2004.

Grunfeld, Helena, Tara Pin, and Dara Seng. (2010). "Challenges in Operationalising the Capability Approach for Evaluating the Contribution of the Cambodian ICT4D Project, iREACH, to Capabilities, Empowerment and Sustainability." Paper presented at the 2010 Conference of the Human Development and Capability Approach Association, University of Jordan, Amman, September 21–23, *2010*.

Hamel, Jean-Yves. (2010). "ICT4D and the Human Development and Capabilities Approach: The Potentials of Information and Communication Technology," Human Development Research Paper 2010/37, United Nations Development Programme. Retrieved from http://hdr.undp.org/en/reports/global/hdr2010/papers/HDRP_2010_37.pdf

Human Development and Capability Association. (2009). "Thematic Groups." Retrieved from http://www.capabilityapproach.com/Thematic.php

Johnstone, Justine. (2007). "Technology as Empowerment," *Ethics and Information Technology*, vol. 9, pp. 73–87. doi: 10.1007/s10676–006–9127-x

Kleine, Dorothea. (2011). "The Capability Approach and the 'Medium of Choice': Steps Towards Conceptualising Information and Communication Technologies for Development," *Ethics and Information Technology*, vol. 13, no. 2, pp. 119–130. doi: 10.1007/s10676–010–9251–5

Madon, Shirin. (2004). "Evaluating the Developmental Impact of E-governance Initiatives: An Exploratory Framework," *Electronic Journal of Information Systems in Developing Countries*, vol. 20, no. 5, 1–13. Retrieved from http://unpan1.un.org/intradoc/groups/public/documents/APCITY/UNPAN021286.pdf

Nussbaum, Martha C. (1992). "Human Functioning and Social Justice: In Defense of Aristotelian Essentialism," *Political Theory*, vol. 20, pp. 202–246. doi: 10.1177/0090591792020002002

Nussbaum, Martha C. (1993). "Non-Relative Virtues: An Aristotelian Approach." In *The Quality of Life*, edited by Nussbaum, Martha C. and Amartya Sen (Oxford: Clarendon Press) 242–269.

Nussbaum, Martha C. (2000). *Women and Human Development: The Capabilities Approach*. Cambridge: Cambridge University Press.

Nussbaum, Martha C. (2006). *Frontiers of Justice*. Cambridge, MA: Belknap Press of Harvard University Press.

Oosterlaken, Ilse. (Under review). "Inappropriate Artefacts, Unjust Design? Human Diversity as a Key Concern in the Capability Approach and Inclusive Design." In *The Capability Approach, Technology and Design*, edited by Oosterlaken, Ilse and Jeroen van den Hoven. Dordrecht: Springer.

Pearson, Noel. (2005). *The Cape York Agenda: "Fundamental Transformation Through Radical Reform."* Cairns, Australia: Cape York Institute for Policy and Leadership. Retrieved from http://www.cyi.org.au/WEBSITE%20uploads/Documents/Cape%20York%20Agenda%20final.pdf

Ratan, Aishwarya L., and Savita Bailur. (2007). "Welfare, Agency and 'ICT for Development.'" In Proceedings of the 2nd IEEE/ACM International Conference on Information and Communication Technologies and Development in Bangalore, India, December 15–16, 2007. Retrieved from http://research.microsoft.com/pubs/78278/O_85_Welfare_agency_ICTD2007.pdf

Sen, Amartya. (1992). *Inequality Re-Examined*. Oxford: Clarendon Press.

Sen, Amartya. (1993). "Capability and Well-Being." In *The Quality of Life*, edited by Nussbaum, Martha C. and Amartya Sen (Oxford: Clarendon Press) 30–53.

Sen, Amartya. (1999). *Development as Freedom*. Oxford: Oxford University Press.

Wolff, Jonathan, and Avner de-Shalit. (2007). *Disadvantage*. Oxford: Oxford University Press.

Zheng, Yingqin, and Geoff Walsham. (2008). "Inequality of What? Social Exclusion in the e-Society as Capability Deprivation," *Information Technology and People*, vol. 21, no. 3, pp. 222–243. doi: 10.1108/09593840810896000

5 Happiness and Meaning in a Technological Age

A Psychological Approach

Michael Steger and Joo Yeon Shin

The quest for happiness and meaning in life can be seen as one of humanity's most enduring concerns. In looking forward to what happiness and meaning may look like in a technological age, it is appealing to look backward first, and consider what happiness and meaning looked like in a pre-technological age. Written on a dozen 2,600 year old clay tablets, the Epic of Gilgamesh is considered to be the world's oldest surviving story. This tale relates a fantastical series of adventures pursued by the king of Uruk, who ruled the Sumerian city 4,700 years ago[1]. Gilgamesh craved fame and adventure, leading to both great achievements and harrowing disappointments. The emotional core of this tale centers on a strange relationship, and a couple of odd quests the titular king pursues. The relationship is between Gilgamesh and a wild man, Enkidu, created by the god Anu to challenge the two-thirds divine Gilgamesh. At first rivals, they become friends.

The first of the central quests is initiated because Gilgamesh and his friend have become lazy and corpulent. Their lives are monotonous, and although rich in pleasure, they are bereft of challenges. Their quest is proposed by Gilgamesh to liven up their overly sedentary and tepid lives and help them achieve greater fame. Gilgamesh determines that the most apt challenge is to cut down a legendary forest and destroy its guardian monster. Enkidu attempts to dissuade him, but eventually agrees to go with Gilgamesh. They defeat the guardian, uproot the sacred trees, and later kill the gigantic Bull of Heaven sent by the spurned goddess, Ishtar. Their vainglorious pursuit of stimulation and acclaim ends, unsurprisingly, in disaster. During this first quest, Enkidu incurs a curse and soon dies in excruciating agony. Gilgamesh builds a monumental gate, and expresses his emotional devastation, "I am going to die!—am I not like Enkidu? Deep sadness penetrates my core, I fear death, and now roam the wilderness . . . " This leads to the second quest, Gilgamesh's pursuit of immortality. Gilgamesh finds a guide and labors to meet a wise soul who chastises him for his quest, telling him that although the gods live forever, "to the fool they gave beer dregs instead of butter," and that his quest would never change the inevitability of his death. Gilgamesh refuses this answer and pushes on. Eventually, of course, this quest also ends in disappointment:

For whom have I labored? For whom have I journeyed?
For whom have I suffered?
I have gained absolutely nothing for myself,
I have only profited the snake, the ground lion!

As he reconciles with his failure to surpass his mortality, Gilgamesh consoles himself by taking the guide who accompanied him back to the great city he constructed. Thus, on the brink of a final realization that his days are numbered, the once-trivial and vain king seems to seek justification for his life in what he has built and accomplished in his lamentably finite life.

The themes in the tale of Gilgamesh are familiar, even to psychologists studying happiness and meaning today. Interpersonal rivalry and friendship, the desire for new achievements and growth, the hungry striving for status and fame, and the thirst for a way to achieve immortality and outlast our mortal days: not only do these themes fill contemporary novels, they all appear in modern theories on achieving happiness and meaning in life. Interestingly, the world's oldest story also provides a hint at one possible role for technology in developing happiness. The first quest, although entailing the tragic loss of Enkidu, culminates with the construction of a magnificent gate for the city of Uruk. The second quest, the failed bid for immortality, ends at that same city's gate, with Gilgamesh extolling the marvels of its construction. In this story, technology can be seen as a way to order and regulate our surroundings, express human ingenuity, and make our mark on the world. Of course, it can also be noted that similar advancements made it possible for Gilgamesh and Enkidu to become weak and crave adventure. Agricultural technology created an easy abundance of food and wine grapes for those in positions of power, and transportation and draft animal technology facilitated the trade that delivered the luxury products that made the idleness of Gilgamesh and Enkidu so enjoyable. There is a tension in the tale of Gilgamesh between deriving happiness from the pleasures of the world and deriving happiness from labor, struggle, and ultimately mattering.

Psychological debate about happiness and flourishing also traces this tension. In this chapter, we propose a new way to look at this debate, by linking pleasure-based happiness to "hot" processing brain systems and by linking mattering-based happiness to "cold" processing brain systems. We first review psychological thought on what people are ultimately striving for, known variously as fulfillment, flourishing, and optimal functioning, and which we will generically refer to as the Good life. In particular, we will show where happiness and meaning best align in the debate on the true nature of the Good life. Following this, we examine how broad categories of technological advancement may appeal in unique ways to "hot" and "cold" ways of experiencing happiness.

PSYCHOLOGY OF THE GOOD LIFE

Psychological studies of the Good life are concerned with understanding ways of achieving well-being, thriving, and flourishing. Such studies have been conducted for decades, and the intellectual history of the Good life is even older. However, research in this area has exploded following the widespread acceptance of the positive psychology movement (Seligman and Csikszentmihalyi 2000). In psychological terms, the Good life is marked by more than a mere absence of illness, psychopathology, pain, or weakness; it also includes the presence of positive experiences and emotions, such as happiness, satisfaction, contentment, and joy, as well as the cultivation of positive human traits and characteristics, such as personal strengths, resilience, meaning, and purpose. Although the general parameters of having few negative experiences and traits combined with having abundant positive traits and experiences is highly agreeable to most psychologists, there is nonetheless considerable debate about which kinds of traits and experiences, the timing of them, their relative versus absolute levels, and so on.

Psychological theories of the Good life have generally conformed to philosophical distinctions that trace their roots back to Aristotle. Aristotle provided the first elaborated consideration of the Good life, contrasting the pursuit of pleasure with the development of one's most virtuous self. As they've been considered by psychologists, these two approaches primarily have been used to support subjective and objective theories of the Good life (Deci and Ryan 2000; Diener 2000; Ryan and Deci 2001; Ryff and Singer 1998; Waterman 1993). In psychology, the subjective tradition has been most comprehensively represented by Diener's (2000) subjective well-being model. However, there are numerous theories in the objective well-being tradition; no single theory has gained dominance.

The subjective tradition focuses on achieving the Good life by having a greater balance of positive emotional experiences than negative experiences and subjectively evaluating life as satisfying (known as subjective well-being, Diener 2000). The term "happiness" is usually included in the subjective tradition. Although certainly someone could read a great deal into the question "are you truly happy?," the psychological literature most often considers happiness to refer to relatively speedy judgments about the satisfactoriness of whatever is happening in the moment. In this typical view, happiness can be caused by almost anything. Beyond saying that the dimensions of emotions and satisfaction are important to assess, the subjective tradition tends not to stipulate that any particular information needs to be considered to form judgments of happiness. Subjective theories have typically referred to happiness as the desired end, but because reasons why people feel happy can vary from person to person, moment to moment, and may range from incredibly profound to completely trivial, happiness is seen less as the end of the road than as a marker of interesting things that are happening within a person. Thus, its aims are typically less focused on what causes people to feel happy than on seeking to understand what people who are experiencing happiness do differently than people who are not. Happiness is in the eye of the beholder and is (by virtue of the operationalization) notable for the abundance of pleasure and absence of pain. As a tavern keeper tries to tell Gilgamesh, apparently anticipating the Greek philosophers Gorgias, Aristippus, and Epicurus:

> *Humans are born, they live, then they die, this is the order that the gods have decreed. But until the end comes, enjoy your life, spend it in happiness, not despair. Savor your food, make each of your days a delight, bathe and anoint yourself, wear bright clothes that are sparkling clean, let music and dancing fill your house, love the child who holds you by the hand, and give your wife pleasure in your embrace. That is the best way for a man to live.*

Few people would argue that happiness should *exclude* these things. However, subjective perspectives are open to criticism because of their very subjectivity. For example, it is at least possible that someone who leads a morally reprehensible life (Hitler's name gets thrown around a lot by people who object to the subjective tradition) can be perfectly happy (e.g., Ryan and Deci 2001). One challenge facing the subjective tradition is to account for the philosophical or conceptual objection to someone like Hitler qualifying as being happy or living the Good life. An additional challenge is to account for the fact that sometimes pain and discomfort are not only inevitable, but desirable, as in the sorrow we feel when we lose a loved one (Ryan and Deci 2001).

Objective theories, which can be traced back to humanistic psychologists such as Erich Fromm (1956), Viktor E. Frankl (1963), and especially Abraham Maslow (1971), attempt to handle the conundrum of Hitler and the necessity of unpleasant feelings to the Good life by articulating particular definitions or qualifications for the Good life, or by prescribing a set of qualities someone would need to have to be considered to be functioning optimally. These theories are often framed in terms of achieving the highest levels of fulfillment, maximizing one's potential, or attaining optimal human functioning (e.g., Keyes, Shmotkin, and Ryff 2002). They prioritize self-awareness, self-direction, intentional activity, and motivational qualities. These theories have typically embraced meaning in life, viewing it as an important by-product of flourishing (Ryan and Deci 2001) or as a desirable outcome in and of itself.

Whereas happiness has typically referred to subjectively pleasing psychological states, meaning in life refers to the sense people make of their lives, the significance and coherence they derive from their life events, and the purposes or missions that drive their behavior over the course of their lifetime (Kashdan and McKnight 2009; Steger, Frazier, Oishi, and Kaler 2006). Meaning in life is unique among psychological variables by virtue of its collusion of cognitive, sense-making qualities and conative, motivational qualities. Research on meaning in life is fairly typical of research in the objective tradition—it focuses quite a bit on how people attain meaning in life, on what kinds of sources and activities yield meaning, and on establishing that meaning is somehow vital to human functioning. From the objective tradition, this makes a great deal of sense because meaning, as an objectively good thing to have, is the end of the road to some extent, and by virtue of it being an "objectively" necessary thing, meaning should be critical to how people function.

Whereas happiness is about pleasure at some basic level, meaning is about finding significance, importance, and purpose in life; it is about mattering. When people feel they matter, they are expressing their belief and perception that they are important and significant to the world and to the lives of other people (Elliott, Kao, and Grant 2004; Rosenberg and McCullough 1981). We see mattering as a primary dimension of the cognitive component of meaning in life, which we term comprehension (Steger 2009; Steger in press). Comprehension captures people's abilities to make sense of their selves, the world around them, and their place in the world. We matter when we can find sense in being alive and see some important ties and relationships between ourselves and other entities in the world (especially people, Baumeister 1991).

Despite the existence of various camps, it is not clear whether these subjective and objective theories represent different forms of the Good life, or whether they simply prioritize different aspects of the human experience. Subjective approaches lend more credence to pleasant states and 'gut feelings,' whereas objective approaches lend more credence to intentional self-development and autonomous endorsement of positive social mores. But this does not mean that the most desirable outcome includes only one or the other (see Kashdan, Biswas-Diener, and King 2008). In fact, it is difficult to imagine a persuasive argument that "the Good life" should only consist of pleasure, but not a sense of mattering, or that it should only be about pleasureless striving and an inert sense that one's life matters. The Good life should have both.

SIMPLIFYING THE SUBJECTIVE-OBJECTIVE DEBATE: THE DUAL-PROCESS MODEL OF THE GOOD LIFE

The conclusion that the Good life should have both pleasure and mattering assumes these are separate, or at least separable, qualities. Depending on how they are operationalized in research, pleasure (as in positive emotions, the absence of negative emotions, happiness, pleasant daily events, etc.) and mattering (meaning in life, self-actualization, life satisfaction, a sense of social belonging, etc.) are fairly strongly correlated (e.g., Compton, Smith, Cornish, and Qualls 1996; Lyubomirsky, King and Diener 2005;

Diener 2000; King, Hicks, Krull, and Baker 2006; Steger, Frazier, Oishi, and Kaler 2006; Steger in press). However, most of these same studies also demonstrate that the relations among these variables do not reach a level that would require them to be considered as equivalent, at least not statistically (e.g., McGregor and Little 1998).

We would like to accept for the time being that pleasure and mattering are distinguishable, although they likely function in close partnership. We are not arguing that pleasure is "subjective" and mattering is "objective"—either one could be either subjective (subjective mattering might be "parenting is the most important job anyone can do") or objective (objective pleasure might be the warm feelings of emotional intimacy)—or that either one is a "better" kind of fulfillment. We also agree that pleasure and mattering are likely to co-occur (Kashdan, Biswas-Diener, and King 2008), and that they should co-occur to attain the Good life.

A psychological account for the Good life therefore should be concerned less with different kinds of experiences and more with how humans are able to experience pleasure and mattering. Using an evolutionary perspective, we propose that people experience these two domains of the Good life through the operations of two separate but related neurological systems. In essence, we are proposing a "dual-process model" of the Good life. We base our dual-process model of the Good life on existing models and on our understanding of the well-being, happiness, and meaning in life literatures. Empirical exploration of this model is yet to be performed.

Dual process models have been proposed to explain a wide variety of human behaviors and attributes, such as memory and learning (e.g., Smith and DeCoster 2000). One process is usually automatic, and the other is conscious or deliberative. That is, we learn or remember in ways that operate both unconsciously and automatically (such as associating the smell of baked sweets with comfort) and consciously and deliberatively (such as associating a shape that looks like some rocks from Stonehenge with the number 3.14). As biological organisms, humans evolved from simpler creatures, and an argument can be made that we have at least one "brain" in our heads. The oldest set of structures in the brain (the hindbrain, which includes the brainstem, medulla, and reticular formation, and the midbrain, which includes structures associated with the limbic system, such as the hippocampus, amygdala, thalamus, and cingulate cortex) can be considered a "hot" system that helps the organism quickly evaluate safety and welfare on the basis of subjective impressions of pleasure and pain, and automatically learn from pleasurable or painful experiences (e.g., Barrett, Tugade, and Engle 2004; Kahneman 2003; Sloman 1996).

Superimposed over this phylogenetically old system of emotional and implicit processes is a newer system—the cerebral cortex—that evolved along with the advanced human cognitive capacities. This "cold" system is used in more rational and deliberative evaluations of how well people are able to make sense of their circumstances and achieve a reassuring degree of predictability. The "hot" system strongly resembles subjective ideas about happiness in that pleasant states are highly prized and influential. The "cold" system strongly resembles objective ideas about happiness in that self-reflective processes and conscious comparisons of one's values, qualities, and accomplishments with socialized standards and goals are seen to drive behavior (see table 5.1).

For the sake of simplifying the debate, we propose that the "hot" system helps us quickly evaluate whether things are "all right" in our lives. If we feel good, things are good. If we feel bad, things are bad. We propose that the "cold" system comes into play when we evaluate how our lives are going overall, in the abstract and longer-term sense. If we make sense of our circumstances and are progressing toward our aspirations, things are good. If not, things are bad. It should be possible for these systems to calculate conflicting estimations of how we are doing in terms of the Good life (e.g., "My grandmother's eulogist is so sexy!" "I am so bored volunteering with these fatally ill children" "Boy, childbirth hurts—this kid better be worth it!"), but usually they play off one another. For example, watching a funny

Table 5.1 A Dual Pprocess Model of Happiness and Meaning

Hot and Cold Good Lives in a Technological Age

	"Hot" route	"Cold" route
Basic system	Quickly sense internal and environmental threats and resources	Deliberatively formulates and assesses hypotheses, abstract representations, and planning
Neurological substrate	Hindbrain, limbic system	Frontal lobes of the cerebral cortex
Related constructs	Pleasure, pain, positive emotions, negative emotions	Meaning in life, intrinsic motivation, life satisfaction, identity and self-worth
Examples	Delicious food	Food that reminds you of a loved one who has passed on
	The nonverbal joy of a parent seeing a baby's first smile	Saving money in a child's college fund
	Masturbating to internet pornography	Having some quiet time with a romantic partner
	A favorite scenic vista	A culturally revered work of art

video clip makes people think their lives are more meaningful (King, Hicks, Krull, and Baker 2006), and people who appraise their lives as more meaningful report higher daily levels of positive emotions (Steger, Kashdan, and Oishi 2008).

THE DUAL PROCESS MODEL AND TECHNOLOGY

Using the dual process model as our lens, we next consider whether technological advances might or might not influence our ability to achieve the Good life. We will examine technological impacts in the five "ways to happiness" identified by the New Economics Foundation (NEF). NEF, founded in 1986 by the leaders of the Other Economic Summit, is an independent think-and-do tank that aims to improve quality of life by promoting innovative solutions that challenge mainstream thinking on economic, environmental, and social issues (The Happy Planet Index 2.0 2009). In 2008, NEF was commissioned by the U.K. government's Foresight Project on Mental Capital and Well-being to develop a set of evidence-based actions to improve personal well-being. They reviewed the interdisciplinary social science research reports of over 400 scientists from across the world. Through this extensive examination of psychological and economic literature, the five evidence-based, everyday actions to improve well-being were identified (see Aked, Marks, Cordon, and Thompson 2008 for more details).

These five ways are: Connect, Be Active, Take Notice, Keep Learning, and Give. These five ways are concerning building good relationships, maintaining physical health and wellness, savoring, curiosity and education, and self-transcendence. It is interesting to note that the tavern keeper from the Epic of Gilgamesh had a 2,700-year head start on many of these topics:

- *Savor your food* [Take Notice]
- *Make each of your days a delight* [Take Notice, Be Active]
- *Bathe and anoint yourself* [Be Active, in a health and hygiene sense]
- *Wear bright clothes that are sparkling clean* [Be Active—health and hygiene]
- *Let music and dancing fill your house* [Be Active, Keep Learning—learning to play an instrument is an example given by the NEF]
- *Love the child who holds you by the hand* [Connect], and
- *Give your wife pleasure in your embrace* [definitely Connect]

We must necessarily limit the number of technological advances we consider here, so we will consider how gadgets (big tvs, evolving media delivery systems), communication (cell phones, computers), medicine (genomic and genetic science, treatments), and parahuman (robotics, artificial life) advances may interact with our hot and cold flourishing systems in each of the five ways to happiness. We also would like to add a "way" to the Good life to NEF's list. After all, their concern was with happiness, which may slant things toward pleasure. We would suggest adding "Believe" to the list, to make reference to the roles of human spiritual and religious beliefs, as well as nonreligious and nonspiritual beliefs and cognitive models that people use to understand and make sense of our existence in the grand scheme of things. Believe may be more closely linked to mattering than pleasure. More in-depth considerations of individual technologies are available elsewhere (e.g., Bittman, Brown, and Wajcman 2009; Disalvo 2010; Fisher and Robinson 2009; Gelles 2009; Kang 2007; Robinson and Martin 2009; Salleh 2000; Singer 2005; Surowieki 2005; Wight and Raley 2009; Yu and Chou 2009).

GADGETS

The variety of stimulations and diversions that gadgets provide people has been rapidly expanding. From three television stations to hundreds, from "Pong" to thousands of video game titles released every year, from an arcane governmental file-sharing system to 30 billion webpages (Boutell.com, n.d.): the amount and specificity of information people can obtain seems practically limitless. The boundlessness of this resource presents competing challenges. On the one hand, never-ending stimulation provides the "hot" system with incentives to direct time toward the gadgets, rather than toward other activities, such as connecting and being active. An example of this might be someone who spends a great amount of time on a single-player video game. On the other hand, people are able to find communities of like-minded people, bonded by amazingly narrow interests (for example, there appear to be 47 members of the Yahoo! Group dedicated to interest in koalas; perhaps not quite as impressive as the tens of millions who play interactive video games like World of Warcraft or Halo), which should enable the "cold" system to make meaningful connections (e.g., Lanthie and Windham 2004).

Gadgets also have enabled the development of tools such as Google Scholar and Wikipedia, which are distributing previously rarefied technical information on an unprecedented scale. Both the "hot" and "cold" systems should work together to support continued learning, both in terms of helping people experience visceral pleasures with people sharing mutual interests (as in World of Warcraft), and also in terms of more deliberative research facilitated by the additional information delivered by gadgets (such as Google Scholar). Similarly, both systems should support giving, as evidenced by the enormous outpouring of financial aid for rescue and recovery efforts following the Haitian earthquake (over half a billion U.S. dollars in the first 15 days (Preston and Wallace 2010). It seems unlikely that gadgets promote a great deal of taking notice through either hot or cold appeals. The "cold" route to taking notice is akin to mindfulness, which is a practiced perspective of being in the moment and nonjudgmentally aware of what is happening (Brown and Ryan 2003; Hanh 1975; Jain, Shapiro, Swanick, Roesch, Mills, Bell, and Schwartz 2007). The key selling point of many gadget advances seems to be the sheer amount of data they can quickly deliver, which is fairly antithetical to savoring a particular stimulus. Finally, as gadgets are capable of more and more astonishing abilities (refrigerators can tell us when to buy more lettuce!), it is possible that they could increase momentary experiences of awe, which may support beliefs through the "hot" system, while also challenging traditional belief systems due to the increasing array of incredible things that humans are directly responsible for.

In sum, Gadgets would seem likely to benefit some ways to happiness, such as to Keep Learning and Give, and possibly to Connect, that may be accomplished through the mediation of gadgets themselves. However, gadgets would seem to appeal most strongly to the "hot" systems that leverage pleasure, which may encourage increasing time spent tinkering with gadgets rather than time in which people could Be Active or Take Notice. The impact on Believe might depend largely on what people believe in and how those beliefs explain the increasingly "miraculous" gadgets around us. Cell phones that instantly stream videoconferencing around the world could be taken as a sign of an active Creator's hand guiding creative developments, or perhaps as a sign of the evolved complexity and ingenuity of the human mind

COMMUNICATION

Given that there is little evidence that mediated communication, such as mobile phones or Internet chatting, diminishes people's satisfaction with interactions (see Surowiecki 2005), communication technology should benefit at least the "hot" route to connecting. However, 'disembodied' communication might derail the "cold" route (Kang 2007), and it seems at least possible that skills in reading the critical nonverbal communication humans evolved to use (e.g., Keltner 2009) might be underdeveloped if email, text messages, phone calls, and chatrooms form the bulk of social interactions. There would not seem to be any direct effects of communication technologies on Being Active, Taking Notice, Keeping Learning, or Believing—any impact on these areas would depend on when, how much, and toward what ends people use the technology, factors that are not embedded in the technology itself. Communication technology would seem to serve giving in ways similar to gadgets, potentially giving opportunities greater publicity and making giving easier.

In sum, communication technologies probably will have net positive impacts on happiness, primarily through making it easier to Connect and Give, with the caveat that it remains to be seen what communication problems could emerge among a population whose predominant forms of communication are electronically mediated.

MEDICINE

The primary impact of medical technologies is—hopefully—on increasing the longevity of human quality of life. This includes both lengthening the life span and also making those living years better by attenuating the impacts of negative health factors, such as chronic illness or injury. As such, medicine should have its primary impact on being active. The "hot" route should respond to any reductions in pain and any gains in functioning accomplished through medical treatments. The "cold" route should respond through expanded opportunities to engage in meaningful activity. For example, if part of the Good life from the objective, Aristotelian perspective is maximizing one's potential (e.g., Ryan and Deci 2001), then being able to devote extended years to musical, athletic, or dance avocations should be a good thing. Reducing pain should also help people take notice, as more pleasant states are associated with a broadening in perspective (e.g., Fredrickson 1998). It is not clear how medical technology could help people cultivate mindfulness, the "cold" route to taking notice. In terms of medical practice, however, it seems equally likely that the seemingly constant perseveration on immortality (e.g., Häyry 2011) could push a nonmindful obsession with the future, or the increasing willingness of medical practitioners to examine spiritually based approaches to health could increase the acceptance and practice of mindfulness.

Medical technology seems unlikely to directly influence connecting or keeping learning going. An easy assumption is that if people live longer, they'll connect and learn longer, but it is not clear whether this benefit would indeed materialize. One possibly ironic influence of medical technology could emerge in giving. If people live longer, particularly after retirement, there is an increased likelihood that they will consume a greater portion of their financial resources, limiting the amount that could be bequeathed upon death. This may be particularly true if medical technology merely keeps us alive, rather than keeping us alive and well (Datko 2010; but see also Bloom, Canning, and Graham 2003). Finally, the human ability to alter the course of life and death would seem to pose a challenge to many people's beliefs, and the clash of medical technology and beliefs may be a major factor in the development of this technology (e.g., stem cell research).

In sum, the assumption being made here is that medical technology will enable people to live longer and better. This should enable greater use of the Be Active, and possibly the Take Notice ways to happiness. However, taking David G. Myers's and Ed Diener's (1995) famous breakdown of the lack of relationship between increased wealth and life satisfaction as a starting point, we attempted to illustrate whether there was a visibly significant relationship between life expectancy and life satisfaction. As can be seen in figure 5.1, the very least that might be inferred from these data are that life expectancy and life satisfaction do not appear to increase in lockstep.

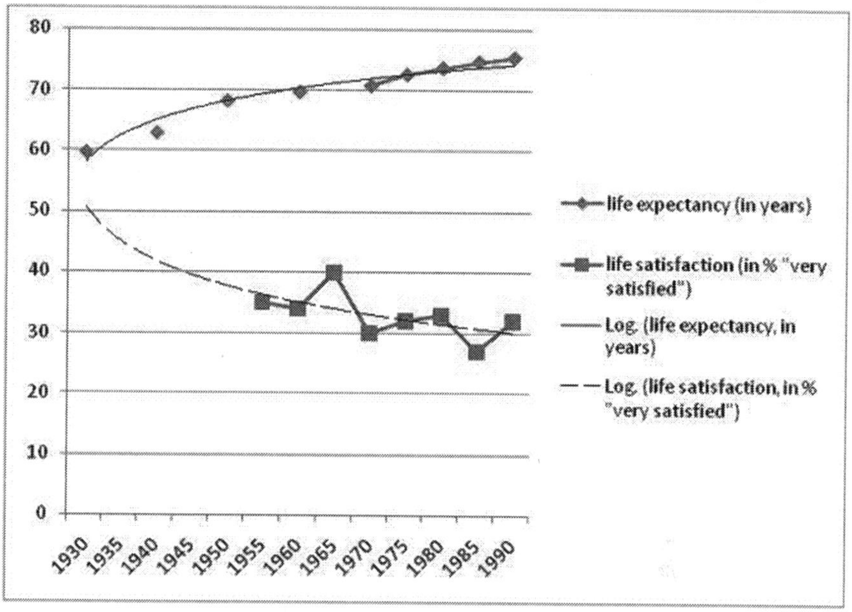

Figure 5.1 Life expectancy and life satisfaction do not appear to increase together. The figure portrays available points as well as the logarithmic trend line for both life expectancy and life satisfaction.
Source: Myers & Diener, 1995; Infoplease, 2007 .

PARAHUMAN

Robots build our cars, vacuum our living rooms, and even (apparently) make great romantic partners (e.g., Iggulden 2008). In addition to artificial life forms, technology is increasingly making its way into the human body. Although that has long been the case, the technology may now be considered to be even better than the real thing, as in the case of Paralympian Oscar Pistorius, who was denied the chance to compete in the Beijing Summer Olympics because his prosthetic lower legs put him at an unfair disadvantage (Associated Press 2008). Further, biological technology makes modifying the form and genetic content of life forms easier all the time, and it is probably only a matter of time before humans try the products of test tube organs and genetic modifications. As technology becomes better able to mimic, and perhaps surpass, the human organism, the clearest implication is for a challenge to fairly widespread beliefs about what it means to be human (Parsons 2010).

The primary challenge is to the "cold" system, which grapples with questions of whether robots deserve legal rights, whether people who are unable to provide legal consent can be technologically modified, and whether a couple comprised of a robot and a human can adopt a human child, and even whether a partially robotic person can compete in professional sports (e.g., Asaro 2006). There may even be a "hot" route detriment to beliefs; just look at photos of Roxxxy, the first sex robot, and see whether it makes you more or less happy (Parsons 2010). At the same time, it should be clear that some people will find ways to connect and be active with such entities. Parahuman technology holds few implications for taking notice, keeping learning, or giving.

In sum, parahuman technology may be a fundamental challenge to what we Believe, and what we hold to be unique, or not, about human existence and our own place in life. Challenges do not always result in negative outcomes, but it seems likely that there will be some transitional discomfort around this issue. At the same time, parahuman technology is like the other categories, the impact is probably less tied to what it is and more reliant on how it is used.

In this chapter, we have proposed a dual process model of two neurological systems that give rise to differing experiences of happiness. Technological changes may variously appeal to the "hot" or "cold" systems, enhancing or hampering people's pursuit of the Good life. At least one consensual effort has suggested people can build the Good life if they Connect, Be Active, Take Notice, Keep Learning, and Give; and we suggest that people are also likely to find benefit if they clarify their outlook on life and their place in it, which we called Believe. Each of the main technological changes we see looming on the horizon interacts somewhat differently with these ways to the Good life, but overall, we were forced to conclude that the specific nature of the technology itself is less likely to matter than whether the use of technology will inhibit or enhance people's abilities to Connect, Be Active, Take Notice, Keep Learning, Give, and Believe.

CONCLUSIONS

Although it is perhaps our species' most ancient tale, the Epic of Gilgamesh offers the lesson of balance between pleasure and mattering. Many of the pivotal moments in the tale are about the opportunities of pleasure—Enkidu is tamed through sexual pleasures enjoyed with a woman, making his friendship with Gilgamesh possible, and the wisest characters in the tale seek to spur Gilgamesh toward immersion in the pleasures of mortal life. Yet, pleasure brings restlessness and emptiness to the king. From the riches of his palace, Gilgamesh seeks something that matters more, and the gods themselves created Enkidu to provide Gilgamesh with a worthy rival. In this story, pleasure seems to bring life and vibrancy, but also stagnation; striving to matter brings friendship and memorable achievements, but also struggle and disappointment. Gilgamesh, who shared with us only the technologies of clothing, masonry, and simple tool-making, nonetheless shares with us the central questions of how to live life. The ancient king, who would not recognize many of the touchstones of our technological age, would nonetheless recognize that technology neither denies nor delivers the Good life, it merely becomes woven into the tapestry against which unfolds the human struggle to enjoy a life that matters.

NOTES

1. The following sources were consulted: The Mesopotamia Reader, maintained by Richard Hines, (http://www.wsu.edu/~dee/MESO/GILG.HTM); text translation, standard Akkadian version, provided at http://www.ancient-texts.org/library/mesopotamian/gilgamesh/.

REFERENCES

Aked, Jody, Nic Marks, Corrina Cordon, and Sam Thompson. (2008). *Five Ways to Well-Being: The Evidence*. London: New Economics Foundation.

Asaro, Peter M. (2006). "What Should We Want from a Robot Ethic?" *International Review of Information Ethics*, vol. 6, pp. 9–16.

Associated Press. (2008, January 14). "'Blade Runner' Barred From Olympics." Retrieved from http://www.cbsnews.com/stories/2008/01/14/sports/main3707924.shtml

Barrett, Lisa F., Michele M. Tugade, and Randall W. Engle. (2004). "Individual Differences in Working Memory Capacity and Dual-Process Theories of the Mind," *Psychological Bulletin*, vol. 130, pp. 553–573.

Baumeister, Roy F. (1991). *Meanings of Life*. New York: Guilford Press.

Bittman, Michael, Judith Brown, and Judy Wajcman. (2009). "The Mobile Phone, Perpetual Contact and Time Pressure," *Work, Employment and Society*, vol. 23, no. 4, pp. 673–691.

Bloom, David E., David Canning, and Bryan Graham. (2003). "Longevity and Life-Cycle Savings," *The Scandinavian Journal of Economics*, vol. 105, no. 3, pp. 319–338.

Boutell.com. (n.d.). "WWW FAQs: How Many Websites Are There?" Retrieved from http://www.boutell.com/newfaq/misc/sizeofweb.html

Brown, Kirk W., and Richard M. Ryan. (2003). "The Benefits of Being Present: Mindfulness and Its Role in Psychological Well-Being," *Journal of Personality and Social Psychology*, vol. 84, no. 4, pp. 822–848.

Compton, William C., Maggie L. Smith, Kim A. Cornish, and Donald L. Qualls. (1996). "Factor Structure of Mental Health Measures," *Journal of Personality and Social Psychology*, vol. 71, pp. 406–413.

Datko, Karen. (2010). "The Living-Too-Long Problem." Retrieved from http://articles.moneycentral.msn.com/SmartSpending/blog/page.aspx?post=1756838

Deci, Edward L., and Richard M. Ryan. (2000). "The "What" and "Why" of Goal Pursuits: Human Needs and the Self-Determination of Behavior," *Psychological Inquiry*, vol. 11, pp. 227–268.

Diener, Ed. (2000). "Subjective Well-Being: The Science of Happiness and a Proposal for a National Index," *American Psychologist*, vol. 55, pp. 34–43.

Disalvo, David. (2010). "Are Social Networks Messing with Your Head?" *Scientific American Mind*, vol. 20, no. 7, pp. 48–55.

Elliott, Gregory C., Suzanna Kao, and Ann-Marie Grant. (2004). "Mattering: Empirical Validation of a Social-Psychological Concept," *Self and Identity*, vol. 3, pp. 339–354.

Fisher, Kimberly, and John P. Robinson. (2009). "Average Weekly Time Spent in 30 Basic Activities Across 17 Countries," *Social Indicators Research*, vol. 93, no. 1, pp. 249–254.

Frankl, Viktor E. (1963). *Man's Search for Meaning: An Introduction to Logotherapy*. New York: Washington Square Press.

Fredrickson, Barbara L. (1998). "What Good Are Positive Emotions?" *Review of General Psychology*, vol. 2, pp. 300–319.

Fromm, Erich. (1956). *The Art of Loving*. New York: Harper and Row.

Gelles, David. (2009). "Immortality 2.0: A Silicon Valley Insider Looks at California's Transhumanist Movement," *The Futurist*, vol. 43, no. 1, pp. 34–41.

Hanh, Thay N. (1975). *The Miracle of Mindfulness*. Boston: Beacon Press.

The Happy Planet Index 2.0. (2009). About NEF. http://www.happyplanetindex.org/info/about-nef.html

Häyry, Matti. (2011). "Considerable Life Extension and Three Views on the Meaning of Life," *Cambridge Quarterly of Healthcare Ethics*, vol. 20, pp. 21–29.

Iggulden, Caroline. (2008, December 11). "Inventor Builds She-3PO Robot. *The Sun*. Retrieved from http://www.thesun.co.uk/sol/homepage/news/article2023392.ece

Infoplease Almanac. (2007). "Life Expectancy at Birth by Race and Sex, 1930–2007." Retrieved from http://www.infoplease.com/ipa/A0005148.html

Jain, Shamina, Shauna L. Shapiro, Summer Swanick, Scott C. Roesch, Paul J. Mills, Iris Bell, and Gary E. Schwartz. (2007). "A Randomized Controlled Trial of Mindfulness Meditation Versus Relaxation Training: Effects on Distress, Positive States of Mind, Rumination, and Distraction," *Annals of Behavioral Medicine*, vol. 33, no. 1, pp. 11–21.

Kahneman, Daniel. (2003). "A Perspective on Judgment and Choice," *American Psychologist*, vol. 58, pp. 697–720.

Kang, Seok. (2007). "Disembodiment in Online Social Interaction: Impact of Online Chat on Social Support and Psychosocial Well-Being," *CyberPsychology and Behavior*, vol. 10, no. 3, pp. 475–477.

Kashdan, Todd B., Robert Biswas-Diener, and Laura A. King. (2008). "Reconsidering Happiness: The Costs of Distinguishing Between Hedonics and Eudaimonia," *Journal of Positive Psychology*, vol. 3, pp. 219–233.

Kashdan, Todd B., and Patrick E. McKnight. (2009). "Origins of Purpose in Life: Refining Our Understanding of a Life Well Lived," *Psychological Topics, vol. 18, pp*. 303–316.

Keltner, Dacher. (2009). *Born to Be Good: The Science of a Meaningful Life*. New York: W. W. Norton.

Keyes, Cory L. M., Dov Shmotkin, and Carol D. Ryff. (2002). "Optimizing Well-Being: The Empirical Encounter of Two Traditions," *Journal of Personality and Social Psychology*, vol. 82, pp. 1007–1022.

King, Laura A., Joshua A. Hicks, Jennifer A. Krull, and Gus A. Baker. (2006). "Positive Affect and the Experience of Meaning in Life," *Journal of Personality and Social Psychology*, vol. 90, pp. 179–196.

Lanthie, Richard P., and Craig Windham. (2004). "Internet Use and College Adjustment: The Moderating Role of Gender," *Computers in Human Behavior*, vol. 20, pp. 591–606.

Lyubomirsky, Sonya, Laura A. King, and Ed Diener. (2005). "The Benefits of Frequent Positive Affect: Does Happiness Lead to Success?" *Psychological Bulletin*, vol. 131, pp. 803–855.

Maslow, Abraham H. (1971). *The Farther Reaches of Human Nature*. New York: Viking.

McGregor, Ian, and Bryan R. Little. (1998). "Personal Projects, Happiness, and Meaning: On Doing Well and Being Yourself," *Journal of Personality and Social Psychology*, vol. 74, pp. 494–512.

Myers, David G., and Ed Diener. (1995). "Who Is Happy?" *Psychological Science, vol. 6,* pp. 10–19.

Parker-Pope, Tara. (2008, May 6). "Psychiatry Handbook Linked to Drug Industry." *New York Times*. Retrieved from http://well.blogs.nytimes.com/2008/05/06/psychiatry-handbook-linked-to-drug-industry/

Parsons, Zack. (January 11, 2010). "A Test Drive of Roxxxy, the World's First Sex Robot." Retrieved from http://www.somethingawful.com/d/news/roxxxxy-love-robot.php

Preston, Carolina, and Nicole Wallace. (2010, January 27). "Donations to Help Haiti Exceed $528-Million, Chronicle Tally Finds," *The Chronicles of Philanthropy.* Retrieved from http://philanthropy.com/article/Donations-to-Aid-Haiti-Exce/63756/

Robinson, John, and Steven Martin. (2009). "IT and Activity Displacement: Behavioral Evidence from the U.S. General Social Survey (GSS)," *Social Indicators Research*, vol. 91, no. 2, pp. 115–139.

Rosenberg, Morris, and Claire McCullough. (1981). "Mattering: Inferred Significance and Mental Health Among Adolescents," *Research in Community and Mental Health*, vol. 2, pp. 163–182.

Ryan, Richard M., and Edward L. Deci. (2001). "On Happiness and Human Potentials: A Review of Research on Hedonic and Eudaimonic Well-Being," *Annual Review of Psychology,* vol. 52, pp. 141–166.

Ryff, Carol D., and Burton Singer. (1998). "The Contours of Positive Human Health," *Psychological Inquiry,* vol. 9, pp. 1–28.

Salleh, Anna. (2000, May 29). "Technology Won't Bring Lasting Happiness, Say Students," *ABC Science Online*. Retrieved from http://www.abc.net.au/science/articles/2000/05/29/131423.htm

Seligman, Michael E. P., and Mihaly Csikszentmihalyi. (2000). "Positive Psychology: An Introduction," *American Psychologist*, vol. 55, pp. 5–14.

Singer, Emily. (2005, December 5). "Don't Worry, Be Happy: Nobel Laureate Eric Kandel Explains How Genetic Research Could Lead to a New Generation of Anti-Anxiety Drugs," *Technology Review.* Retrieved from http://www.technologyreview.com/biomedicine/15960/?a=f

Sloman, Steven A. (1996). "The Empirical Case for Two Systems of Reasoning," *Psychological Bulletin*, vol. 119, pp. 3–22.

Smith, Eliot R., and Jamie DeCoster. (2000). "Dual-Process Models in Social and Cognitive Psychology: Conceptual Integration and Links to Underlying Memory Systems," *Personality and Social Psychology Review,* vol. 4, pp. 108–131.

Steger, Michael F. (2009). "Meaning in Life." In *Oxford Handbook of Positive Psychology,* edited by Lopez, S. J. (2nd ed.) (Oxford: Oxford University Press) pp. 679–687.

Steger, Michael F. (in press). "Experiencing Meaning in Life: Optimal Functioning at the Nexus of Spirituality, Psychopathology, and Well-Being." In *The Human Quest for Meaning,* edited by Wong, P. T. P. and P. S. Fry (2nd ed.). Mahwah, NJ: Lawrence Erlbaum.

Steger, Michael F., Patricia Frazier, Shigehiro Oishi, and Matthew Kaler. (2006). "The Meaning in Life Questionnaire: Assessing the Presence of and Search for Meaning in Life," *Journal of Counseling Psychology,* vol. 53, pp. 80–93.

Steger, Michael F., Todd B. Kashdan, and Shigehiro Oishi. (2008). "Being Good by Doing Good: Daily Eudaimonic Activity and Well-Being," *Journal of Research in Personality*, vol. 42, pp. 22–42.

Surowiecki, James. (2005, January). "Technology and Happiness: Why Getting More Gadgets Won't Necessarily Increase Our Well-Being," *Technology Review.* Retrieved from http://www.technologyreview.com/Biotech/14091/

Waterman, Alan S. (1993). "Two Conceptions of Happiness: Contrasts of Personal Expressiveness (Eudaimonia) and Hedonic Enjoyment," *Journal of Personality and Social Psychology*, vol. 64, pp. 678–691.

Wight, Vanessa R., and Sara B. Raley. (2009). "When Home Becomes Work: Work and Leisure Patterns of Home-Based Workers," *Social Indicators Research*, vol. 93, no. 1, pp. 197–202.

Yu, Sen C., and Chien Chou. (2009). "Does Authentic Happiness Exist in Cyberspace? Implications for Understanding and Guiding College Students' Internet Attitudes and Behaviors," *British Journal of Educational Technology*, vol. 40, pp. 1135–1138.

6 The Ambivalence of the Good Life

Happiness, Economics, Technology, and Relational Goods[1]

Luigino Bruni

Just imagine a city without noisy and quarrelsome apartments, in which each family has its own house acoustically and visually isolated from others such that no neighbor can disturb another; where the few remaining skyscrapers are constructed in such a way so as to avoid all encounters on the stairs or on the landings; where office and workplace communications are solely via e-mail, or by Skype for the more delicate decisions; where all formerly common spaces, from parks to neighborhoods, have been parceled up and privatized, and everyone protects his or her own little piece of the city; where by simple e-mail we can order groceries with home delivery without having to get out and lose precious time; where the interactive media have become so sophisticated that we feel in the company of others all day long, although we spend more time alone in front of the computer and the television; where even college courses are delivered at home via the Internet, with highly trained virtual instructors that can follow us personally from anywhere in the world, with no need whatsoever for face-to-face encounters.

An 'ideal' city: conflicts have been eliminated because the precondition for conflict, that is, insisting on maintaining a common ground, a *communitas*, has itself been eliminated. This stylized scenario is very close to the reality that is taking shape in the cities we are envisioning and planning in our market and technological societies. Yes, market and technology, because the "marriage" between market and technology is the driving and main force responsible of the big change in the nature of the interpersonal relationships that is occurring in our postmodern globalized society. This chapter will attempt to offer a few explanations why such a scenario is taking shape, perhaps offering as well a few reflections for those who, like me, are quite concerned over such a prospect that is ambivalent as all main cultural revolutions in human history.

An image and an insight are the source of this chapter (and of my research in the last years: Bruni 2011): the image is that of "Jacob's struggle with the angel," as narrated in Genesis (ch. 32), where Jacob is blessed *and* wounded by the other, and the corresponding insight is the unbreakable link between 'wound' and 'blessing' in every authentic human relationship. Sooner or later, human beings understand that to experience the "blessing" that is bound up in relationships with the others, we must accept its "wound" as well. We come to understand that we cannot enjoy life without going through the risky and dangerous territory of the life in common; taking any escape route from this struggle inevitably leads to a joyless human condition.

This is the central idea that sets the course of this chapter, which attempts to engage a dialogue between economics and the struggle with the wound and the blessing of the other. The good life is ambivalent by nature, because the life in common is, at the same time, the only *oikos* where the social animal can flourish *and* the arena of conflict and death. Market society and technology are main protagonists of this ambivalence.

Economic science in late modernity, with its promise of life in common without sacrifice, represents a great escape from the contagion of a personal relationship with the other. Today, precisely for this reason, the humanism of the market economy, despite having produced many benefits for civilization, is largely responsible, as is technology, for the bleak and lonely drift of modern market societies. Market and technology are, in fact, the great mediators of face-to-face human relationships, on the basis of the promise that thanks to such mediations, we can finally build up a painless and peaceful society, mediating encounters with others with whom we interact without contention or wound. In effect, this is increasingly how we 'encounter' each other in anonymous postmodern markets. But, if it is true that being human begins with gratuitousness, ever a risky and potentially painful experience in interpersonal relationships, perhaps we are going beyond the bounds of what it means to be human. The deception, however, is that this innocuous, painless encounter with the other is also an encounter that cannot lead to a fully human life, either for the individual or for society, as contemporary studies on the 'paradox of happiness,' the topic of this chapter, increasingly confirm. Today we are paying for this great illusion of modernity in the currency of happiness, and it is time that someone calls its bluff.

The common thread of the chapter is an all-around reflection on human relationships, primarily those that are peer, or face-to-face. I am convinced that a society without mediators cannot be a civil society; yet, a society that takes recourse only in mediators to regulate human relations is even less so. Much of the discourse of this chapter will develop between these terms 'without' and 'only.'

Finally, although in what follows I will primarily highlight the problematic aspects tied to the 'thirdness' of mediation of market and technology, we must not forget its potentially civilizing role: The protection from injury by the other, assured by the mediated and decentralized system of prices and the mediation of law, can play a positive and civilizing function, above all in those societies in which the market is underdeveloped and the experience of equality and freedom is ever under threat. There is, however, a tipping point, a threshold, beyond which the anonymous relationships of markets produce lawlessness, loneliness, and the loss of bonds of identity with others: my impression is that opulent Western societies have already passed beyond this threshold, which delimits what it means to be human.

Within this scenario, let us now enter into the happiness-economics-sociality narrative.

HAPPINESS AND ECONOMICS

First in the United States, then in Europe, and now somewhat around the world, happiness has come to the fore in economics since people's happiness began to be measured by means of questionnaires, then compared to traditional economic indicators, such as income, wealth, unemployment, and so forth. Today, we can say that the theme of happiness is quietly but decisively changing our understanding of the economics of well-being. What emerged even from the earliest studies of the phenomenon known as the paradox of happiness in economics is the nonexistent or minimal correlation between income and people's subjective happiness, or between economic well-being and well-being in general. Methodological innovations by social psychologist Hadley Cantril in the 1960s were followed by the introduction of the paradox of happiness by American economist and demographer Richard Easterlin in 1974. In 1976, Tibor Scitovsky, a prominent American economist, published his book *The Joyless Economy*, which introduced unhappiness in contemporary society as a serious research subject for economics. Scitovsky's book was part of a larger movement in economics in the 1970s in which it became infused by other disciplines and complicated by dimensions of human choice. Other important works include Fred Hirsch (1976) on 'positional goods' and *Rational Fools* by Amartya Sen (1977).

From this brief review, let us turn back to our claim that human relations are central to happiness and that these can only be truly enjoyed when concomitant risks and pains are accepted as well. If it is true that there is no blessing without wound, then there is no good life, happiness, or *eudaimonia* without a risky and potentially tragic encounter with the other. We will begin from a curious fact concerning the emergence of economic science that has contributed in no small measure to the bleak drift of our era. Modern economics first emerged in Italy as associated with happiness, and only a few years later in Scotland was it associated with wealth (Bruni 2006). Indeed, we find the word 'happiness' in the first major school of European economic thought in Naples. In particular, we find the ideas of public happiness, reciprocity, relationships, and trust, which today constitute the theoretical platform of what is called the 'civil economy.'[2]

It may be useful then to begin from the view of this ancient, yet modern, Neapolitan tradition on 'public happiness,' summarily reconstructing the thought of the 18th-century Neapolitan economist-philosopher Antonio Genovesi, the leader of this school of thought. Over the course of the chapter, we will see that the promise of happiness during this first stage of the civil economy was translated two centuries later into the economy 'without injury,' and thus 'without joy' as well, which characterizes our era.

Antonio Genovesi is in perfect agreement with Adam Smith on many points, including the civilizing role of the market, his antifeudal polemic, and above all, the distinction between the motivations for actions (which can be virtuous or vicious) and their civil consequences (which are those of greatest interest to the social scientist). The Latin category of 'public happiness' is directly tied to this anthropological and relational view, which in a certain sense summarizes the Neapolitan economic tradition—so much so that many authors use the expressions 'civil economy' and 'public happiness' interchangeably to refer not just to the Neapolitan but to the Italian school of thought. For Genovesi, happiness is by nature constitutively relational; the paradoxical nature of happiness and the good life is an evident characteristic of his thought that one encounters frequently in the Italian tradition.

There is a passage that is central to understanding this aspect of the idea of happiness in Genovesi, a summary and a natural terminus of a theory of happiness that we have constructed step by step to this point:

> You toil for your own interests, and no person could act otherwise than for his or her happiness; such a one would be less than human. However, do not desire to make others miserable; rather, if you can, and when you can, expressly consider how to make others happy. The more one works for one's own interest, so much the more, if one be not mad, must one be virtuous. It is a law of the universe that one cannot make oneself happy without making others happy. (Genovesi 1963 pp. 449)

The paradox consists in stating that happiness emerges from making others happy. Also for this tradition, happiness has a paradoxical nature in that it is constitutively relational: one cannot live a 'good life' unless with and thanks to others (making 'others happy'). But for precisely this reason, we do not have full control over happiness. A human being needs reciprocity to fulfill himself or herself, but to have it, one must make the leap of gratuitousness—which, however, may not be reciprocated. In any case, without gratuitousness, genuine reciprocity does not develop, nor society with it.

RELATIONSHIPS AND WELL-BEING

With a leap of over two centuries, we find today that the fundamental aspects of the discourse on Genovesi and the civil economy are absolutely current; they eloquently explain the absence of joy in an economy that has chosen to exclude a perilous and gratuitous relationship with the other. The decisive role that interpersonal relations occupy in the development of a good life is now widely recognized in the contemporary social sciences. Research around the so-called paradox of happiness shows that the quality of genuine relational life is what carries the most weight, even compared to income, in people's estimation of subjective well-being.[3]

In particular, there is much empirical and experimental evidence in psychological studies on the subject. Daniel Kahneman and colleagues (2004) have shown the importance of relationality in the happiness of human beings. The subjects partaking in their study reported a greater self-evaluation of their own well-being when their activities were carried out in the company of other people. Two economists, Stephan Meier and Alois Stutzer (2004), showed a robust correlation between voluntary activities and subjective well-being. Furthermore, when psychologists analyze the characteristics of people who consider themselves more satisfied, it emerges without exception that these people have meaningful and positive interpersonal relationships (Diener and Seligman 2002).

Psychological studies thus offer us much data that confirm the importance of relationality for people's happiness and satisfaction. Some scholars have defined relatedness as a primary need essential to well-being (Deci and Ryan 2001). In particular, scientists who consider themselves part of the eudaimonistic school of thought, of Aristotelian inspiration, hold that there is a universal nexus between the quality of human relationships and subjective well-being: "The evidence supports that the bond between relatedness and subjective well-being is complex. The studies suggest that, of all the factors that influence happiness, relatedness is very near the top of the list. Furthermore, solitude is significantly correlated in a negative manner with affection and life satisfaction" (Deci and Ryan 2001, p.154).[4]

Finally, in research conducted with a colleague in the Department of Political Economics of the University of Milano-Bicocca (Bruni and Stanca 2008) a strong correlation emerged between the time that a person spends in activity with a relational component (with friends, family, or in volunteer work) and the self-evaluation of subjective well-being (or happiness), a correlation that remains robust even when controlling many other variables (age, geographic region, education, culture, etc.). From the reviewed literature, it seems that empirical studies give a clear and universal message: people's happiness depends in large measure on genuine relationality.

THE PARADOX OF AFFLUENT UNHAPPINESS

The economic works published by Cantril, Easterlin, and Scitovsky in the 1960s and 1970s showed them to be pioneers in decrying the unhappiness in contemporary economies, known today as the 'paradox of happiness.' By 'paradox of happiness' is meant a plurality of data that, beyond the various successive theories and interpretations, nevertheless substantially agree on one point: once per capita income has exceeded a certain threshold that permits one to live 'decently,' it is no longer an important factor in people's subjective happiness or, in any case, it is far less important than other factors, such as relational and family life or health.

Today the explanation for the lack of correlation largely revolves around the metaphor of the treadmill: an increase in income brings with it an increase in something else, exactly as in a treadmill, on which one runs but actually remains stationary because the treadmill moves the opposite direction under one's feet. Kahneman and colleagues (2004) distinguish between two types of treadmill effect: the hedonic treadmill and the satisfaction treadmill. The hedonic treadmill sources from the theory of adaptation level: when we have a low income, we use, for example, a utilitarian car, which gives us a 'hedonic' (or pleasure) level of, say, 5; when our income increases we purchase a new luxury car that, after having caused an increase in well-being for a few months (to a level of, say, 7), soon falls back to the same level of well-being of the utilitarian car (5) as the effect of a psychological mechanism of adaptation. When income increases, it so happens that this improvement in material conditions causes people to continuously demand more intense pleasures to maintain the same level of satisfaction. The satisfaction treadmill—which normally is added to the hedonic treadmill—thus works in such a way that subjective happiness (the self-evaluation of one's own happiness) remains constant despite the increase in objective happiness (the quality of goods consumed). Thus, in the example of the car, with the higher salary my aspirations about the ideal car increase, and this makes it such that I will experience the same level of satisfaction before the increase in income (although I am objectively more comfortable in the new car). Advertising is a powerful mechanism that leverages precisely these 'treadmills.'

It is interesting to note that in the realm of material goods, adaptation and aspirations have a near total effect: increases in comfort are nearly completely absorbed after a more or less brief time. These effects thus determine a 'destruction of wealth,' or better yet, a nonefficient use of the same. Why? Because there are other, noneconomic environments in which adaptation and aspirations play a lesser role, such as in family, emotive, and civil environments.

A third explanation of the paradox of happiness, the one most widely accepted by economists, places emphasis on the 'positional' effects of income and consumption. This hypothesis—which historically goes back to Thorstein Veblen's work on 'conspicuous consumption' (1899), and more recently on 'relative consumption' by James Duesenberry (1949)—consists of the insight that the pleasure we draw from consumption depends primarily on the relative value of the consumption itself, that is, how the absolute level of our consumption differs from that of others with whom we normally compare ourselves. Human beings, it seems, value their things through the eyes of others. The interesting point is that these 'positional' mechanisms lead to failures of economic rationality—or how economics has traditionally understood it. This is the well-known dynamic of so-called 'negative externalities'; consumption by others 'pollutes' my well-being, a case in which market competition is not civilly beneficial but destructive of individual and collective resources. That is why some authors, Richard Layard (2005) and Robert H. Frank (1985), in particular, propose using taxation to reduce consumption of such conspicuous or relational goods, just as demerit goods are taxed (e.g., distilled spirits) because they are considered socially harmful and addictive.

HAPPINESS AND BEYOND

At this point, we must ask a crucial question: which happiness and which sociality do contemporary scholars have in mind when trying to explain the paradox of happiness in economics? In general, the previously mentioned theories suffer, in varying degrees, from a high degree of anthropological reductionism. In my view, however, the error is in thinking that envy and rivalry are the fundamental anthropological characteristics that explain human happiness. Current prevailing theories seem to me to be good explanations of unhappiness and frustration, but I am not convinced that they are positive theories of human happiness, nor that human happiness is parsed wholly along the lines of envy and competition. We hardly consider our lives happy or fulfilled because we consume more than our co-workers.

That does not mean that current studies on happiness do not represent an important innovation. Although scholars of happiness and behavioral economics (e.g., Daniel Kahneman or Robert Sugden) may be able to make a few advances with respect to the 'official' neoclassical economics, I do not think we can yet declare ourselves satisfied; more should be done in areas that include the relational dimension within the analysis of happiness. On the one hand, the economics of happiness, in empirically studying human feelings, is not closed to the feelings that arise from intrinsic motivations and social and relational needs. In this approach, relational goods are ranked on the basis of the happiness and pleasure they bring to the individual. On the other hand, however, the approach of the psychological economists is still hedonistic (based on pleasure) and not eudaimonistic (based on happiness, in the Aristotelian sense of 'good life'): in fact, a real distinction is not made between pleasure and happiness, and the entire evaluation of happiness is entrusted to subjects who must self-assess their own level of subjective well-being. Where then are the problems, one might ask?

There are at least two. We might call the first the "Scitovsky effect."[5] If the perceived pleasure of a good is entrusted solely to self-evaluation, we must then keep in mind that there is a strong tendency to exchange creative goods for comfort goods; these may appear equivalent in terms of immediate pleasure, although in the long term, the 'objective' happiness acquired from the two types of goods is quite different, as we will discuss later. We systematically commit 'shortsighted' errors in consumption. The second we might call the "Kahneman effect": there are important cognitive errors in the process of self-evaluation of subjective happiness. People adapt to goods, and, due to treadmill effects, in short order they no longer subjectively perceive an improvement in happiness, even though objectively they are better off. This adaptation produces an excessive (that is, inefficient) tendency to change and to acquire new goods that do not objectively improve our standard of living, but which subjectively 'deceive' us.

Relational goods (spouse, children, friends, and so forth), however, are also subject to an 'adaptation effect' (Frey and Stutzer 2002). Can we, however, entrust the evaluation of the quality of our lives, of our happiness, solely to subjective perception? I think not, precisely because the adaptation effect can carry so much weight; if the principal or sole indicator of happiness becomes subjective self-evaluation, we risk making many mistakes. Subjective happiness is certainly important, but it alone is not sufficient to evaluate the goodness of life: We cannot entrust the evaluation of our well-being solely to self-evaluation.

We must recognize that once we ascertain that the gross domestic product (GDP), which measures the wealth of a country, is insufficient as an indicator of well-being, we cannot simply replace it with an alternative indicator of subjective happiness: there are indispensable civil values for 'human flourishing' (or *eudaimonia*) that are not easily translated into either GDP or individual perceptions of happiness. Amartya Sen's criticism deserves special attention. Sen (2009) is not the only contemporary important thinker to criticize happiness as the basis of an ethical theory. Sen criticizes Bentham's utilitarian vision of happiness, that is, the maximization of social and individual pleasure, as it is incompatible with his theory of capabilities. Sen criticizes also—albeit not explicitly—the much broader idea of happiness that we find in Aristotle (*eudaimonia*) and in Mill.

Sen tells us that even broadening the limits of the concept happiness to include the idea of fulfillment or human flourishing, which for the Aristotelian tradition is the supreme and ultimate good, happiness is not the only variable that counts in life, or the most important one. Why? The basic reason is simple. If, by reinterpreting Aristotle, happiness were to become all that has value in a person's life, then happiness would become a useless concept for social analysis, because knowing that all good and true things trace back to happiness leaves us without any criterion for specific choices, which frequently feature a conflict between several good situations. For Sen, then, one should keep happiness within the boundaries of common sense, including utilitarian happiness, considering it as but one dimension of life that in certain instances can give way to others. I personally have worked for several years around the idea of Aristotelian happiness (Bruni 2010), showing that many of the criticisms of the studies on happiness are actually applicable to Bentham's, rather than Aristotle's, view. I must acknowledge, however, that now I agree with Sen; while enriching and complicating the idea of happiness, it is more useful to limit happiness and compare it with other dimensions of the good life when analyzing people's specific choices.

As an example, consider a person (a woman, as Sen often does) who, for religious or ideological reasons, is convinced that being subject to her husband is part of her vocation, and she is content and happy in that state. Imagine that an impartial outsider to that situation freely suggests a path of awareness and freedom to her. If she begins a path of liberation and increases her capabilities, certainly in the early stages, and perhaps even in later stages, her happiness will decrease (if interpreted reductionistically), but her freedom will increase. Can we then say that the woman is happier, while possibly suffering spiritually and perhaps physically? I do not know; we can, however, say with certainty that she is more free, and thus prefer to trigger the process of her becoming aware, even though we know that there will be a cost in terms of happiness. We could certainly include freedom, in an Aristotelian sense, in the concept of happiness, but, again, in so doing, the concept of happiness would become so broad as to be useless.

To sum up: Modern democracies need more indicators of well-being, in addition to GDP, because any *reductio ad unum* always places democracy and freedom in danger because, paraphrasing Plato, happiness is one, but happinesses are many. What then does the study of happiness add to the understanding of social reality and well-being? If happiness is just a new name for utility, then why disturb (and perhaps devalue) this ancient word? If, with respect to classic utility, happiness were to add only the empirical dimension of self-evaluation, then it would be sufficient to speak of subjective utility, or self-evaluated utility, rather than happiness. Jeremy Bentham transformed hedonic happiness into utility; perhaps today we are called to do more. If, instead of utility, we use the word happiness today with a new and truly relevant meaning for civil life, then we must take the Aristotelian tradition seriously (happiness as *eudaimonia*), and with it neglected categories such as that of 'relational good.' In the next sections, we will attempt to further develop this discourse in search of a "positive" theory of happiness that is centered on relationality.

RELATIONSHIPS: THE GOODS THAT STANDARD ECONOMICS DOES NOT SEE

As we have seen, the happiness that is discussed in current economic theories is far removed from the classic idea of happiness, which was tightly tied to the virtues and quite distinct from pleasure. It is on this point that the ideas of happiness for ancients and moderns profoundly differ. Happiness for modern economists in fact coincides with hedonistic pleasure, with no reference whatever to the constitutive fragility of the good life. The paradoxes of happiness, however, are decrying the deception of such a happiness-pleasure concept. If the extension of the market reduces the quality of interpersonal relationships, and if economic ties erode other social ties, then goods can become evils, bringing not well-being but ill-being. The relationship between happiness and aspects of family life is a highly significant empirical fact in this regard (cf. table 6.1).

One of the indicators that signals a crisis in neoclassical economic theory today is its inability to account for intrinsically motivated human relationships. In particular, in conventional economic science there is no room for noninstrumental relationships, thus, there is no room for relationships that are now called 'relational goods.' The reason for this inability is readily stated: relational goods emerge (as we will soon see) from noninstrumental motivations (or gratuitousness), but neoclassic economic theory treats goods as means, never as ends in themselves. Contemporary economics sees only individuals who choose, ignoring the relationships they establish. Relationships are considered only in subsequent analysis, and consequently they can only be instrumental. If in economics we want to study noninstrumental relationships, and we must, then we need to have the right tools: A science is required that is immediately relational (without, however, becoming all-encompassing!). The problem, therefore, is not so much individualism, but rather solipsism and 'methodological narcissism.'

I will introduce a vision of happiness that is directly related to the classical Aristotelian and Thomistic tradition, the basic anthropological assumption of which is the necessity of noninstrumental relationships for a good

Table 6.1 Effects on Happiness

	Index of Reduction in Happiness
Income	
One–third Reduction in Household Income	1
Work	
Unemployed (rather than employed)	3
Uncertain work (rather than certain)	1.5
Inflation increases by 10%	0.5
Family status	
Separated (rather than married	4.5
Widow/Widower (rather than married)	2
Health	
Personal health worsens by 1 point (on a scale of 1-5)	3

Source: Layard (2005).

or happy life. This is why the key concept of such a theory of happiness, or human flourishing, is that of 'relational goods.'

RELATIONAL GOODS

The category of relational goods was introduced into the theoretical debate nearly simultaneously by four authors: the philosopher Martha Nussbaum (1986), the sociologist Pierpaolo Donati (1986), and the economists Benedetto Gui (1987) and Carole Uhlaner (1989).

Benedetto Gui defined relational goods as "non-material goods, which are not services that are consumed individually, but are tied to interpersonal relations" (1987, p. 37). Carole Uhlaner defined them as goods that "can only be 'possessed' by mutual agreement that they exist after appropriate joint actions have been taken by a person and non-arbitrary others" (1989, p. 254). Relational goods are, therefore, goods that cannot be produced, consumed, or acquired by a single individual, because they depend on interaction with others and are enjoyed only if shared with others. According to Uhlaner, "goods which arise in exchanges where anyone could anonymously supply one or both sides of the bargain are not relational" (1989, p. 255).

Actually, relational goods can be better understood as a *third* category of goods, neither private nor public. Given its brief history, the concept of relational goods still lacks a shared definition among the few economists who use them in their models. Gui (2002) proposes to analyze every form of interaction as a particular productive process that he calls *encounter.* It is the relationship itself that constitutes the "good" in relational goods. The dimension of reciprocity is, therefore, foundational here. What's more, the *identity* of the other person is essential for the value and sometimes the existence itself of the relational good.

The key-characteristics of the interpersonal relationships that can be described as "relational goods" are therefore: (i) *motivation dependent* (unlikely the standard market goods, in relational goods the "why" one enters the relationship is essential for the existence and value of them), and (ii) *identity dependent* (they are not anonymous).

In Bruni and Stanca (2008), one of the first empirical papers on relational goods, membership of a voluntary organization is associated with a statistically significant increase in life satisfaction, and the increase is somewhat larger and more significant (than in simple membership). It is interesting to observe that the effect of volunteering for life satisfaction is *quantitatively the same as that of moving up by one decile in the income scale.* All our results are robust to the use of an alternative definition of the relational indicators. All data show that the quality of intrinsic relationality is very important in human happiness—as Aristotle said.

PRIMARY AND SECONDARY RELATIONSHIPS

It can be useful toward a theory of relational goods to distinguish two large families of relational goods, which we can call, following Pierpaulo Donati (in the naming convention, not the content), 'primary' and 'nonprimary' (or secondary) relational goods.[6] It is, in fact, easy to see that a relational good formed in a commercial encounter for a haircut is quite different from a relational good formed between a mother and daughter or between two friends.

A nonprimary relational good can then be defined as an outcome that is summed with the other components produced by the interaction-encounter. In this case, the relational good is an addendum to a summation: as an example, we can represent the value of an encounter with the expression P_1+P_2+RG, where just the third term (RG) is the relational good, and its possible zero (or negative) value does not change the value of the other two.[7] In the case of nonprimary relational goods, the value of the good in play is not nullified by the absence of a relational good. Thus, although a given encounter generates neither a relational good or bad, it can still produce economic goods and utility for the agents involved in the encounter. To cite a second example, if I go to a doctor who is technically skilled but with whom no relational good is created, the economic good 'medical visit' still exists and retains its value (although lower with respect to the same visit to a doctor with whom a relational good emerges). Other examples of nonprimary relational goods are a legal consultation or coffee with friends in a neighborhood cafe.

In the case of a primary relational good, however, the relational component of the encounter cannot be null without destroying the good itself, thus zeroing out its value. For this reason, the primary relational good is with a more complex function, which can take the form of a production function. In particular, the 'primary relational good' component could be formalized as a parameter that acts as a multiplier of the entire summation in which we find the nonrelational components of the interaction as terms (such as the exchange of information or various types of material favors): $RG(P_1+P_2)$.

In a primary relational good, there are always other nonrelational components generated by an encounter. In the case of a primary relational good, however, the role of the relationality is essential (substance), because the other nonrelational components (accidents) can produce utility (in a case in which utility is absent, the summation of the latter produces no effect in terms of well-being). In other words, if within the environment of classic primary relational goods, such as familial relationships or deep friendships, the affective and communicative components are lacking in the relationship, the value of the consumed good is not simply modified (as in the case of nonprimary relational goods), but the whole relationship loses value. If the emotional bond is lost between a couple, the material components of the interaction no longer have any value. Moreover, unlike nonprimary relational goods, primary ones cannot be substituted or compensated with an equivalent monetary value; they cannot be converted into money without profoundly altering their nature.

"L'ENFER C'EST LES AUTRE"

We have arrived at the key step. In light of these considerations, how then can we explain the paradox of happiness if we include relational goods in our discourse, particularly primary ones? When income and consumption levels increase, it can and does happen that what is earned from income in terms of well-being is less than what is lost due to the diminishment of relational goods. It is reasonable to suppose that the overall effect of income on happiness is positive for low levels of income, but it becomes negative after exceeding a certain threshold.

Let's consider a simple logic model. The income increase acts on happiness in two ways: (a) directly, because having more income increases my opportunities of choice in the market, greater opportunities (of health, education, comfort, and so forth) that of themselves, normally, would increase subjective happiness; (b) indirectly, because the increase in income interferes with other areas of life, symbolized by the X, that include relational goods (especially the primary ones), but in general, all the activities that source from gratuitousness and intrinsic motivations, including one's interior life and prayer (see Figure 6.1).

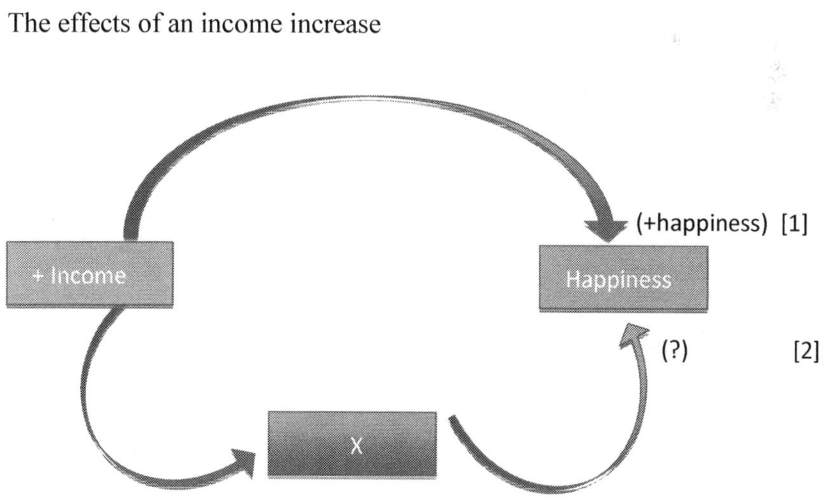

Figure 6.1 The relative weight of these two 'effects' varies with the growth in income and the sum (1+2) can become negative beyond a critical point.

In other words, the increase in income generally comes from an increase in working hours. The increase in hours changes the allocation of resources between instrumental and intrinsically motivated activities. Instrumental activities associated with the increase in income have an impact on happiness, which, however, varies with the variance in income increase: In effect, income is very important when it is low, but it counts much less when it is high. Unfortunately, higher income, which in itself carries little weight on happiness, carries a great deal of weight on unhappiness, because it reduces the time for gratuitousness and for intrinsically motivated activities.

This is why we can imagine a happiness (H) that depends on income (I) and on the value of X, that is. $H_a = f(I_a, X_{ab})$. The pedix of X is "ab," in order to express that in many social activities, such as relational goods, we depend on others. We can thus trace following graph in Figure 6.2.

Up to a certain critical threshold, income and relationships based on gratuitousness can be seen as complementary goods; afterward they become rivals, and income can displace noninstrumental relationships. At this point, a question naturally arises: Why do rational beings who learn from their mistakes not understand that if they are beyond the critical zone, they would be better off with less income and more gratuitous relationality? Why not go backward on the curve to stop at its peak?

I will limit myself to a couple of considerations related to role of technology.[8] In contemporary advanced economies, in the context of relational goods, we are increasingly witnessing the low-cost production and consumption of real surrogates, which we might call pseudo-relational goods. A particularly eloquent example is represented by the consumption

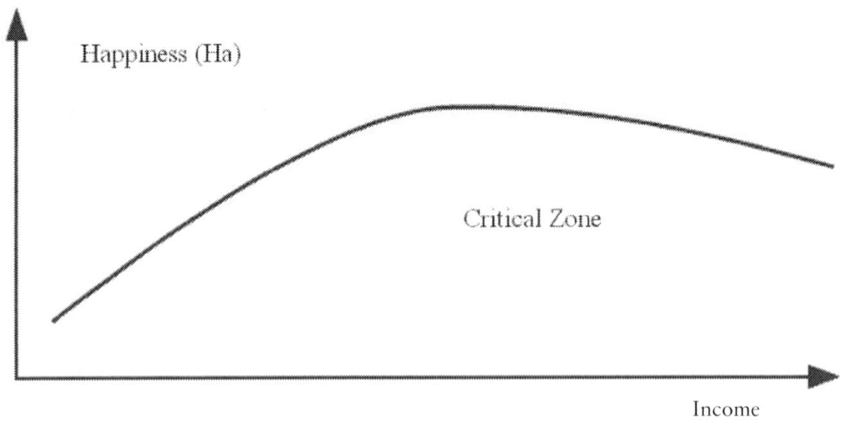

Figure 6.2 Income/Happiness relationship. Beyond a critical threshold, the relationship between income and happiness can change from virtuous to vicious, because it begins to displace gratuitousness.

of simulated relationships in the last few years, such as those offered with greater frequency on television or the Internet. The recent explosion of the so-called reality shows in many countries of the Western world represents one of the macroscopic symptoms of this trend. From the point of view of the viewers, in fact, these are truly simulated relationships (pseudo-relational goods).

The Internet sites dedicated to friendships and romantic relationships constitute a recent example of pseudo-relational consumption. The virtual relationship simulates a real one, replicating some of its essential attributes previously mentioned: the identity of the other counts to some extent (in that as a consumer I choose which television programs to watch and those with whom I interact in a chatroom); there is also simultaneity (in chatting we produce and consume the good through our interaction). It lacks only the painful fragility of flesh and blood relationships, the 'wound' that is the key element of our entire discourse.

In a previously cited study based on the *World Values Survey* (Bruni and Stanca 2008), we identified a significant effect of relational goods on life satisfaction,[9] as well as a crowding-out effect by television consumption (a pseudo-relational good) on actual relationships.[10] These results emerged on the basis of an empirical analysis in which the indicators used had the characteristics of identity and genuineness that are associated with relational goods. A first set of indicators referred to active engagement in individual volunteer activities. A second set of indicators had to do with the time spent in the areas of specific social groups, such as family, friends, colleagues, and so forth.[11]

Television and the Internet are the products of a technological market that not only takes time and energy away from relationships with others, but that, as we have seen, sell pseudo-relationships with others at an infinitely lower price and risk; with a television program or a relationship with a virtual friend we enter and exit with a click, whereas the investment and the risk in a friendship or marriage are infinitely greater. If I lose the capacity to distinguish true relational goods from the false ones (which is a question of culture), and I see only the highly asymmetric risks and costs and not the different returns in terms of happiness, the day is not far off in which commodities will entirely take the place of goods, including relational goods, commodities that will offer ever new substitutes to replace even relational goods.

Furthermore, economic and technological development works in two directions, both of which are relevant to our discourse. First, technology tends to reduce the cost of standard market goods, while it does not do likewise with relational goods; the 'technology,' costs and risks of relational goods are more or less the same over the last millennia. As a consequence, the relative cost of relational goods thus tends to increase in countries with advanced technology, such as those that are experiencing Easterlin's paradox of happiness. Cultivating meaningful relationships with others is very costly in modern market economies, because we find substitutes for relational goods at tremendously lower costs.

This is the real risk that is concealed today in television and new technologies: They appear as new forms of relationship that promise happiness without injury. I risk nothing by spending three hours in front of the television, while I risk much more investing that same time in relationships with others, because if the others do not respond, not only do I not get anything for that investment, I get a relational evil. Unfortunately, the false relational goods of television and the Internet are not very helpful in the crucial moments of life. We need not come to the point of dreaming of a premodern or antitechnological world, and this certainly is not the tone of the chapter; it is sufficient to reflect on the risk that this absence of injury can mean over time for a life made up solely of virtual relationships. This, however, is the charm of a beautiful apple that can be poisoned, or at least noxious.

There is yet another element. The growing efficiency of markets thanks to technology creates a strong tendency to separate relational goods from standard consumption goods. As another example, to consume musical goods a few decades ago, it was necessary to go to the theatre, the opera, a ballroom, or even a party with friends; there was a need for community, with its risks. In a community context, the musical good cannot be distinguished from the relational good: it was necessary to listen to music with others. Today the techno-market permits the unlinking of the two goods, and music consumption is separated from relational goods. An analogous discourse is true for all recreational and cultural goods, but we observe the tendency to dissociate relational goods from individually consumed goods, to varying degrees, in all forms of consumption. One might ask: Where then is the problem? Is this not an increase in freedom of choice and thus of well-being? I can still go dancing or go to the stadium if I want to; if I don't want to, I can listen to music alone or watch the game on TV. We are freer, thus happier, according to standard economics. Unfortunately, things are different than that, as data show. The logic of prices and costs applied to human relationships and the transformation of values into prices produces social, ethical, and spiritual costs in ourselves and others for which we do not know how to account; we only need look closely at our existence to see just how much authentic and meaningful relationships are being impoverished.

Relational goods, to the extent that they consist of relationships, can only be enjoyed reciprocally: here lies all their charm as well as their vulnerability. Relational goods depend on the response of the other: if the other does not respond, the relational good is not created, or worse, becomes a 'relational bad.' The contradiction of the good life associated with injury has traversed the entire trajectory of Western thought up to modernity, which—thanks to markets—sought to resolve the fragility of the good life by abdicating a fully civil life, hoping in this way to avoid the paradoxes due to its vulnerability, hoping to avoid encountering the other who can wound us.

This is a sad fate indeed, represented by the genius of Jean-Paul Sartre in the pièce théatrale No Exit, which represents Hell as a closed room where the three characters cannot meet because they are afraid of becoming lost. It is a tragic but effective symbol of the contemporary human condition, of the condemnation and necessity of others, summed up in the terrible cry, which by itself sums up the entire discourse that we have attempted to this point: "*L'enfer c'est les autres*" (Hell is the others).

If a society severs the bond that links people to each other, or abrades it so much that it reduces it to the single thin thread of the contractual nexus of the market, and does not realize it, it is entering a mortal crisis. Much of the crisis of Western culture and its possible passing will inevitably be played out on the field of markets.

CONCLUSION

We have come to the end of this discourse on the relationship between economics, happiness, and relational goods, that is, at the end, a study on sociality, with some references to technology, a key ally of the market's ethos.

The vast sociological tradition (from Max Weber, to Ferdinand Tönnies, to Georg Simmel), to say nothing of the modern philosophical tradition (Georg Wilhelm Friedrich Hegel would suffice), has grasped, and in a certain manner discovered, with extreme clarity the duplicity and the ambivalence of life together, or the 'sociable unsociableness' of humanity (Immanuel Kant): It has discovered the 'not,' the negative, that is hidden in the other, who is like me but who is not me, a tragic negative that lurks at the heart of life together in communitas. But, rather than traversing that obscurity, that negative, it has sought escape routes. A major escape route was the invention of the market. One cannot but be amazed that, while modern philosophy came to realize over the course of a few decades—from Hegel to Friedrich Nietzsche—that flight from the 'not' of the other led to nothingness and nihilism, modern economics has been living in its infancy for two and a half centuries: no other age like our age of globalization has been so convinced that it can resolve the contradiction of sociality by trusting in the alchemy of contracts and markets.[12] The great crisis of contemporary economics is concealed in this excessively prolonged infancy, eloquently symbolized by the paradox of unhappiness.

This chapter, however, and in general the research project of which this work is a part, not only decries a great illusion of which society and the humanism of the market are the victims, it is also an attempt to indicate exit routes from the paradoxes we have indicated, even if from afar. These exit routes are the many economic experiences that have been open to gratuitousness, in which economic life is experienced that does not flee from the grievous and terrifying encounter with the other. These seek instead to imagine the economic dimension as a well-rounded human area, thus rejecting the major dichotomies of modernity: economy/society, gift/contract, and eros/agape. The civil economy tradition wants to fearlessly traverse these tensions in search of a happier and more human civil life, without however naively denying the difficulty and the deadly risk that such a process carries with it.

Two Final Considerations

All human communities, whether business, political, communal, or family, are places of life and death, of blessing and wound. I am convinced that a good life together plays out in knowing how to identify the critical point of mediation (by the state and the market, in particular): no city could function without rules, contracts, and justice, which are the highest mediation and the indispensable thirdness that all common civil and democratic life vitally requires.[13] But if the mediations of personal encounters exceed a critical point, life in common becomes dismal: If to avoid conflict we design (returning to the imagery painted in the introduction) condominial regulations, workplaces, and cities that prevent our crossing paths in the halls, stairways, common spaces, and the town square (and the decline in public spaces in our cities is worrying), then the cure becomes far worse than the disease. For example, a good policy is one that can mediate reciprocity without preventing people from encountering each other out of fear, otherwise the 'embrace' of the other is lost. And without hugs we die. We cannot give in to the fear of injury and avoid this embrace, although risking even mortal injury, because from that open wound caused by the other, and which we cause the other, flows the only blessing that makes human life worth living.

Finally, we must also recognize the significant positive benefits to life that the humanism of the techno-market has produced. Even if our tone has primarily been critical, the relationality of the contract and of markets undoubtedly contains positive and civilizing elements, which are the triumph and the result of civilization. The principle benefits that the relationality of the market produces are symbolized by two values of modernity: equality and liberty, especially individual liberty. There is, however, a third principle forgotten by modernity: fraternity. The liberty and equality promised by the market required sacrificing fraternity, because their achievements were effected by the expulsion of the fraternal relationship from the public sphere. Liberty and equality can remain—and historically have frequently remained—experiences of *immunity*; this is not true of fraternity that, unlike the other two principles, is a link, a bond, and, as any other bond, is at the same time a bind *and* a relationship. It is the need/fear of fraternity at the heart of the ambivalence of the good life. But—and this is the insight that in a certain sense inspired this chapter—without fraternity life cannot flourish; there is neither happiness nor full humanness.

Certainly, life is neither happy nor fully human when liberty and equality are lacking, but the great illusion of market humanism has been to think that one might preserve something authentically human while removing the relationship of fraternity, with its tragic burden of sorrow and suffering. The great challenge of postmodernity will thus be to hold these three principles together while devising and building a three-dimensional humanism. The paradox of 'affluent misery,' and the other paradoxes that we have considered, basically tell us the high price we are paying for the sacrifice of fraternity, the forgotten principle of modernity.

And here we must return to the image of the "struggle between Jacob and the angel" in the ford of the river Jabbok, with which we opened our discourse. The story is inserted into the overall picture of Jacob's return to the land of his fathers after his exile with his uncle Laban to escape his swindled brother Esau. To fully understand the meaning of the angel's—or the mysterious being's—blessing to Jacob, we must begin with the experience of wounded fraternity between Jacob and his twin brother Esau. Genesis (ch. 27) recounts the blessing that Jacob deceitfully snatched from his aged and blind father Isaac, illegitimately taking it from Esau. Jacob—with the complicity of his mother Rebecca—disguised himself as Esau and obtained the blessing from his father. The wound that Jacob received from the angel should be understood in relation to that much deeper fraternal wound.[14]

Contemporary market society has sacrificed fraternity for liberty, here as well in the deceit of promising ourselves a common life without suffering and gratuitous encounters. The result, however, has not been the elimination of suffering and sorrow from common life, but rather their multiplication. The market society has, in fact, created structures that wound and mechanisms that exclude the men, women, boys, and girls of many countries from the market and politics, and from their mediations; here I cannot but think of Africa, where the deadly wounds of powerful politicians and of the market—often far more inhuman than the social structures of those cultures—have been added to the wounds of traditional communities. Common life includes wounds: if we do not face them head on, we must find scapegoats for all the wounds in the world, 'black holes' that absorb all the negativity from which we avert our eyes, but which we continue to produce. In these black holes of history, the wounds of others multiply, become infected, and never become blessings. The challenge is demanding but absolutely vital for the quality of human life in the third millennium.

NOTES

1. The author thanks Benedetto Gui, Pier Luigi Porta, Vittorio Pelligra, Alessandra Smerilli, Robert Sugden, and Stefano Zamagni for precious suggestions, and "Econometica" (Milan) for financial support.

2. For historical and theoretical perspectives on the civil economy, see Bruni and Zamagni (2007).
3. On this topic cf. Bruni and Porta (2005).
4. Regarding the relationship between deep interpersonal relationships and health, cf. also Ryff and Singer (2000).
5. Tibor Scitovsky was a prominent American economist who, with the publication of *The Joyless Economy* in 1976, argued that the economic science of that time was too caught up in the objective measure of wealth to believe that the lack of joy could be a research topic for a serious economist. That denunciation of unhappiness was prophetic though, and only today is its full importance understood.
6. Together with Luca Zarri, I introduced this distinction (2007). I want to emphasize that the distinction between primary and nonprimary relational goods is essentially a distinction of degree, not of substance.
7. I am aware that I am making some drastic simplifications here that hold up only if we remain on a completely abstract level. In reality, as we know, the value of a relational good always modifies the value of other components, and vice-versa; if, for example, the emotional relationship between an instructor and a student does not work, objective learning is affected. So we always take these formulas as parables, which must be translated when we pass from formula to reality.
8. For an in-depth analysis, see Bruni (2006).
9. This result is in line with that obtained by Stephan Meier and Alois Stutzer (2004).
10. Both results are quite strong with respect to the use of alternative indicators of relationality and esteem by instrumental variables.
11. To fully grasp the scope and the significance of the previously mentioned effects, it is important to keep in mind that television consumption constitutes the most important free time activity in the world, with an average daily television consumption of 217 minutes in Western Europe and 290 minutes in the United States.
12. In my view, a sign of nihilism in contemporary economics is the theory of decline, which, at its core, is a declaration of a deadly crisis in the Western model, which is constructed around the myth of growth and development.
13. A specific discourse should be developed on the form of mediation represented by the media that, although a different type of mediation, have several critical elements in common with the forms of mediation analyzed here (we referred to television and the Internet, and their adulteration of the true relational goods).
14. "Make me the kind of appetizing dish I like and bring it to me to eat and I shall give you my special blessing before I die" (Genesis 27:5). It is interesting to note a message concealed in the Hebrew language in which that story was written. The Semitic root of the word blessing, *barach* (of the root *brk*), refers in fact, according to some interpreters, to a person's thigh, to Jacob's very thigh wounded in the struggle.

REFERENCES

Bruni, Luigino. (2006). *Civil Happiness: Economics and Human Flourishing in Historical Perspective*. London: Routledge.

Bruni, Luigino. (2010). "The Happiness of Sociality: Economics and Eudaimonia: A Necessary Encounter," *Rationality and Society*, vol. 22, no. 4, pp. 383–406.

Bruni, Luigino. (2011). *The Wound and the Blessing. Economics, Relationships, and Happiness.* New York: New City Press.

Bruni, Luigino, and Pier Luigi Porta. (Eds.). (2005). *Economics and Happiness. Framings the Analysis.* Oxford: Oxford University Press.

Bruni, Luigino, and Luca Stanca. (2008). "Watching Alone: Happiness, Relational Goods and Television," *Journal of Economic Behavior and Organization,* vol. 65, pp. 506–528.

Bruni, Luigino, and Stefano Zamagni. (2007). *Civil Economy.* Oxford: Peter Lang.

Bruni, Luigino, and Luca Zarri. (2007). *La grande illusione. False relazioni e felicità nelle economie di mercato contemporanee* (Working Paper no. 39). Aiccon [Associazione italiana per la promozione della cooperazione e del non profit]: Forlì, Italy.

Cantril, Hadley. (1965). *The Pattern of Human Concerns.* New Brunswick, NJ: Rutgers University Press.

Deci, Edward L., and Richard M. Ryan. (2001). "On Happiness and Human Potentials: A Review of Research on Hedonic and Eudaimonic Well-Being," *Annual Review of Psychology*, vol. 52, pp. 141–166.

Diener, Ed, and Martin E. P. Seligman. (2002). "Very Happy People," *Psychological Science*, vol. 13, pp. 81–84.

Donati, Pierpaolo. (1986). *Introduzione alla sociologia relazionale.* Milan: Franco Angeli.

Duesenberry, James. (1949). *Income, Saving and the Theory of Consumer Behaviour.* Cambridge, MA: Harvard University Press.

Easterlin, Richard. (1974). "Does Economic Growth Improve the Human Lot? Some Empirical Evidence." In *Nation and Households in Economic Growth: Essays in Honor of Moses Abromowitz*, edited by Davis, P. A. and M. W. Reder. New York: Academic Press.

Frank, Robert. (1985). *Choosing the Right Pond.* New York: Oxford University Press.

Frey, Bruno S., and Alois Stutzer. (2002). *Happiness in Economics.* Princeton, NJ: Princeton University Press.

Genovesi, Antonio. (1963). *Autobiografia e lettere.* Milan: Feltrinelli.

Gui, Benedetto. (1987). "Eléments pour une définition d'économie communautaire," *Notes et documents*, vols. 19–20, pp. 32–42.

Gui, Benedetto. (2002). "Più che scambi incontri. La teoria economica alle prese con i fenomeni relazionali," *Sacco e Zamagni*, pp. 15–66.

Gui, Benedetto. (2005). 'From Transactions to Encounters: The Joint Generation of Relational Goods and Conventional Values.' In *Economics and Social Interactions: Accounting for Interpersonal Relations,* edited by Benedetto Gui and Robert Sugden (Cambridge: Cambridge University Press) pp. 23–51.

Kahneman, Daniel, Alan Krueger, David Schkade, Norbert Schwarz, and Arthur Stone. (2004). "A Survey Method for Characterizing Daily Life Experience: The Day Reconstruction Method (DRM)," *Science,* vol. 306, pp. 1776–1780.

Layard, Richard. (2005). *Happiness. A New Science.* New York: Penguin Books.

Meier, Stephan, and Alois Stutzer. (2004). "Is Volunteering Rewarding in Itself?" (IZA Discussion Paper no. 1045). Bonn: Institute for the Study of Labor (IZA).

Nussbaum, Martha. C. (2001). *The Fragility of Goodness: Luck and Ethics in Greek Tragedy and Philosophy.* Cambridge, MA: Cambridge University Press. (Originally published 1986)

Ryff, Carol D., and Burton Singer. (2000). "Interpersonal Flourishing: A Positive Health Agenda for the New Millennium," *Personality and Social Psychology Review*, vol. 4, pp. 30–44.

Scitovsky, Tibor. (1976). *The Joyless Economy: An Inquiry into Human Satisfaction and Consumer Dissatisfaction*. Oxford: Oxford University Press.
Sen, Amartya. (1977). "Rational Fools: A Critique of the Behavioural Foundations of Economic Theory," *Philosophy and Public Affairs*, vol. 6, no. 4, pp. 317–344.
Sen, Amartya. (2009). *The Idea of Justice*. Cambridge, MA: Harvard University Press.
Uhlaner, Carole J. (1989). Relational Goods and Participation: Incorporating Sociality into a Theory of Rational Action," *Public Choice*, vol. 62, pp. 253–285.
Veblen, Thorstein. (1998). *The Theory of the Leisure Class*. New York: Prometheus Books. (Originally published 1899)

7 Desire-Satisfactionism and Technology

Anton Tupa

According to desire-satisfaction theories of the good life, roughly, what is good for a person is the satisfaction of a relevant set of desires. 'Desires' can be understood in a broad sense, and the set of desires can include all pro-attitudes, ranging from slight inclinations to complex plans and projects. A desire is satisfied if the desired state of affairs is the case. Alternatively, one might say that a desire is satisfied if a proposition appropriately expressing the desired state of affairs is true. According to desire-satisfactionism, roughly, one has a good life to the extent to which a relevant set of pro-attitudes and the world match up.

In this chapter, I critically discuss contemporary theories of desire-satisfactionism, the main objections to them, and how such objections can be overcome. In doing so, I will be exploring lines of significant and promising development, with a focus on what it would take to have a fully complete and specified desire-satisfaction theory of the good life. Also, I will be developing a desire-satisfactionist perspective on technology. I will be assessing some of the relations between technology and desire, as well as normative issues relating to desire-satisfaction in the use of technology.

THE NATURE OF DESIRE-SATISFACTIONIST THEORIES

Desire-satisfaction theories have many attractive features, two of which are especially noteworthy. First, they make what is good for one subjective in an important way. What is good for a person, according to these theories, depends crucially on that person's particular constitution or will. Second, desire-satisfaction theories mediate between two extremes: they mediate between, on the one hand, having the good life entirely depend on mental states, and on the other, entirely depend on non-mental states of the world. Desire-satisfaction accounts treat some mental states—desires—as important. Moreover, because desire-satisfaction does not typically involve only mental states, these theories treat states of the "outside" world as essential. Thus, desire-satisfaction accounts of the good life occupy a middle ground, taking on virtues of each side, while avoiding the troubling characteristics of the extremes.

'Desire-satisfaction theories' picks out a *family* of theories. Each member of the family shares the characteristics previously outlined. However, desire-satisfaction theories can vary in what desires they make relevant to the good life. For example, on a simplistic version, any of one's desires can be relevant to the goodness of one's life. Some restrictions or alterations to the simplistic version are often thought necessary. Many recent works about desire-satisfaction accounts of well-being involve debates over the criteria of the set of desires the satisfaction of which enhance well-being. For simplicity, call the set of desires the satisfaction that enhances well-being the "relevant set." It would seem a testament to the *prima facie* plausibility of desire-satisfaction accounts generally that much of the literature on desire-satisfactionism involves criticizing certain specifications of the relevant set in contrast with some other.

A first draft of a desire-satisfaction account of the good life is as follows: State of affairs *S* is good for a person *A* at time t_1 only if *A* at t_1 intrinsically desires *S*. Call this the basic theory. First, note that according to the basic theory, *A* must actually desire a state of affairs for it to enhance the goodness in his life. Actual states of affairs are good because they are the satisfaction of actual states of mind.[1] Second, the desire and the relevant state of affairs must be concurrent. If *A* were to cease desiring *S*, then *S* would no longer be good for *A*.[2] Third, only intrinsic desires are in the relevant set.[3] If instrumental desires were to be in the relevant set, then an odd sort of double-counting (or triple-counting, and so on) would occur for the goodness of *A*'s life.[4] Last, although not explicitly stated, the basic theory may make the extent of goodness of *S* for *A* determined by the strength or intensity of *A*'s desire for it.

DESIRE-SATISFACTIONISM: CRITICISMS AND RESPONSES

Is the basic account of a desire-satisfaction theory of the good life acceptable? Richard Brandt (1979) has criticized a similarly basic account of the rationality of desires. Strictly speaking, Brandt is not concerned with desire-satisfaction accounts of welfare, but his arguments about the rationality of desires have been influential for desire-satisfaction account of welfare. Derek Parfit (1986), James Griffin (1989), and others have criticized a basic account of well-being along lines very similar to Brandt's. These lines of criticism have been very influential, but they have come under scrutiny in recent literature.

Brandt argues that a basic account of rationality of desires that treats all present, actual and intrinsic desires as rational is implausible. He has four principal examples of present, actual, and intrinsic desires that are not rational: (1) desires formed from false beliefs, (2) desires formed by cultural conditioning, (3) desires based on generalizations from atypical examples, and (4) exaggerated intensity of desire caused by early deprivation. In each of these examples, Brandt explains how it can be the case that a desire can be intrinsic and yet still somehow malformed. Brandt suggests that if a person would lose the desire upon receiving all relevant and available information (after undergoing a kind of cognitive psychotherapy), then that desire is not rational (1979, pp. 10–14). The natural conclusion that follows is that the basic account of rationality of desires is wrong. Moreover, what Brandt says about the basic account of rationality of desires plausibly extends to the basic desire-satisfaction account of the good life.

Brandt's examples, or examples similar to them, have come under scrutiny by Chris Heathwood (2005) and Mark C. Murphy (2002). Heathwood's ambitious defense of the basic theory sorts the supposed counter-examples into two sorts: those that belong in the relevant set but, contrary to popular opinion, are not problematic for the basic theory, and those that belong in the relevant set and are indeed problematic for the basic theory, but not in a way that is uniquely problematic for the basic theory. Let us explore one of Heathwood's lines of argument.

Take the problem of ill-informed desires. Heathwood picks up on a supposedly problematic example from Thomas L. Carson (2000). Imagine that a person is thirsty and desires to quench her thirst within the next hour. Her desire is intrinsic (e.g., her desire is not derived from a desire to avoid dehydration, sickness, etc.). She sees a river and sees the opportunity to quench her thirst within the hour. However, the river is polluted and quenching her thirst will make her sick. She drinks from the river and becomes sick.

Heathwood argues that such a desire is in the relevant set for the basic theory—that is to say, satisfying the desire is good for her in a sense—but that the example is not genuinely a problem for the basic theory. Heathwood's defense begins with a distinction between something's being intrinsically bad for a person and all-things-considered bad for that person. Something is intrinsically bad for a person if, in isolation to other of his desires, it in itself makes him worse off. Something is all-things-considered bad for a person if, in taking all of her desires into account (including the specific desire which she satisfies and others), it is bad for her. Heathwood suggests that in the example of the thirsty person, the satisfaction of her desire to quench her thirst within the next hour is indeed intrinsically good for her. However, it is not all-things-considered good for her, according to Heathwood, because several other and stronger of her intrinsic desires are frustrated (such as her desire not to be sick).

Heathwood's argument has an initial air of plausibility because indeed the thirsty person is made better off for a short while before she gets sick. Although Heathwood does not use the terminology, this could be said to be an example of a "swamped desire-satisfaction"—a desire is satisfied, but the state of affairs that obtains frustrates other, countervailing desires. Heathwood is willing to call such a desire "defective" in a sense, but not *intrinsically* defective. So the example of the thirsty drinker is not a counter-example to the basic theory, according to Heathwood.

Are all ill-informed desire-satisfactions of the swamped desire-satisfaction variety?[5] Certainly there are ill-informed desire-satisfactions that do not have the feature of intuitively making a person's life better for a short while before the effects of the resulting state of affairs set in, thereby frustrating other, stronger desires. For example, imagine that a person desires to eat pistachio ice cream, but he has never tried it before. Suppose that his desire is intrinsic (it is not derived from a desire to eat something cold on a hot day, etc.). Imagine that he takes a bite, finds it disgusting, then demands a cone of chocolate, and yells at the shop employee for even having such a disgusting flavor for sale. In such an example, it is hard to see how there is any welfare-enhancing desire-satisfaction, either intrinsic or all-things-considered.[6]

A proponent of the basic desire theory could offer a line of defense against the pistachio ice cream example. The proponent would suggest that all that the result in the pistachio example shows is that the man did not really intrinsically want to eat pistachio ice cream after all. The proponent suggests that his intrinsic desire was never to eat pistachio ice cream: it was only ever to taste it. His desire to eat the ice cream, according to this line of defense, was derived from his desire to taste it: He did taste it and, insofar as that desire is satisfied, his welfare is enhanced. This is yet another example of a swamped desire-satisfaction, so the argument goes. Call this strategy the 'rediscription' strategy.

The rediscription strategy works reasonably well for at least some examples, especially those involving the senses. However, it still is hard to comprehend the benefit the person receives by having his desire to taste pistachio ice cream satisfied. Moreover, the rediscription strategy has an implausible implication: that any desire that is extinguished upon the acquisition of information cannot have been intrinsic. To be sure, at least some desires that are extinguished with some relevantly related experience. For example, if a woman hopes to kiss someone for the first time and then does so, her desire for that first kiss must be extinguished. But not all examples of extinguished desires are relevantly like that.

It is intuitive that people can intrinsically desire states of affairs that they do not fully or significantly comprehend. In examples that do not involve the senses, such as a desire for a certain career path, the problem for the basic account becomes more forceful. If someone wants to win an election to become a public official, yet does not fully understand what this job entails, and if the job does not fit with her personality, then being elected will be bad for her, or at least not good for her. It is hard to redescribe her desire to be elected in some other way, such that there is a swamped desire-satisfaction. Thus, it seems that there are still some problematic examples of ill-informed desires for the basic theory.

To be sure, the brief previous discussion does not do justice to the full panoply of Heathwood's arguments. Moreover, it sets aside altogether

Murphy's arguments (which disagree with more basic assumptions of this chapter) that address the issue of ill-informed desires. Nevertheless, many people have found there to be a profound problem with the basic account of desire-satisfactionism. So long as it is possible for one to have an intrinsic desire for a state of affairs one does not fully comprehend, and one loses this desire with the acquisition of information, then there is a *prima facie* case that there are problematic ill-informed desires.

What are the conditions in which a desire can be excluded from the relevant set? Why is it that their satisfaction is at least sometimes not rational, welfare-enhancing, or goodness-enhancing for the desiring person? Brandt suggests that desires that are lost after a process of cognitive psychotherapy are not rational (1979, p. 113). First, Brandt is not asserting that all desires that are malformed are thereby not rational. Rather, he is asserting only that those desires that are lost after vividly receiving the best available information are not rational. Thus, there is room for a desire to be "misbegotten" and yet still be rational—provided that the agent retains it after learning of its origin. Here is an explanation of what is going on in Brandt's examples: a person's desires can distort her *personality* in an important way. This explanation is inspired by Peter Railton's account of the nonmoral good (1986a, 1986b). Moreover, this line of response is picked up by Anton Tupa (2006). If a personality is thought of as a set of dispositions, which taken with a belief set, to desire certain states of affairs, then false beliefs might result in desires that distort a personality, but a sufficiently rich set of true beliefs does not result in a desire set that distorts a personality.

Where to go from here? There are two reasonable lines of response. Here is (1): restrict the actual desire-set by some further necessary condition to make the relevant set. One *could* require that the desire be formed with a robust set of true beliefs. This, however, goes against a lesson from Brandt's examples, because it is possible for a desire to be malformed, yet nevertheless rational. A better additional necessary condition is to require that one's actual present desires pass a counterfactual test—they *would* survive scrutiny with sufficiently rich information. A plausible draft of (1) for accounts of the good life is as follows: *S* is good for *A* at t_1 only if *A* at t_1 intrinsically desires *S*, and at t_1 *A* is constituted in such a way that *A* would desire *S* with sufficiently robust information.

Alternatively, one can treat the relevant set as being constituted entirely by some or all of the desires one *would* have *were* one to be sufficiently informed. Call this approach (2). Note that along this line, there is no requirement that a person actually presently desires a state of affairs for that state of affairs to be good for her. This has been the more popular response to the failure of basic desire-satisfaction accounts. According to (2), roughly: *S* is good for *A* at t_1 only if *A* at t_1 would intrinsically desire *S*, were *A* to have sufficiently robust information. In the works cited above, Griffin (1989) and Parfit (1986) explore and develop variations of (2), even if they do not ultimately accept any variation of (2), whereas version (1) might be seen as an adaptation of Brandt's account.

It is unclear how much information should be given to the person's hypothetical counterpart for either (1) or (2). Brandt suggests the best information according to the science of the day (1979, p. 112). However, one might argue that Brandt's account gets it wrong both in terms of quality and quantity: that the person should have as much correct information objectively as is possible. How about requiring—the hypothetical counterpart: know all there is to know—to be fully informed? Several notable lines of argument have arisen in response to a full information requirement.

Railton (1986a, 1986b) points out that desire-satisfaction accounts such as (2) have problems because if we look to the possible scenario in which *A* is fully informed, then some of *A*'s other desires may change as well. For example, if a person is fully informed, it is entirely possible that he will not desire to know more. However, the less informed counterpart could intrinsically desire to know more and, intuitively, the satisfaction of that desire would be good for him. Railton proposes instead that the relevant set be composed of the desires an ideally informed person would want his less informed counterpart to want. This seems roughly to solve the sort of problem Railton introduces.[7]

Connie Rosati's (1995) worry is that a fully informed person would be so different from his actual counterpart that the desires of the fully informed person would not be relevant to his actual counterpart's well-being or goodness. One promising line of response to Rosati's objection is to appeal to the idea of a *personality* as previously discussed. If a personality is a set of dispositions to desire states of affairs given a belief set, then a normal, imperfect person and his idealized counterpart can share a personality—they very likely have different belief and desire sets, but they share the basic dispositions that give rise to these. So a common personality is what makes the fully informed person's advice pertain to the less informed person's life. Admittedly, this is not a complete response to Rosati's objections, but it does seem to be a promising line of response.

Sobel's (1994) criticism is that if one person has a certain kind of experience or knowledge, and that rules out having another kind of experience or knowledge, then it is impossible to be fully informed in the right way. For example, if *A* has his first kiss ever with *B*, *A* cannot have his first kiss ever with *C*. So it would seem that some experiences and accompanying beliefs rule out others. Moreover, Sobel is right to point out that one cannot easily shift the position to get around this objection, such as having *A* kiss *B*, then have amnesia, then kiss *C*, and so on, then later remember all of them.

However, there are several responses to Sobel's objection. First, Sobel's standards for what it takes to be fully informed are very high. In the example provided, he takes it that there is a "firstness" built into such experiences. It is far from clear that this is correct. Second, even if Sobel is right about it being impossible for someone to be fully informed, having less information might still do—after all, it is logically possible to know all true propositions and have a great deal of other knowledge, and this is an enormous amount of information. Third, Sobel requires that hypothetical fully informed people "stand" states of affairs side by side and compare them (such as with first kisses), but perhaps this is not necessary. Perhaps the strength of desire could be determined and the corresponding good that results from each of the kisses could then be determined without comparing the exact experiences. Thus, there are some promising lines of defense for version (2) against some of the more important criticisms in the literature.[8]

Up to this point, the concern has been to get right the information and counterfactual conditions of the relevant desire set. Each version is stated as providing necessary but not sufficient conditions for the relevant set. What other restrictions ought there be on the relevant set of desires for an account of the good life? The answer to this question proceeds largely independently of whether (1) or (2) or even the basic theory is correct.

Consider a case from Parfit (1986, p. 494). Imagine that *A* meets *B* and interacts with *B* briefly. *A* comes to learn that *B* has an illness, and *A* desires intrinsically that *B* get well. Suppose furthermore that *A* and *B* part ways never to learn anything of each other again. In such a case, is *A*'s desire that *B* get well in the relevant set? Parfit and many others have thought not. Call this the problem from irrelevant desires. One further restriction on the relevant set can be as follows: *S* is good for *A* at t_1 only if at t_1 *S* sufficiently involves *A*. *S* has to involve *A* in order for the desire for *S* to be in *A*'s relevant set. Detailed accounts of the conditions of 'sufficient involvement' are proposed by Mark Carl Overvold (1980) and Tupa (2006). Alternatively, Mark Lukas (2010) has argued that desire-satisfaction theories can "bite the bullet" on the problem of irrelevant desires.

Last, some have worried about desires for perverse things—such as a person's desire to be worse off than she is. Interestingly, Stephen Darwall (2002) has proposed a "rational care" account of welfare that has promise in solving the problem of perverse desires. Darwall's theory, simplified somewhat, is that something is good for a person by virtue of being something it makes sense for anyone caring about her to want for her for her sake. If a desire stems from hatred, indifference, or concern for someone else, then the desire cannot be in the relevant set. Darwall's approach is controversial, because there is a risk that caring for someone is desiring that he have a high level of well-being. Nevertheless, Darwall's rational care requirement has strong prospects for solving the problem of perverse desires.[9]

A near-final draft of version (1) then is: *S* is good for *A* at t_1 if and only if (i) *A* at t_1 intrinsically desires *S*, (ii) at t_1 *A* is constituted in such a way that *A* would desire *S* with sufficiently robust information, and (iii) *S* sufficiently involves *A*. A near-final draft of version (2) then is: *S* is good for *A* at t_1 if and only if and because (i) *A* at t_1 is constituted in such a way that, were *A* to have sufficiently robust information, *A* would intrinsically desire *S* for *A* as *A* is, and (ii) *S* sufficiently involves *A*. Recall that the difference between

(1) and (2) is that (1) requires that the person desires the state of affairs, yet the actual desire must pass a counterfactual test, while (2) does not require that the person have an actual desire for the state of affairs, and so is fully counterfactual. Versions (1) and (2) can add a requirement that the relevant desire should stem from care for the subject.

DESIRE-SATISFACTIONISM AND PERSONHOOD

Much of the previous section is an explanation of the internal debates between various kinds of desire-satisfaction theories. Because of these internal debates, one might somehow mistakenly infer that there is something wrong with the family of desire-satisfaction accounts of the good life. This conclusion would be a mistake, or at least premature. First, desire-satisfaction theories have been discussed in deep analytical detail for only a short duration—the family is a young family in the realm of philosophy. In such a case, deep disagreements about which version is best are natural and expected. Second, some of the problems addressed in the internal debates affect other families of theories of well-being too, yet there is not as deep a parallel literature on the debates. For example, some hedonists will say that all of a person's pleasure enhances that person's well-being; yet other hedonists will say that only pleasure that relates to a person's life enhances a person's well-being.

Wayne Sumner (1996), for example, argues that the mental state of life-satisfaction, if it passes a part-counterfactual test, enhances welfare. This debate within the family of mental-state theories mirrors the debate over previously mentioned irrelevant desires. Moreover, to get around certain objections to hedonism, some mental-state theorists such as Sumner adopt an information-based counterfactual test that one's mental state must pass in order to be in the relevant set. This mirrors the debate over the information condition that desire-theories face to some extent.

Third, the internal debates among desire-satisfaction accounts stem from concern about how best to capture the concepts central to subjective accounts of the good life. Concepts such as 'subjective,' 'desire,' and 'good for' are at the heart of the internal debates among desire-satisfaction theories. There is a lot of agreement that these concepts are important and that the best account of well-being or the good life must capture them, but there is disagreement on how best to capture them. Thus, the disagreements are much less basic than one might initially think. Fourth, there are strong responses that the desire-satisfaction theorist can make to many, if not all, of the objections. This chapter contains a number of those responses to potential objections.

The central motivation, arguably, for the various lines of defense to the previously mentioned objections is the idea that desires in a person's relevant set must fit the nature of the person. Recall the concept of personality, as it has been used in this chapter, is that of the set of dispositions, which taken with a belief set, generates a set of desires. An information condition is probably necessary to avoid distortions of the personality, or so it has been argued (this is the argument over ill-informed desires). The concept of 'fit with a person' also ties in with another of the previous debates—that the desires in the relevant set must in some way be fit for the person. Some further restrictions or qualifications would seem to be required for the desires to fit with a person as well (this is the debate over irrelevant desires and over perverse desires). Despite the fact that some desire-satisfaction theories are somewhat sophisticated, the core motivation is the concept of the person, reflected both in the concept of personality, understood in the technical way described in this chapter, and in the concept of fit with a person.

DESIRE AND TECHNOLOGY

The interplay between desire-satisfaction theories and technology is dynamic. First, technology can provide information that can help one to *discover* what one would desire with more robust information. Second, technology might *create* new desires, which in some cases might seem to work against common intuitions. Third, technology might help to *satisfy* one's desires, and thereby make one's life better.

Start with the first point, which is not so much an argumentative point as it is a point of clarification about the nature of desire-satisfactionism thus developed. The desire-satisfaction theories thus developed are about the truth conditions of the good life, and they are not meant to be decision procedures one engages in to discover what is good for one. Thus, there is no absolute rule according to which any piece of information about a state of affairs increases the likelihood that an actual desire is in the relevant set according to a nonbasic desire-satisfactionism. Nevertheless, a persuasive general rule of thumb is that the more one knows about a state of affairs, the more likely a desire for it will be in the relevant set.

Technology can, but does not always, enhance one's belief-set in this regard. It goes without saying that the Internet makes more information available, yet exposure to the Internet might not make one more informed in the relevant sense for desire-satisfactionism. Anton Vedder (2008) addresses the reliability of information on the Internet. Nonexperts, argues Vedder, have a difficult task in assessing the reliability of the content of information on the Internet. Sometimes, people are more likely to seek out and believe information that already conforms to their belief-set. One way to overcome this natural bias is to seek out information with a pedigree. More traditional media, such as books, newspapers, and journals often offer the nonexpert a pedigree accompanying the information to help establish reliability of that information. Such pedigree is often missing on Internet content, argues Vedder. Thus, Vedder sees two sorts of reliability problems for information on the Internet: one is a content problem and one is a pedigree problem. Vedder suggests, among other things, that a credibility-conferring system on the Internet should be developed and that Internet consumers should be made more aware of the differing levels of reliability of information on the Internet.

Another related point on the role of technology in helping ordinary people obtain a desire set coextensive with the relevant set is that of virtual reality. While the virtual reality technologies one sees in movies might be far off, there are functional virtual worlds in which people gain information. Philip Brey (2008) gives a nuanced account of the morality of virtual worlds, including an ethical analysis of the standard of accuracy of virtual information. Such an analysis is helpful for the desire-satisfaction theorist in understanding what it is to be fully informed about what things are like (in addition to having all propositional knowledge). As previously noted, Sobel's concern is that it is impossible for one to be fully informed, because some knowledge of what something is like precludes knowledge of what other things are like. Virtual reality addresses part of Sobel's concern in that it makes clearer the extent of knowledge people can gain about what things are like. Even if in a technical way a person's having some information precludes that person from having other information, virtual reality makes clearer the extent of information one can obtain and how such information could make a person's desire-set overlap to a greater degree with her ideal desire-set.

Studies of information on the Internet also reveal a presupposition of versions of ideal desire-satisfactionism and show that further inquiry is required. Brandt has a concern for the effect of cognitive psychotherapy on a person (1979, pp. 110–115). Brandt imagines that a person undergoing cognitive psychotherapy would be exposed repeatedly to the best scientific information of the day by a method that makes the information most vivid to the subject. Brandt's concern is that exposure to information will not necessarily make it clear why the information is relevant. Seemingly, those who favor ideal-desire theories since Brandt set aside the concern about relevance, presumably imagining that an ideal counterpart to an actual person could be informed as if by miracle. Is such an assumption warranted?

David Shenk (1997) argues that the Internet provides an information glut that has the unexpected effect of making our decisions in many cases worse. Shenk discusses such phenomena as overconfidence and impaired judgment that accompany substantial increases in the amount of information available. Shenk concludes that it would be better in many cases to limit the information one receives. Although Shenk's work does not refute ideal desire-satisfactionism, if his arguments about the Internet are applied to full-information theories, they do renew Brandt's concern about the problem of relevance for such theories. Shenk's work is a reminder that there is more to information than mere belief formation; there is also information processing and comprehension. This is an underexplored area of research in ideal desire-satisfactionism, and it stands in need of further consideration.

The second point, that technology creates new desires, is guided by neither absolute rules about whether the creation of new desires is good, nor guided by rules of thumb. Some instances of desire creation from technologies will not implicate a controversial aspect of desire-satisfactionism. Take an example of a desire to listen to a new kind of music. In such an example, desire-satisfaction theories explain how such technologies affect well-being in a plausible way—indeed, probably in a better way than do most other accounts of well-being. In other sorts of examples, however, there is a concern that a desire created by a technology is suspect or troubling.

Take for example the selection of offspring either on the basis of sex selection or on the basis of avoiding disease. Anita Silvers, Dan Brock, and others make spirited contributions to a collection of essays on this issue (Wasserman, Bickenbach, and Wachbroit 2005). New technologies make prenatal testing for disease possible and might create a desire in prospective parents not present before, or at least only weakly present before the rise of the relevant technologies due to its being beyond anyone's control. Desire-satisfactionism, thus far developed in this chapter, treats such desires as in the relevant set. Some might find this counter-intuitive. Although the matter is controversial, some might wish to restrict the relevant set of desires in a way not yet discussed. Desire-satisfactionism is often understood as a theory of prudential but not moral good for a person. As a theory of prudential good, the concern is not what is morally right or wrong, but what is in a person's self-interest. In light of this, some desire-satisfactionists would be willing to "bite the bullet" on the implication of technology creating desires.

Technologies can also satisfy desires. Begin with the distinction between treatment and enhancement. Some will argue that at least in some cases, use of a technology for treatment is acceptable for a sick person, but that an enhancement is unacceptable for a healthy person. Desire-satisfactionism, thus explained, does not treat the distinction between treatment and enhancement as relevant to the good life. Although this chapter does not explore the allocation of resources, desire-satisfactionism treats the satisfaction of the relevant desires to be good for a person, and there has been no restriction thus far placed on the relevant set that distinguishes between treatment and enhancement. Almost no philosopher or scientist gives wholehearted and unqualified approval or disapproval of use of technologies to satisfy desires and enhance quality of life. For example, Leon Kass (2003) is cautiously pessimistic about the moral prospects of using some new technologies to enhance lives. On the other hand, some writers have offered qualified approval of use of technologies, such as implantable devices in the brain (Clausen 2009), or in use of products that come from technologies, such as cognitive-enhancing drugs (Greely 2008) (if such drugs are not themselves technologies).

Of course, there is much more to be said about the interplay between desire-satisfactionism and technology, some of which will be taken up later in this volume. Moreover, there is much more to be said about which version of desire-satisfactionism is the most plausible. This chapter has included an account of the most contentious lines of debate over desire-satisfactionism and a sketch of the various ways in which desire-satisfactionism and technology interact in philosophically important ways.

NOTES

1. It is possible to construe a basic account as treating desire-satisfaction, rather than states of affairs, as good for one, but set this aside.
2. There are some complex cases for the concurrence requirement. For example, a person might have a present desire for a future state of affairs. Here, the structure is: *A* at t_1 desires *S* to obtain at t_2. In this example, the desire is at a given time, while the content of the desire is indexed to a future time. It is unclear how the concurrence requirement would treat such an example. Set this aside.
3. A desire is intrinsic if the state of affairs is desires in itself, and not for some further state of affairs. A desire is instrumental if the state of affairs is desired to get some other state of affairs.
4. For a provocative argument on the grounds of individuation of desires that cuts across the intrinsic-desire requirement, see Murphy (2009).
5. To be sure, many of a person's ill-informed desire satisfactions are not at all problematical. If the example of the thirsty drinker were modified to include a *pure* river from which the person drinks, then no problem ensues. The issue though is where there are ill-informed desire-satisfactions that are not swamped desire-satisfactions in a way that causes problems for the basic theory.
6. The pistachio ice cream example would be more analogous to the thirsty drinker example if the ice cream eater were to like pistachio ice cream in general, but would be allergic to some ingredient in this particular ice cream.
7. An alternative expression of Railton's insight is that what is relevant is what the ideal advisor would want for his less informed counterpart. This alternative expression puts the formulation in terms of desired states of affairs, rather than desired desires.
8. Interestingly, option (1)—the part counterfactual version—does not necessarily require full information. In the counterfactual state, perhaps *A* only needs to understand *S* in a sufficiently robust way to pass the test (in order to make the desire a member of the relevant set).
9. Heathwood (2005) has a different line of argument in response to the problem of perverse desires—a line of argument according to which the problem of perverse desires is a kind of paradox of desire, and thus not a special problem for desire-satisfaction accounts of well-being.

REFERENCES

Brandt, Richard. (1979). *A Theory of the Good and the Right*. Oxford: Clarendon Press.

Brey, Philip. (2008). "Virtual Reality and Computer Simulation." In *Handbook of Information and Computer Ethics*, edited by Himma, Kenneth Einar and Herman T. Tavani. Hoboken, NJ: John Wiley.

Carson, Thomas L. (2000). *Value and the Good Life*. Notre Dame: University of Notre Dame Press.

Clausen, Jens. (2009). "Man, Machine and in Between," *Nature*, vol. 457, pp. 1080–1081.

Darwall, Stephen. (2002). *Welfare and Rational Care*. Princeton, NJ: Princeton University Press.

Greely, Henry. (2008). "Towards Responsible Use of Cognitive-Enhancing Drugs by the Healthy," *Nature*, vol. 456, pp. 702–705.

Griffin, James. (1989). *Well-Being: Its Meaning, Measurement and Moral Importance*. Oxford: Oxford University Press.

Heathwood, Chris. (2005). "The Problem of Defective Desires," *Australasian Journal of Philosophy*, vol. 83, no. 4, pp. 487–504.

Kass, Leon. (2003). *Beyond Therapy: Biotechnology and the Pursuit of Happiness*. Report of the President's Council on Bioethics, 2003. New York: Harper Perennial.

Lukas, Mark. (2010). "Desire Satisfactionism and the Problem of Irrelevant Desires" *Journal of Ethics & Social Philosophy*, vol. 4, no. 2, pp. 1–24.

Murphy, Mark. (2009). "The Simple Desire-Fulfillment Theory," *Nous*, vol. 33, no. 2, pp. 247–272.

Overvold, Mark Carl. (1980). "Self-Interest and the Concept of Self-Sacrifice," *Canadian Journal of Philosophy*, vol. 10, pp. 105–118.

Parfit, Derek. (1986). *Reasons and Persons*. Oxford: Oxford University Press.

Railton, Peter. (1986a). "Facts and Values," *Philosophical Topics*, vol. 14, pp. 5–31.

Railton, Peter. (1986b). "Moral Realism," *The Philosophical Review*, vol. 95, pp. 163–207.

Rosati, Connie. (1995). "Persons, Perspectives and Full Information Accounts of the Good," *Ethics*, vol. 105, pp. 296–325.

Shenk, David. (1997). *Data Smog: Surviving the Information Glut*. New York: Harper Collins.

Sobel, David. (1994). "Full Information Accounts of Well-Being," *Ethics*, vol. 104, pp. 784–810.

Sumner, Wayne. (1996). *Happiness, Welfare and Morality*. Oxford: Oxford University Press.

Tupa, Anton. (2006). *Development and Defense of a Desire-Satisfaction Conception of Well-Being*. Doctoral dissertation, University of Florida.

Vedder, Anton. (2008). "Responsibilities for Information on the Internet." In *The Handbook of Information and Computer Ethics*, edited by Himma, Kenneth Einar and Herman T. Tavani. Hoboken, NJ: John Wiley.

Wasserman, David, Jerome Bickenbach, and Robert Wachbroit. (Eds.). (2005). *Quality of Life and Human Difference: Genetic Testing, Health Care, and Disability*. New York: Cambridge University Press.

Part III

Consumer Products and Well-Being

8 Consuming Happiness

Lindsey Patterson and Robert Biswas-Diener

Among the questions that have long excited and perplexed philosophers, policymakers, religious scholars, and psychologists is the issue of whether happiness is an "inside-out" or "outside-in" phenomenon. That is, can we attribute individual happiness to objective life circumstances or to subjective attitudes and interpretations of those circumstances? A cursory examination of the topic yields an answer of "both." People's emotions and satisfaction are affected by the objective circumstances of their lives (Veenhoven and Hagerty 2006; Diener, Suh, Lucas, and Smith 1999), but their subjective experience of daily events is also important to happiness (Lyubomirsky, Sheldon, and Schkade 2005). Interestingly, many people intuitively focus on addressing externalities rather than attending to their own attitudes and aspirations in their individual pursuit of happiness. This is particularly true of the consumption of material goods, both basic and luxury. Most people assume that fulfilling basic needs, status needs, comfort needs, and experience needs are important elements of happiness. To fulfill these needs, people need to consume products and services that are grounded in the material world. This chapter reviews the potential benefits and pitfalls of consumption behaviors in the pursuit of happiness. It is argued that although it is important to satisfy physical, social, and psychological needs, the consumption of material items may only have short-term benefits as individuals adapt to their purchases and begin to desire more. However, experiential purchases and purchases that engage the consumer in enjoyable activities can have both short- and long-term positive impacts.

MASLOW AND THE FULFILLMENT OF NEEDS

In his seminal theory of motivation, Abraham Maslow (1954) proposed that people need to fulfill basic physical needs as well as social and psychological needs. At the heart of this theory is the understanding that people are physical creatures in need of food, shelter, thermoregulation, and other physiological fulfillments. To fulfill these needs, individuals often acquire various basic material resources. There is some support for the idea that basic needs are fulfilled through material resources and that people at lower economic levels are highly concerned with material sufficiency. In a study by one of us (Robert Biswas-Diener), for instance, Maasai tribal people living a materially simple lifestyle overwhelmingly mention "sturdier roofs," "more money," and "have an automobile" when asked what element of their lives they would change if they could (Biswas-Diener, Vitterso, and Diener 2005). Furthermore, the consumption of material resources is sometimes necessary to fulfill basic needs.

As people satisfy basic and even higher level needs, the consumption of material resources may enhance well-being and life satisfaction. A recent study of demographically representative samples from more than 130 countries (Diener, Ng, Harter, and Arora 2010) found that income and the possession of luxury conveniences were associated with higher rates of life evaluation but not necessarily increased positive affect. Once basic needs are met, individuals are more likely to be able to satisfy higher level needs. In a follow-up study to that by Diener and colleagues (2010), Diener and Tay (in press) found that the richest decile of the sample in the Gallup World Poll was almost nine times more likely to have met psychosocial needs, such as receiving respect or feeling a sense of mastery. Therefore, to the extent that people consume for the purpose of fulfilling physical, social, and psychological needs, material consumption may be an essential and beneficial endeavor.

On the other hand, there are several reasons to believe that a focus solely on material consumption as a strategy for need fulfillment is not the optimal strategy for achieving happiness. Among these reasons is the finding that, although the rich and poor appear to differ on life satisfaction, they do not appear to be different on levels of positive versus negative affect (Diener and Tay, in press). In addition, a number of researchers have found diminishing marginal utility of income on happiness, such that increasing income affects life satisfaction at diminishing rates of return (Diener, Ng, and Tov 2009; Inglehart and Klingemann 2000).

A further caveat is to be found in the research suggesting that a heavy focus on the acquisition of material goods may distract people from other valuable aspects of their lives, such as their closest social relationships (Solberg, Diener, and Robinson 2004). It could be that psychosocial needs are as "basic" as physical needs, and some scholars have called for a minor revision in Maslovian theory to the effect that individuals ought to invest in nonphysical outcomes, such as social relations, in concert with physical needs (Sheldon, Elliot, Kim, and Kasser 2001). Additionally, Biswas-Diener (2008) argues that psychological processes, such as aspirations and adaptation, are important elements of the relationship between physical consumption and happiness.

ADAPTATION AND CONSUMERISM

Although material items can have benefits if consumed to satisfy physical, social, or psychological needs, the benefits of consumption may be short lived. Moreover, consumers have the tendency to adapt to the positive benefits received after a purchase and can even begin to desire more. Researchers from a range of fields, including psychology, economics, political science, and marketing, have consistently linked the "hedonic treadmill" to consumerism (Binswanger 2006; Bruni and Stanca 2005; Edelstein 2008; McFadden 2006; Raghunathan and Irwin 2001).

The hedonic treadmill was first introduced by Philip Brinkman and, and Donald T. Campbell (1971) and suggested that emotions fluctuate according to daily events but are centered on a stable neutral set point. Moreover, every individual has the capability of experiencing any positive or negative emotion at a given point in time; however, affect is typically set at some innate neutral position. Therefore, the experience of emotion is generally neutral until a stimulus triggers an emotional reaction that, for a limited time, changes the emotion experienced. After a short period of time, however, the individual often becomes accustomed to the stimulus (i.e., the purchase), and her emotions return to the neutral set point. The idea that people adapt to events and circumstances is suggestive of a psychological architecture that buffers people from life's ills and allows them to take risks and accommodate change that might be associated with material or social dividends.

Since its introduction in 1971, some changes to the model have been suggested. Principally, Ed Diener, Richard E. Lucas, and Christie N Scollon (2006) suggested five major revisions, which are as follows: (1) neutrality is not the set point, instead the set point is positive for most individuals; (2) there are individual differences in the set point; (3) well-being is not a single construct, so there are multiple potential set points for every individual; (4) the set point can change in response to external factors that are perceived to be significant; and (5) there are individual differences in the impact of all external factors.

The main idea connecting the hedonic treadmill to consumerism is that people continually purchase material items that they are socialized to desire because they receive some immediate, positive benefit (Bruni and Stanca 2005). People will adapt to most of these purchases relatively quickly and will almost never be satisfied with their purchase for a couple of reasons: (1) they will adjust to it and take it for granted or (2) the market will expand, introducing new products promised to be even better (Bruni and Stanca 2005). Essentially, individuals seek happiness where it rarely pays off because there is always something more to want. However, this does not mean that individuals are never satisfied or even happy with their purchase. The crucial question is whether or not a purchase is perceived by the consumer as actually able to change the set point for that individual.

In answering the previous question, a distinction should be made between immediate and long-term change in a set point. In terms of an immediate change, researchers have found that consumption can increase happiness for a short period of time (Bruni and Stanca 2005; Raghunathan and Irwin, 2001). The immediate satisfaction of a purchase is, however, contingent on various circumstances. Research has shown that immediate happiness following a purchase can depend on the context in which the product is situated, agreement about the purchase from others, the number of product options available for purchase, and the number of purchases made by others (Binswanger 2006; Bruni and Stanca 2005; Raghunathan and Corfman 2006; Raghunathan and Irwin 2001). If the purchase is immediately able to impact happiness, oftentimes its effects are short lived. As in Brinkman and Campbell's original hedonic treadmill model, individuals adapt to the limited sensation caused by the purchase, especially if it is broadly categorized (e.g., clothing as opposed to dress shirt; Redden 2008). Although it is not surprising that some purchases do not significantly impact an individual's well-being, is there any type of purchase that can impact happiness in the long run?

There has been very little research on the long-term impacts of purchases; however, it is conceivable that some purchases could be perceived as sufficiently meaningful and affect well-being. In the U.S., there is much market emphasis on time-saving products (e.g., microwave, washer and dryer, car). What consumers are led to believe is that these products will save time in doing daily chores, giving them more leisure time. However, it has been suggested that even time-saving products can have a limited positive impact and cause a negative outcome in the long run (Binswanger 2006). Moreover, after purchasing a supposedly time-saving product, individuals can experience the "rebound effect," or the act of engaging in the activity that allowed for more time more often than before the product was purchased (Binswanger 2006). To illustrate this effect, Binswanger used automobiles as an example. Cars are purchased and used for the convenience and purpose of saved time, but because cars get people to their destination faster than other, older modes of transportation, people are able to travel farther distances, maintaining and oftentimes increasing the amount of time spent traveling. In the end, it appears that people may put too much stock in material purchases as a source of happiness because they underestimate the extent to which they will adjust to the purchase over time.

EXPERIENTIAL PURCHASES AND HAPPINESS

From existing research literature on the topic, it seems that consumerism has, at best, only a limited positive impact on well-being. However, it is conceivable that a major purchase could have a long-term positive impact on happiness. Although research has indicated that materialism has various negative effects on well-being, it has consistently been found that experiential purchases can have an enduring, positive impact on happiness (Carter and Gilovich 2010; Howell and Hill 2009; Van Boven 2005; Van Boven, Campbell, and Gilovich 2010; Van Boven and Gilovich 2003). Experiential purchases are those that are made with the intention of creating or achieving a perceived significant life experience (Van Boven and Gilovich, 2003). Common examples of experiential purchases are travel and education.

Although some purchases can be most obviously categorized as experiential (e.g., the purchase of a study abroad experience), there are many tangible, material purchases that may also be considered experiential. For example, a bicycle could be perceived as experiential if purchased to enjoy sunny weekends outdoors or a fancy dinner could be experiential if it is the centerpiece of a party or celebration. Furthermore, it is difficult to dichotomously define all purchases as either material or experiential. In fact, research has suggested that, although many items are consistently identified as one or the other, there is some variation among individuals in differentiating an experiential from a material purchase (Van Boven, Campbell, and Gilovich 2010). Moreover, an individual's perception of the purchase (i.e., material or experiential) is a moderating factor on long-term happiness (Carter and Gilovich 2010). Despite the relative difficulty of defining some purchases, most research focuses on those purchases that are either clearly defined as one or the other or on purchases that the individuals self-categorize, avoiding many potential confounds in the conceptualization of the two types of purchases.

There is sufficient evidence to support the notion that experiential purchases positively affect well-being, both immediately as well as longitudinally. For instance, Leaf Van Boven and Thomas Gilovich (2003) found that a diverse sample of participants was more likely to report being happy with an experiential purchase compared to a material purchase. They also found that participants reported experiential purchases as better financial investments than material items (Van Boven and Gilovich 2003). What then makes experientialism more beneficial to well-being than materialism?

In answering this question, Van Boven (2005) suggested three reasons why experiential purchases can positively impact long-term happiness: (1) people adapt less to experiential purchases than to material purchases, (2) experiential purchases are not as readily compared to similar alternatives, and (3) social relationships are often cultivated during or as a result of experiential purchases. In support of Van Boven's (2005) first point, that the results of an experiential purchase do not dwindle, research has shown that people typically choose to make an experiential purchase when thinking about future happiness, and that happiness can grow significantly more over time as a result of an experiential, as opposed to a material, purchase (Carter and Gilovich 2010; Van Boven and Gilovich 2003).

Evidence also indicated that material purchases are more difficult to make than experiential purchases because there are more alternatives or options to choose from, that experiential purchases are not typically comparable to alternatives, and that comparisons to similar alternative purchases negatively impact well-being and satisfaction with a material but not an experiential purchase (Carter and Gilovich 2010). These findings suggest that experiential purchases produce unique life experiences that can be appreciated without wondering if a different option would have been better.

Finally, researchers have found that people who endorse experientialism report higher levels of social relatedness than do those who make more material purchases. Van Boven and his colleagues (2010) found that people prefer discussing experiential purchases to material purchases and judge their social partners more favorably if they engage in experientialism compared to materialism. These findings suggest that social relationships are another benefit to experientialism, positively influencing the well-being of the experiential purchaser.

Although the broad notion that experiential purchases positively impact well-being has generally remained unquestioned (Carter and Gilovich 2010; Howell and Hill 2009; Van Boven 2005; Van Boven, Campbell, and Gilovich 2010), researchers have identified some caveats to this rule. First, evidence has suggested that individuals from low socioeconomic status and with little education may value material over experiential purchases and benefit more from material possessions (Van Boven, Campbell, and Gilovich 2010; Van Boven and Gilovich 2003). In demographic groups that struggle to satisfy basic needs, it is likely that material items are more useful than a life experience.

A second caveat to the idea that experiential purchases are more beneficial to well-being is that the outcome of the experience matters. It has been found that some experiences that resulted in negative outcomes can have neutral or even negative impacts on well-being over time (Nicolao, Irwin, and Goodman 2009). Finally, motivation of the purchase can affect individual happiness as well as others' perceptions of the purchase and the purchaser. Extrinsically motivated purchases, often associated with materialism, have been found to have detrimental effects on projected well-being, associated with negative appraisals of the purchase by others, and can lead to stigmatization, whereas intrinsically motivated purchases, typically associated with experientialism, have been associated with perceived happiness by others and positive appraisals of personality (Van Boven, Campbell, and Gilovich 2010). Although some purchases may seem to clearly be either experiential or material, it is important to consider the motivation (i.e., intrinsic or extrinsic) for making the purchase because motivation has the potential to influence the purchase outcome on happiness.

RECREATION, HOBBIES, AND ENTERTAINMENT AND HAPPINESS

Mihaly Csikszentmihalyi (1999) has suggested that material items can lead to happiness up to a certain point. However, consumption of material goods may resemble a quadratic function, where too much material consumption and emphasis on material acquisition can lead to unhappiness quickly (Csikszentmihalyi 1999; 2000). Perhaps happiness through materialism can be optimized—that is, it can result in a better emotional return on investment—if the purchase affords the buyer an opportunity to engage in an enjoyable leisure activity. Moreover, it is conceivable that some material purchases lead to happiness through recreation (e.g., sports), hobbies (e.g., composing music), and entertainment (e.g., Broadway show). In support of this notion, evidence has suggested that happiness is related to leisure activities (Csikszentmihalyi and Hunter 2003). Additionally, Luo Lu and Michael Argyle (1994) found that satisfaction with leisure activity was positively associated with happiness, where leisure satisfaction was characterized as control of the activity and mentally and socially absorbing.

Moreover, it seems that psychological engagement in the activity may be the key to potential material happiness. For example, Csikszentmihalyi and Jeremy Hunter (2003) found that the challenge to skill ratio, as a measure of engagement during a leisure activity, helped determine adolescents' level of happiness. Furthermore, Csikszentmihalyi (1999; 2000) suggested that material purchases can lead to happiness if they help to induce a state of flow or autotelic experience, which is a psychological and physical state of complete involvement, concentration, and enjoyment. Csikszentmihalyi and colleagues (2003; 1991) found that people typically report the highest levels of happiness while experiencing flow during leisure activities. Furthermore, happiness is not determined by the material circumstance but by the involvement and enjoyment in the activity that was provided by the material purchase (Sheldon and Lyubomirsky 2006). In this way, material items can lead to happiness to a certain extent, if they afford the purchaser the ability to engage in leisure activities, such as recreation, hobbies, and entertainment.

CONCLUSION

When addressing the question of whether happiness is an outside-in or inside-out phenomenon, it would be a mistake to dismiss the impact of the material world on our life satisfaction. There is overwhelming evidence that income, especially at the national level, is associated with higher rates of satisfaction. Further, it may be insensitive to suggest that happiness is only a state of mind to those living in objective poverty. An anecdote from one of our (Biswas-Diener) trips to India illustrates this point nicely: While interviewing a young mother in a shanty in a slum in South Kolkata, she was asked "if you could change one thing about your life, what would you change?" She could have mentioned "world peace" or "better relationships." Instead, she said she would like an oscillating fan because she was too hot. To the extent that we can control our circumstances through material consumption—whether that means buying comfort items as individuals or investing in roads as societies—we can create situational benefactors of happiness.

This important take-home message must be tempered with the caveat that overemphasizing the role of consumer goods in happiness may be psychologically toxic and environmentally unsustainable. People likely overestimate the extent to which goods will produce happiness and likely invest disproportionately more energy into their acquisition that is sensible from a happiness-returns perspective. With this in mind, we make several important recommendations to consumers. First, it is important to realize that material comfort should not be emphasized more than the cultivation of social relationships or the pursuit of meaning at work. This is especially important for people living in the relative affluence of industrialized countries. For these individuals, rising aspirations and adaptation may conspire to create somewhat fewer happiness dividends relative to investing in relationships. This leads us to recommend a "train your appetites" approach to consumption in which people carefully consider the difference between *wanting* and actually *liking* an item. Second, care should be taken to purchase experiences or physical objects that are the most likely to lead to meaningful interactions or enjoyable activities. To be effective in this strategy requires a certain degree of self-reflection such that an individual recognizes the extent to which a particular object is consonant with personal values and which might reasonably be used in ongoing, novel, or social ways.

Finally, it is important to recognize that there is much that we do not know about how consumption and happiness relate to one another. For example, we know relatively little about how to define "basic" in basic needs. Is "air conditioning" a basic need, for instance? And, if so, how do we define how basic air conditioning needs to be? Is a hand fan basic, or an electric fan? An air conditioning unit attached to a single room in a home? In addition, we do not fully understand the interplay of what defines an experiential versus a material purchase. In both cases—and in the case of basic needs—we know that the individual's point of view (the inside-out perspective, if you will) is an important determinant of whether the consumption of goods leads to happiness. It is not simply the object, but what we make of it, that informs our happiness.

REFERENCES

Binswanger, Mathias. (2006). "Does Income Growth Fail to Make Us Happier? Searching for the Treadmills Behind the Paradox of Happiness," *The Journal of Socio-Economics*, vol. 35, pp. 366–381.

Biswas-Diener, Robert. (2008). "Material Wealth and Subjective Well-Being." In *The Science of Subjective Well-Being*, edited by Eid, M. & R. Larsen (New York: Guilford Press) pp. 307–322.

Biswas-Diener, Robert, Joar Vitterso, and Ed Diener. (2005). "Most People Are Pretty Happy but There Is Cultural Variation: The Inughuit, the Maasai and the Amish," *Journal of Happiness Studies*, vol. 6, pp. 205–226.

Brickman, Philip, and Donald T. Campbell. (1971). "Hedonic Relativism and Planning the Good Society." In *Adaptation Level Theory: A Symposium*, edited by M. H. Appley (New York: Academic Press) pp. 287–302.

Bruni, Luigino, and Luca Stanca. (2005). *Income Aspirations, Television, and Happiness: Evidence from the World Values Survey* (Working Paper no. 89) Milan: University of Milano-Bicocca.

Carter, Travis J., and Thomas Gilovich. (2010). "The Relative Relativity of Material and Experiential Purchases," *Journal of Personality and Social Psychology*, vol. 98, no. 1, pp. 146–159.

Csikszentmihalyi, Mihaly. (1999). "If We Are So Rich, Why Aren't We Happy?" *American Psychologist*, vol. 54, no. 10, pp. 821–827.

Csikszentmihalyi, Mihaly. (2000). "The Costs and Benefits of Consuming," *The Journal of Consumer Research*, vol. 27, no. 2, pp. 267–272.

Csikszentmihalyi, Mihaly, and Jeremy Hunter. (2003). "Happiness in Everyday Life: The Uses of Experience Sampling," *Journal of Happiness Studies*, vol. 4, no. 2, pp. 185–199.

Csikszentmihalyi, Mihaly, and Maria M. Wong. (1991). "The Situational and Personal Correlates of Happiness: A Cross-National Comparison." In *Subjective Well-Being: An Interdisciplinary Perspective*, edited by Strack, F., M. Argyle, and N. Schwarz (New York: Pergamon Press) pp. 193–212.

Diener, Ed, Richard E. Lucas, and Christie N. Scollon. (2006). "Beyond the Hedonic Treadmill: Revising the Adaptation Theory of Well-Being," *American Psychologist*, vol. 61, no. 4, pp. 305–314.

Diener, Ed, Weiting Ng, James Harter, and Raksha Arora. (2010). "Wealth and Happiness Across the World: Material Prosperity Predicts Life Evaluation, Whereas Psychosocial Prosperity Predicts Positive Feeling," *Journal of Personality and Social Psychology*, vol. 99, pp. 52–61.

Diener, Ed, Weiting Ng, and Will Tov. (2009). "Balance in Life and Declining Marginal Utility of Diverse Resources," *Applied Research in Quality of Life*, vol. 3, pp. 277–291.

Diener, Ed, Eunkook M. Suh, Richard E. Lucas, and Heidi L. Smith. (1999). "Subjective Well-Being: Three Decades of Progress," *Psychological Bulletin*, vol. 125, pp. 276–302.

Diener, Ed, and Louis Tay. (in press). "Needs and Subjective Well-Being Around the World," *Journal of Personality and Social Psychology*.

Edelstein, Jean. (2008). "The End of Aspiration?" *Public Policy Research*, vol. 14, no. 4, pp. 248–251.

Howell, Ryan T., and Graham Hill. (2009). "The Mediators of Experiential Purchases: Determining the Impact of Psychological Needs Satisfaction and Social Comparison," *The Journal of Positive Psychology*, vol. 4, no. 6, pp. 511–522.

Inglehart, Ronald, and Hans-Dieter Klingemann. (2000). "Genes, Culture, Democracy and Happiness." In *Culture and Subjective Well-Being*, edited by Diener, E. & E. M. Suh (Cambridge, MA: MIT Press) 165–184.

Lu, Luo, and Michael Argyle. (1994). "Leisure Satisfaction and Happiness as a Function of Leisure Activity," *Kaohsiung Journal of Medical Sciences*, vol. 10, no. 2, pp. 89–96.

Lyubomirsky, Sonya, Kennon M. Sheldon, and David Schkade. (2005). "Pursuing Happiness: The Architecture of Sustainable Change," *Review of General Psychology*, vol. 9, pp. 111–131.

Maslow, Abraham. (1954). *Motivation and Personality*. New York: Harper.

McFadden, Daniel. (2006). "Free Markets and Fettered Consumers," *American Economic Review*, vol. 96, no. 1, pp. 3–29.

Nicolao, Leonardo, Julie R. Irwin, and Joseph K. Goodman. (2009). "Happiness for Sale: Do Experiential Purchases Make Consumers Happier than Material Purchases?" *Journal of Consumer Research*, vol. 36, no. 2, pp. 188–198.

Raghunathan, Rajagopal, and Kim Corfman. (2006). "Is Happiness Shared Doubled and Sadness Shared Halved? Social Influence on Enjoyment of Hedonic Experiences," *Journal of Marketing Research*, vol. 43, no. 3, pp. 386–394.

Raghunathan, Rajagopal, and Julie R. Irwin. (2001). "Walking the Hedonic Treadmill: Default Contrast and Mood-Based Assimilation in Judgments of Predicted Happiness with a Target Product, *The Journal of Consumer Research*, vol. 28, no. 3, pp. 355–368.

Redden, Joseph P. (2008). "Reducing Satiation: The role of Categorization Level," *Journal of Consumer Research*, vol. 34, no. 2, pp. 624–634.

Sheldon, Kennon M., Andrew J. Elliot, Youngmee Kim, and Tim Kasser. (2001). "What's Satisfying About Satisfying Events? Comparing Ten Candidate Psychological Needs," *Journal of Personality and Social Psychology*, vol. 80, pp. 325–339.

Sheldon, Kennon M., and Sonya Lyubomirsky. (2006). "Achieving Sustainable Gains in Happiness: Change Your Actions, Not Your Circumstances," *Journal of Happiness Studies*, vol. 7, no. 1, pp. 55–86.

Solberg, Emily. C, Ed Diener, and Michael Robinson. (2004). "Why Are Materialists Less Satisfied?" In *Psychology and Consumer Culture: The Struggle for a Good Life in a Materialistic World*, edited by Kasser, T. & A. D. Kanner (Washington, DC: American Psychological Association) pp. 29–48.

Van Boven, Leaf. (2005). "Experientialism, Materialism, and the Pursuit of Happiness," *Review of General Psychology*, vol. 9, no. 2, pp. 132–142.

Van Boven, Leaf, Margaret C. Campbell, and Thomas Gilovich. (2010). "Stigmatizing Materialism: On Stereotypes and Impression of Materialistic and Experiential Pursuits," *Personality and Social Psychology Bulletin*, vol. 36, no. 4, pp. 551–563.

Van Boven, Leaf, and Thomas Gilovich. (2003). "To Do or To Have? That Is the Question," *Journal of Personality and Social Psychology*, vol. 85, no. 6, pp. 1193–1202.

Veenhoven, Ruut, and Michael Hagerty. (2006). "Rising Happiness in Nations, 1946–2004: A Reply to Easterlin," *Social Indicators Research*, vol. 79, pp. 421–436.

9 Thinking Through Consumption and Technology

Pak Hang Wong

A few years ago, Zygmunt Bauman published a book polemically titled *Does Ethics Have a Chance in a World of Consumers?* (Bauman 2008). At first glance, Bauman's question is not too different from that of others who have studied the ethics of consumption and related consumerist attitudes, behaviors, and practices (e.g., Borgmann 2000; Cafaro 2001). There is, however, one subtlety in Bauman's formulation of the question that separates it from similar pursuits. Instead of analyzing consumption and the related consumerist lifestyle with a specific ethical theory, Bauman invites us first to uncover the existential conditions engendered by consumer society, and then to examine the possibility of ethics under those existential conditions.

Bauman's approach, I think, offers us a different and significant alternative to normative analysis. It is different, as I have pointed out, in that it begins with a unique set of existential conditions of consumer society but not with a normative standard fixed by a particular ethical theory. And, it is significant because if the set of existential conditions engendered by consumer society repudiates the normative standard of a given ethical theory, then arguing against consumption and the related consumerist lifestyle *with* the theory *under* those existential conditions is unlikely to be successful unless the argument is simultaneously directed at consumer society itself.[1]

Because many technologies are designed and manufactured for people in a consumer society and their functions and uses are an integral part of the consumerist experience, the same line of thought is equally applicable to the normative analysis of technology. Certainly, philosophers have already offered insightful analyses of the impacts of technology on people's well-being with special reference to consumerist lifestyles (e.g., Borgmann 1984; 1992; 2000; Chapter 10, by Edward Spence, and Chapter 11, by Valerie Tiberius, in this volume). However, they often start with a theory of the good life and follow with an application of their theory to evaluate the technology in question. What is missing in their analyses, therefore, is the set of existential conditions of consumer society. It is important because those existential conditions necessarily (dis)favor some accounts of the good life, which can go against a preferred view of the good life. In short, uncovering the set of existential conditions in consumer society is as important to normative analysis of technology as to normative analysis of consumption and consumer lifestyle. The aim of this chapter is to demonstrate one way to examine technology through the set of existential conditions in consumer society and, at the same time, to argue for its importance in *normative* analyses of technology.

BAUMAN AND THE CONSUMERIST SYNDROME

Consumption, consumerism, and consumer society are topics being studied by researchers of different disciplines and from multiple perspectives (for an overview of the research, see Arnould and Thompson 2005). In this chapter, I shall look at *social theory of consumption* (also known as *sociology of consumption*), as it is most relevant for the current purpose. Social theory of consumption aims to examine the processes and meanings of consumption and the constitutive characteristics of consumer society. It is unique in that its focus is not only on consumption and consumerist lifestyle *per se*. Instead, consumption and consumerist lifestyle are examined from a macroscopic perspective and through various changes in sociocultural, epistemological, and institutional dimensions of a society (e.g., Baudrillard 1998; Campbell 1987; Slater 1997; Featherstone 2007; Bauman 2005, 2007, and 2008). Accordingly, consumer society is to be understood as a specific way to organize a society (thus, *consumer* society), and consumerism is to be understood as a specific way of life in *that type of society*.

Bauman, one of the most prominent theorists of consumer society, has conceptualized consumerism as a unique way of *being-in-the-world* that constitutes its own "cognitive frames and tacit assumptions" (Blackshaw 2008, p. 117). He has argued that consumerism is constitutive of people's "assumptions of the ways of the world and the ways of walking them, visions of happiness and ways to pursue them [. . .]." (Bauman, in Rojek 2004, p. 293) So understood, consumerism favors some ways of life over the others and informs its people *what a good life is* and *how to live it*. In this respect, Bauman's analysis is particularly useful for illustrating the outlook of the good life in consumer society and its influences on people.[2] In the following, I shall look at his analysis of consumerism and consumerist values (or, as Bauman has labeled them, the *consumerist syndrome*).

Bauman's analysis is based on a comparison between consumerism with the way of life in the 'society of producers' (i.e., the type of society in an earlier stage of modernity that is exemplified by mass production of goods for mass consumption).[3] In the 'society of producers,' where mundane life is guided by a strong work ethic, work is valued as an end in itself. Work ethic's emphasis on work as an end in itself has not only affirmed the importance of work-related virtues (e.g., diligence, self-discipline, etc.), it has also devalued qualities that are presumed to distract people's fulfillment of their work duty, such as pleasure and (instant) gratification (Weber 2001/1930). For the significance of work in the 'society of producers,' the type of work a person has also becomes the designator of how good a person's life is in the 'society of producers.' Because people's life is structured by their work, and work is characteristically *immobile* in the 'society of producers,' a good life can only afford to be a gradual and long-term project in that type of society. Any gradual and long-term projects, however, require *stability*. It is the drive for stability, Bauman suggested, that fosters the desire for a "reliable, trustworthy, orderly, regular, transparent, and [. . .] durable, time-resistant and secure setting" (Bauman 2007, p. 29).

As a society moves toward becoming a consumer society, the work ethic and the outlook of the good life it supports are replaced by another ethic and a different outlook of the good life. It is so because consumer society, unlike its predecessor, "engages its member [primarily] in their capacity as consumers" (Bauman 2005, p. 24). Or, put differently, consumer society is organized around the "social logic" of consumption (Baudrillard 1998). The move toward consumer society is helped and intensified by various sociocultural, epistemological, and institutional changes (e.g., a new time-space structure, heightened self-reflexivity, the decline of nation-states, globalization, etc.) (e.g., Harvey 1989; Giddens 1990; Beck 1992; Bauman 2000). Bauman has called these processes and changes a liquefaction of modernity. 'Liquefaction' or 'liquidity' is his choice of metaphor to capture the status of the current stage of modernity, which is characterized by an emphasis on *mobility* and *inconstancy*[4] (Bauman 2000).

The processes of liquefaction, especially the heightening sense of (self-) reflexivity and the increasing interconnectedness with globalization, have undermined preexisting sources of authority (e.g., scientific rationality) and generated unprecedented uncertainty, and the uncertain world of liquid modernity is inherently hostile to any gradual and long-term projects, as their outcome cannot be affirmed with confidence in the face of uncertainty without any solid forms of authority. Bauman has pointed out that gradual and long-term projects can hardly maintain their sensibility in an uncertain environment, and, thus, are often substituted by something immediate and short-term (Bauman 2007, p. 31).

In this respect, consumerism—for its emphasis on *instant* gratification—has supplied a ready-made answer to the uncertainty of liquid modernity. Moreover, as Campbell (1987) has argued, the act of consumption associated with consumerism is characterized by its *pleasure*-seeking nature, where pleasure is not to be understood as sensory pleasure but something *in* and *of* emotional experiences achieved through a person's power of creative imagination (Campbell 1987, pp. 65–76). So construed, consumerism can also be seen as an inner protective mechanism against the uncertainty of liquid modernity.

The significance of instant gratification is also reinforced by the new meaning of time in liquid modernity. Bauman has noted that time is *pointillist* in liquid modernity, that is—time becomes a collection of self-contained units, or what he called 'eternal instants' (Bauman 2007, p. 32). Because of the disconnectedness of each 'eternal instant,' an activity is only relevant for *now*. Consumption, or any activities for that matter, has to be done in and for the present if it is to be relevant at all.

However, immediacy is not the only value prioritized by consumer society of liquid modernity, social theorists of consumption have pointed out that *novelty* (or *newness*) is another cornerstone of consumer society. For instance, Bauman has argued that the necessity of novelty is built in the very existence and continuity of consumer society itself, as its survival is rested on a state of perennial incomplete satisfaction or nonsatisfaction of its members (Bauman 2007, pp. 46–47). Novelty is essential to maintain such a state, for it embodies the promise of unexplored experiences, which keeps people in "a state of constantly seething, never wilting excitation [. . .] and in a state of suspicion and disaffection [that give rise to the desires, needs or wants to find] a way out of disaffection" (Bauman 2005, p. 26).

In short, consumer society is propelled to prioritize *immediacy* and *novelty* due to various sociocultural, epistemological, and institutional changes, and work ethic and its outlook of the good life in the 'society of producers' are made obsolete in consumer society. As Bauman has fittingly described:

> the consumerist syndrome has degraded duration and elevated transience. It lifts the value of novelty above that of lastingness. [. . .] It has sharply shortened the timespan separating not just the want from its fulfilment [. . .], but also the birth moment of the want from the moment of its demise, as well as the realization of the usefulness and desirability of possessions from the perception of them as useless and in need of rejection. Among the objects of human desire, it has put the act of appropriation, to be quickly followed by waste disposal, in the place once accorded to the acquisition of possessions meant to be durable and to their lasting enjoyment. (Bauman 2007, pp. 85–86)

Yet, it is important to point out that people's freedom of choice is necessary if immediacy and novelty are values to be realized. For people to take advantage of immediacy, they must be *free* to make choices at any moment. Similarly, for people to experience novelty (or newness), they must have the options to choose from and, more important, they must be able to choose from them *freely*. In this respect, the freedom of choices a person enjoys has also determined how far the person can benefit from immediacy and novelty. Freedom of choices, therefore, becomes the most important parameter for shaping the outlook of the good life in consumer society. This is perhaps why Bauman has claimed that "[f]reedom to choose sets the stratification ladder of consumer society and so also the frame in which its members, the consumers, inscribe their life aspirations—a frame that defines the direction of efforts towards self-improvement and encloses the image of a 'good life'" [5] (Bauman 2005, p. 31).

TECHNOLOGY IN CONSUMER SOCIETY, TECHNOLOGY OF CONSUMER SOCIETY

I have, through Bauman's analysis, outlined the existential conditions of consumer society and attempted to show that immediacy and novelty are prioritized in response to those existential conditions and become the core values in consumer society. Such an account not only allows us to better understand the peculiarities of consumer society, it also allows us to better understand technology in the context of consumer society. As I have pointed out in the beginning, insofar as technologies are something designed and manufactured for consumption, they too can be seen as an expression of the peculiarities of consumer society. In other words, technologies *in* consumer society can also be conceptualized as technologies *of* consumer society in that they reflect and reinforce the values inherent to consumer society (i.e., immediacy and novelty).

A paradigm example here is information technology. Consider, for instance, social media technology, such as social networking sites (e.g., MySpace, Facebook, Google+, etc.), instant messengers (Windows Live Messenger, Google Talk, etc.) and (micro)blogging services (WordPress, Blogger, Twitter, Tumblr, etc.). Their design and functions have embodied the essence of immediacy and novelty. The uniqueness of social media technology is that they allow and encourage real-time inputs and responses from their users. Social networking sites such as Facebook and Google+ ask their users "What is on your mind?" and "Share what's new" at their front page and call those 'updates,' instant messengers are promoted as 'instant' messengers, and Twitter has advertised to let its users to get "[i] nstant updates from your friends, industry experts favorite celebrities, and what's happening around the world."

Their emphasis on real-time inputs and responses has propelled their users to react as soon as they start using them. Similarly, the main purpose of these platforms is to enable their users to share and receive *new* information (i.e., updates), either personal or nonpersonal, with and from other users. As a matter of fact, the proper functioning of social media technology is contingent on the *new* information provided by the users, without which social media technology essentially becomes obsolete. In other words, one of the main driving forces behind social media technology is the continuous novel experiences they can offer their users. In this respect, social media technology is a prime example of a technology of consumer society.

The significance of immediacy and novelty is also epitomized in the current debate of the virtues and vices of the Internet. Recent critics of the Internet have argued that the Internet has flattened our mind and society.

For example, Nicholas Carr (2008; 2010) has argued that the Internet has made deep reading and deep thinking difficult if not impossible, and the shallowness engendered by the Internet (and information technology in general) is a result of overabundance of information available on the Internet and their accessibility made possible by information technology. Carr's argument, in effect, is an argument against the immediacy enabled by information technology that renders contemplation impossible; and, it is also an argument against the endless search for novel information via hyperlinks that themselves become distractions to people's mind.

In light of this, it is interesting to see that the proponents of the Internet and information technology have also appealed to immediacy and novelty in arguing for the virtues of the Internet. For instance, Charles Leadbeater (2008), Clay Shirky (2010), and Steven Johnson (2010) have all argued that information technology, by shortening the distance in time and space, has empowered new forms of collaboration through the Internet. In short, it is the immediacy in this new way of communication enabled by information technology that speeds up generations of new ideas crucial to the progress of the society.

Either way, it should be clear that immediacy and novelty are central to both technology and the discussion of the benefits and harms of technology, at least in the case of information technology as I have attempted to demonstrate. In this sense, technology in consumer society is not only something consumed by the people, but it is also constitutive of the consumerist experience itself.

FROM THE CRITIQUE OF TECHNOLOGY TO THE CRITIQUE OF CONSUMER SOCIETY

So far, my analysis has been primarily interpretive. I have offered another way to understand technology with respect to the existential conditions of consumer society. I want to turn now to an implication of this approach on *normative* analysis of technology. Particularly, I want to argue against the sufficiency and effectiveness of a specific pattern of argument *against* technology, which can be found in both popular and academic discourses. The pattern of argument I shall criticize has the following form:

(1) A technology *T* promotes *immediacy* and/or *novelty* (or other values purportedly reducible to either of them).
(2) Novelty and immediacy is bad/undesirable (either morally or prudentially).
(C) *T* is bad/undesirable (either morally or prudentially).

Here, it is not my intention to argue that technology *does not* promote the consumerist syndrome of immediacy and novelty, nor it is my aim to argue *in favor* of them. My criticism of this pattern of argument is that even if (1) and

(2) are true and (C) follows from them, this pattern of argument is ultimately insufficient and ineffective because it fails to acknowledge the fact that technology is situated in a consumer society and it is a reflection and reinforcement of the consumer syndrome I have attempted to demonstrated.

An example of this type of argument is proposed most clearly by Albert Borgmann (1984) in his critique of modern technology through his 'device paradigm.' He argued that modern technology has concealed the meaningfulness of our lifeworld by rendering things "instantaneous, ubiquitous, safe and easy" (Borgmann 1984, p. 41). There are, of course, numerous ways to interpret Borgmann's argument. However, it should not be too difficult to see the role of immediacy figures in it, as the concealment of meaningfulness is a result of the diminished (temporal) distance between human being and nature made possible by modern technology. In this sense, Borgmann's argument against modern technology is similar to Carr's argument against the Internet in that both of them have highlighted the vices of *immediate* accessibility.

Yet, as I have attempted to show that the preference for immediacy and novelty arises from the existential conditions of consumer society, it is inescapable so long as people are a member of a consumer society. And, the embodiment of immediacy and novelty in technology is also a reflection and reinforcement of the existential conditions itself. Hence, a sufficient and effective argument against immediacy and novelty cannot succeed as an argument against technology *per se* but can only succeed as an argument against consumer society *in general*. In short, any arguments formulated in isolation from the fact that technology is situated in a consumer society are at best incomplete. More important, however, is that it is often unclear from which perspective the argument against immediacy and novelty is coming from. Particularly, if the argument is made from within the viewpoint of consumer society, it is then bounded to be limited by the set of existential conditions of consumer society itself. In other words, the argument merely presents a predicament, namely immediacy and novelty are undesirable, but there is no escape because they are inherent to consumer society.

In other words, any sufficient and effective arguments against immediacy and novelty, whether they are seen as qualities embodied in technology or as core values in consumer society, require us to go beyond consumer society and transcend its peculiar set of existential conditions. Hence, they require us to rethink the notion of consumer society itself.

RENEWING 'CONSUMER' SOCIETY, RENEWING THE 'CONSUMERIST' GOOD LIFE

To be fair to Borgmann's critique of modern technology, he does attempt to 'go beyond' consumer society and 'transcend' its peculiar set of existential conditions via his notion of 'focal things and practices,' such as running and family dinners for (re-)engaging ourselves with the meaningfulness of our lifeworld (Borgmann 1984, pp. 196–226). However, Borgmann's focal things and practices are deeply *anti*-consumerist and they are difficult to be acknowledged in full by people who are *already* a member of consumer society. Indeed, to fully acknowledge his focal things and practices appears to require members of consumer society to revert their present lifestyle and retreat from (or reform) their society in entirety.

Although this option is possible in theory, it is only practically plausible for a few heroic individuals who are prepared to live *outside* their society that is currently organized around a social logic of consumption. The problem of Borgmann's and other similar anti-consumerist proposals, such as Voluntary Simplicity (e.g., Elgin 1993; for a study of the notion of 'voluntary simplicity' see Doherty and Etzioni 2003) and the slow food movement (e.g., Petrini 2007), is that they have rejected consumer society in its entirety and failed to recognize the positive aspects of consumer society. In closing this chapter, I want to suggest other options that truly go beyond consumer society and transcend its peculiar set of existential conditions but without rejecting consumer society as a whole.

Recently, a new way has been presented to conceptualize the notion of 'consumer,' through *political consumerism* (see, e.g., Micheletti 2003; Micheletti, Føllesdal, and Stolle 2004). According to political consumerism, consumers are not *passive recipients* of goods and services, but they are *active participants* of the society through the act of consumption. In other words, consumption is an active way to shape the society around us. In this respect, political consumerism introduces a new form of (sociocultural, political, and moral) responsibility to the notion of 'consumer.' At the same time, by reconceptualizing consumers as active sociocultural, political, and moral actors (and, consumption as a sociocultural, political, and moral act in itself), it requires us to rethink the idea of consumer society itself. More significantly, however, is that by infusing normative considerations into the act of consumption, political consumerism provides resources *within* consumer society to question the preference of immediacy and novelty. In this sense, people do not need to escape from consumer society, instead, they need to transform it from within.

Political consumerism, of course, is not the only game in town. Other researchers (e.g., Blättel-Mink and Hellman 2009; Ritzer and Jurgenson 2010) have explored the notions of 'prosumer' and 'prosumption' in the place of 'consumer' and 'consumption.' And, they have attempted to argue against social theorists' biased focus on consumption in the present research and to reemphasize the intimate links between production and consumption, especially in the context of information technology. In doing so, they have provided new meaning to the act of consumption (or prosumption), which at the same time introduced new role to consumers (or prosumers) and new way to organize a society (i.e., prosumer society). From this perspective, the notions of 'prosumer' and 'prosumption' have also offered another way to rethink consumer society from within.

What is in common to both political consumerism and the notions of 'prosumer' and 'prosumption' is that they both offer a new way to conceptualize the notion of 'consumer.' Particularly, they have both reconceptualized consumers as active actors, either in sociocultural, political, moral spheres or in the economic sphere. This is significant not only for how we consume technology in consumer society (e.g., political consumers need to consider various sociocultural, political, and moral consequences of their purchase and uses of technology); it is also significant to how technology is being designed and manufactured. After all, technology is also a reflection and reinforcement of the values prioritized by society.

CONCLUSION

In this chapter, I have illustrated one way in which social theory of consumption can be relevant to normative analysis of technology. Particularly, I have argued that a specific pattern of argument against technology commonly found in both popular and academic discourses is inadequate, and it has to be supplemented by an analysis of consumer society. In other words, any critiques of technology (and of consumer society) need to take into considerations the broader perspective in which the objects of critique are situated. In this respect, philosophers and other researchers who are interested in normative issues on technology in consumer society can benefit enormously from social theory of consumption.

NOTES

1. Of course, the scope of Bauman's approach is not only limited to moral rightness and wrongness of consumption and related consumerists attitudes, behaviors and practices as a narrow understanding of the term 'ethics' will imply. It will also cover their prudential goodness and badness, particularly their potential impacts on people's good life.
2. Since Bauman has written prolifically on the topic, it is beyond my scope to provide a comprehensive study of his works here. Rather, I shall only outline the key points in Bauman's analysis of consumer society.
3. This type society corresponds to the stage of modernity Bauman labeled as 'solid' modernity, which is exemplified by a Fordist model of production. Solid modernity is so-labeled because the Fordist model requires a "combination of huge factory buildings, heavy machinery and massive labour force," and thus is characterized by "bulk and size, and, for that reason, also with [spatial] boundaries" (Bauman 2000, 57–58).
4. Bauman's 'liquid' modernity and his discussions of various issues associated with it (e.g., globalization, individualization, the new meaning of space and time, the decline of community, etc.) is broadly consistence with Anthony Giddens's (1990) notion of 'late modernity' and Ulrich Beck's (1992) notion of 'second modernity.'
5. It is useful to be reminded that Bauman's analysis—and this section, which is based on his analysis—is unquestionably overgeneralized and

overhomogenized. In short, there is not a radical discontinuation from the society of producers to the society of consumers. Nor, is there a radical distinction between people-as-producers and people-as-consumers. Despite these flaws, these society-types and ways-of-being-in-the-world remain useful, I think, insofar as they render complexities more comprehensible and, thus, illuminate what is otherwise hidden by the complexities.

REFERENCES

Arnould, Eric J., and Craig J. Thompson. (2005). "Consumer Culture Theory (CCT): Twenty Years of Research," *Journal of Consumer Research*, vol. 31, no. 4, pp. 868–882.

Baudrillard, Jean. (1998). *The Consumer Society: Myths and Structures*. London: Sage.

Bauman, Zygmunt. (2000). *Liquid Modernity*. Cambridge: Polity Press.

Bauman, Zygmunt. (2005). *Work, Consumerism and the New Poor* (2nd ed.). Maidenhead: Open University Press.

Bauman, Zygmunt. (2007).*Consuming Life*. Cambridge: Polity Press.

Bauman, Zygmunt. (2008). *Does Ethics Have a Chance in a World of Consumers?* Cambridge, MA: Harvard University Press.

Beck, Ulrich. (1992). *Risk Society: Towards a New Modernity*. London: Sage.

Blackshaw, Tony. (2008). "Bauman on Consumerism—Living the Market-Mediated Life." In *The Sociology of Zygmunt Bauman*, edited by Jacbsen, M. H. and P. Poder (Aldershot, UK: Ashgate) pp. 117–135.

Blättel-Mink, Birgit, and Kai-Uwe Hellmann. (2010). *Prosumer Revisited: Zur Aktualität einer Debatte*. Wiesbaden, Germany: VS Verlag für Sozialwissenschaften.

Borgmann, Albert. (1984). *Technology and the Character of Contemporary Life*. Chicago: University of Chicago Press.

Borgmann, Albert. (1992). *Crossing the Postmodern Divide*. Chicago: University of Chicago Press.

Borgmann, Albert. (2000). "The Moral Complexion of Consumption," *Journal of Consumer Research*, vol. 26, no. 4, pp. 418–422.

Cafaro, Philip. (2001). "Economic Consumption, Pleasure, and the Good Life," *Journal of Social Philosophy*, vol. 32 no. 4, pp. 471–486.

Campbell, Colin. (1987). *The Romantic Ethic and the Spirit of Modern Consumerism*. Oxford: Blackwell.

Carr, Nicholas. (2008, July/August). "Is Google Making Us Stupid," *The Atlantic*. Retrieved from http://www.theatlantic.com/doc/200807/google

Carr, Nicholas. (2010). *The Shallows: What the Internet Is Doing to Our Brains*. New York: W. W. Norton.

Doherty, Daniel and Amitai Etzioni. (Eds.). (2003). *Voluntary Simplicity: Responding to Consumer Culture*. Lanham, MD: Rowman and Littlefield.

Elgin, Duane. (1993).*Voluntary Simplicity: Toward a Way of Life That is Outwardly Simple, Inwardly Rich* (Rev. ed.). New York: Quill.

Featherstone, Mike. (2007). *Consumer Culture and Postmodernism*. London: Sage.

Giddens, Anthony. (1990). *The Consequences of Modernity*. Stanford: Stanford University Press.

Harvey, David. (1989). *The Condition of Postmodernity: An Enquiry into the Origins of Cultural Change*. London: Blackwell.

Johnson, Steven. (2010). *Where Good Ideas Come From: The Natural History of Innovation*. New York: Riverhead Books.

Leadbeater, Charles. (2008). *We-Think*. London: Profile.
Micheletti, Michele. (2003). *Political Virtue and Shopping: Individuals, Consumerism, and Collective Action*. New York: Palgrave Macmillan.
Micheletti, Michele, Andreas Føllesdal, and Dietlind Stolle. (Eds.). (2004). *Politics, Products, and Markets: Exploring Political Consumerism Past and Present*. London: Transaction.
Petrini, Carlo. (2007). *Slow Food Nation: Why Our Food Should Be Good, Clean And Fair*. New York: Rizzoli Ex Libris.
Ritzer, George, and Nathan Jurgenson. (2010). "Production, Consumption, Prosumption: The Nature of Capitalism in the Age of the Digital 'Prosumer,'" *Journal of Consumer Culture*, 10, 2010, 13–36.
Rojek, Chris. (2004). "The Consumerist Syndrome in Contemporary Society: An interview with Zygmunt Bauman," *Journal of Consumer Culture*, vol. 4, pp. 291–312.
Shirky, Clay. (2010). *Cognitive Surplus: Creativity and Generosity in a Connected Age*. New York: Penguin Press.
Slater, Don. (1997). *Consumer Culture and Modernity*. Oxford: Polity Press.
Weber, Max. (2001). *The Protestant Ethic and the Spirit of Capitalism*. London: Routledge. (Originally published 1930)

10 Consumption and Sustainability

A Neo-Epicurean Approach to a Sustainable Good Life in a Technological Age[1]

Edward Spence

> Empty is that philosopher's argument by which no human suffering is therapeutically treated. For just as there is no use in a medical art that does not cast out the sicknesses of the body, so too there is no use in philosophy, unless it casts out the suffering of the soul.—
>
> Epicurus

This chapter explores how Epicurus' philosophy of pleasure provides a way of understanding the problem of sustainability in relation to the notion of a *good life* in a technological age, as well as offering a framework for its solution. I will with some caution explain the problem of sustainability as the problem of sustaining our current level of lifestyles, principally in terms of consumption, both as individuals and as societies worldwide, in terms of at least five variables: (1) ever-diminishing natural resources; (2) increase of pollution to the biosphere, including an increase of greenhouse gases in the atmosphere; (3) increase of the consumption of energy per capita in the form of fossil fuels to maintain our level of lifestyles; (4) increase in world population, which impacts on variables 1, 2, and 3; and (5) climate change and global warming for which most if not all the variables of 1 to 4 appear to be, on current scientific evidence, contributing causal factors.

One of the main claims of this chapter is that any solution to sustainability must also address the crucial and related issue of what constitutes *a sustainable good life*. For in the past, the issue of sustainability and that of a good life (what constitutes a good life generally conceived in philosophical, social, political, and psychological terms) could have been perceived as two conceptually distinct and practical issues that could have been dealt with independently of each other. However, that is no longer the case.

It is true to say that for most people and governments the world over, at least in democratic states, to have a good life is of paramount importance. The issue of climate change and environmental sustainability is also of great importance. However, although of equal importance, the two seem to be in tension and, at times, to pull in opposite directions. This is partly because our market economies encourage and support consumption-driven hedonistic lifestyles that for the medium and long term at least, seem unsustainable. So because having a good life as well as a sustainable environment in which to live it are of comparable importance and value, we need to address the problem of how to live good lives that are also sustainable.[2] The main objective of this chapter, therefore, is to propose a theory of a *sustainable good life* that is at once conceptually sound and practical.

To contextualize the problem of a sustainable good life, I will use Australia as a case study. First, because I am restricted by space in discussing several cases in any meaningful detail; and second, because Australia, like Buridan's donkey (Zupko 2011), seems torn between maintaining its present high standard of the good life (a recent survey by the Organisation for Economic Co-Operation and Development named Australia the happiest nation in the world), as well as deciding what to do about sustainability. The Australian prime minister's latest proposed carbon tax, which offers an interim solution to the problem of sustainability, has polarized the nation, with the carbon tax proposal seemingly heading for an impasse.

Since at least the Kyoto Protocol ratified only lately by the Australian government (by the ex-Prime Minister Kevin Rudd), the problem of sustainability has been the central focus of social, scientific, and political debate. A recent report on climate change (Garnaut Climate Change Review 2011) and its impact on the environment by Professor Ross Garnaut has received wide media coverage. It outlines the critical risks and dangers facing Australia, its neighboring countries, and more generally the rest of the world, if appropriate action is not taken to correct the predicted, if rather uncertain, damaging consequences of climate change on the natural environment.

Interestingly, Garnaut is an economist, and his main proposed solution to the problem of sustainability is an Emission Trading Scheme (ETS). The proposal was its introduction from 2010, under which businesses would be issued with permits for greenhouse gas emissions up to a set limit. Under a transitional arrangement to remain operative until 2012, the proposal was that there would be a low fixed price for these permits set by an independent regulator. The ETS would be tied to the government's plans to reduce emissions of greenhouse gases in line with the Kyoto deadline of 2012. After 2012, the market will determine the permit price for greenhouse gas emissions.[3] Now in 2011, the ETS still remains a proposal.

I mentioned just now that the choice of an economist in the form of Ross Garnaut to produce a report on the problem of climate change, which also has by extension a direct relevance to the problem of a sustainable good life, is particularly interesting. By that I mean that insofar as we live in a political system of a free-market economy, at least in the Western democracies, any solution to the problem of sustainability, as it specifically relates to the notion of a good life, must take into account and be sensitive to the free-market economic forces that create, shape, contextualize, and guide our choices both as citizens and consumers. And insofar as the problem of a sustainable good life is at least partly an economic problem, its solution must also be partly an economic solution.

However, because the problem of sustainability specifically affects every individual on the planet and not merely large business corporations and governments, the responsibility for a solution to that problem lies with every single individual as citizen and consumer and cannot be passed on to institutional and commercial conglomerates alone. Hence, individuals have a primary ethical responsibility to be concerned with and be active in finding and partaking in solutions to the problem of sustainability. However, just as in practical moral philosophy, moral prescriptions must both be justified and practically motivating before they can be accepted as making legitimate normative claims on people. Hence, any proposed solutions to the problem of sustainability must be both rationally justified and practically motivating to be capable of earning legitimate acceptance and compliance by those who are individually and collectively concerned and affected by those solutions.

If I am at least partly right in offering this initial albeit simplistic account of the problem of sustainability, there are in principle at least two main approaches to solving the problem of sustainability and that of the interrelated and equally important problem of a sustainable good life. One is an externalist approach, the other an internalist approach. I define the externalist approach as one that seeks to find an external solution to the problem of sustainability primarily through science and technology and the internalist approach as one that seeks to find a solution to the problem of a sustainable good life through targeting the consumers themselves both individually and collectively as societies. Specifically, the internalist approach targets the consumers by targeting their hedonic desires. In particular, the desires that people seek to satisfy in pursuing lifestyles of pleasure and a good life as broadly understood. Epicurean philosophy, which locates the good and well-being of individuals primarily in pleasure, conceived by Epicureans as the highest good conducive to the attainment of *eudaimonia*, well-being, or happiness, is particularly well-suited in offering a framework with regard to an internalist solution. This is particularly so because the free-market economic system in which most people in capitalist or mixed-capitalist systems, such as the one in China, live is theoretically and practically motivated and at least partly based on the individual's preferences and desires.

Hence, solutions to the problem of a sustainable good life within the context of free market economies able to specifically target the preferences and desires of individuals, such as the Neo-Epicurean framework proposed in this chapter, seem particularly promising. It is important to note, however, that the externalist and internalist solutions are not incompatible and can, as my argument will support, be combined in offering practical solutions to the problem of a sustainable good life. This is not surprising given that externalist solutions are primarily technologically based solutions, and technology itself through various multiple appliances, including household electrical and electronic appliances, are also orientated toward individuals.

Before proceeding in examining an Epicurean approach to an internalist solution to the problem of a sustainable good life, let me first outline at least three essential conditions that any solution to that problem, either internalist or externalist, must meet.

The Hedonic Condition

Any theory offering a solution to the problem of a sustainable good life must be *capable* of appealing to and addressing the hedonic desires of actual rational agents, at least *instrumentally* rational agents. That is, agents who are motivated to seek the most effective means to achieve desirable goals. This is essential in providing adequate rational and psychological motivation capable of motivating actual agents to at least feel predisposed to act favorably toward a solution to the problem of a sustainable good life. Any theory either internalist or externalist that cannot address or appeal to the agent's desires or preferences, at least in principle, would lack motivation and would thus be impractical in offering a solution to the problem of a sustainable good life.

The Ethical Condition

As we are rationally required to act ethically toward others on the basis of any number of ethical theories (including consequentialist, contract, deontic, and virtue theories) that, if not individually at least collectively, render that ethical requirement persuasive if not overriding, then any solution to the problem of a sustainable good life must also be consistent with ethical principles generally conceived. It must negatively not cause unjustified harm to people (it must not be malevolent), and it must moreover whenever possible positively produce good for people (it must be beneficent).

The Eudaimonic Condition

Insofar as instrumentally rational agents desire a good life as a means to the attainment of self-fulfillment or well-being or happiness, which I shall collectively refer to here as *eudaimonia*, any theory that offers a solution to the problem of a sustainable good life must also address and meet the eudaimonic condition. For if it cannot meet it and insofar as people value their own eudaimonia as one of the most desirable and valuable ends of their lives (as suggested by Aristotle) that theory is impractical and therefore inadequate in appealing to the instrumental and prudential considerations of actual agents in decision making.[4] Based on those three conditions, the chapter will postulate and seek support for two interrelated theses.

(A) Any solution to the problem of a sustainable good life must be consistent and allow for a conception of a sustainable good life, generally conceived. A solution to the problem of a sustainable good life, be it externalist or internalist, that was not capable of promoting a reasonable conception of a good life, one that most rational agents can upon informed reflection accept and adopt as reasonable, would lack adequate motivation in persuading individuals and societies to rationally adopt such a solution and would thus be impractical. Purely externalist solutions to the problem of a sustainable good life that fail to address agents' desires directly would prove impractical in just this way and thus prove unsuccessful. Therefore, it is necessary that any practical solution to the problem of a sustainable good life must address peoples' desires both individually and collectively as societies, as these are directly relevant to any conception of a good life.

(B) Furthermore, a solution to the problem of a sustainable good life must also be capable of addressing and meeting all the three conditions previously outlined, namely, the hedonic, the ethical, and the eudaimonic. For purely hedonistic solutions to that problem that, for example, fail to meet the ethical condition or the eudaimonic condition (for example, some hedonic lifestyles might prove self-defeating in not being conducive to the attainment of eudaimonia) would be objectionable and thus unacceptable, once more proving impractical. Similarly, solutions to the problem of a sustainable good life that addressed the ethical condition but failed to address the hedonic or the eudaimonic condition would also prove impractical in lacking adequate motivation.

EPICURUS' PHILOSOPHY OF PLEASURE

Although in agreement with Plato, Aristotle, and the Stoics that an ethical lifestyle based on a virtuous character is necessary for happiness, Epicurus unlike them identified happiness (for ease of reference I will use the terms happiness, well-being, self-fulfillment, and *eudaimonia* interchangeably) not with virtue but with pleasure. Epicurean happiness in a nutshell aims at relief or freedom from anxiety through the enjoyment of pleasures and avoidance of pains.

Epicurus like Aristotle thought of happiness as the highest good valued for its own sake and not for the sake of anything else. Unlike Aristotle, however, he identified happiness with pleasure, as pleasure according to Epicurus is the highest good that everyone pursues for its own sake and not for the sake of anything else.

Two Types of Pleasure

According to Epicurus there are two main types of pleasure: "moving" pleasures and "static" pleasures. Moving pleasures are those experienced in the process of satisfying a desire, e.g., eating a plate of pasta when one is hungry. However, when one has satisfied a moving pleasure, one also experiences a "static" pleasure that comes with the state of satiety that follows when one is no longer in need or want. Such static pleasures, the pleasure one feels when one's moving pleasures have been fulfilled, are the best pleasures because they relieve one of pain and anxiety; the pain of hunger, thirst, or sexual desire, for example (O'Keefe 2001).

Freedom from Fear and Anxiety

According to Epicurus, it is freedom from pain and anxiety that measures the relative merits of different activities. The best activities are those that produce the greatest pleasures, and the greatest pleasures are those that relieve the most pain and anxiety; those pleasures that are most conducive to peace of mind. Those are the pleasures that should be preferred if one is prudent and wise. Freedom from pain and anxiety that leads to *ataraxia* (the state of being unperturbed) or tranquility is therefore the basis of Epicurus' hedonic calculus.

Two Strategies in Dealing with Desire

If pleasure is desire-satisfaction (getting want you want or need) and pain is desire-frustration (not getting what you want or need), there are two strategies that one can implement with any given desire: one can either try and fulfill the desire or try and eliminate it. Insofar as some desires are difficult to fulfill, Epicurus advocates the second strategy: eliminating desires that are hard to fulfill so one is left with a core of desires that can easily be satisfied with minimal effort (O'Keefe 2001).

Three Types of Desire

Epicurus identifies three different types of desire: *natural and necessary desires*, *natural but not necessary desires*, and *vain and empty desires*. Examples of natural and necessary desires include desires for food, water, and shelter among others. According to Epicurus, these desires that are necessary for life are easy to satisfy, difficult to eliminate, and bring great pleasure when satisfied. People, in general, should try to satisfy these desires, as they are "naturally limited"; for when one is hungry or thirsty, for example, a limited quantity of food or water will suffice to satisfy one's hunger or thirst.

Interestingly, some natural desires such as the sexual desire may only be necessary and thus difficult to avoid for the species but not for the individual. So whereas sexual activity although natural is not necessary for an individual, it is both natural and necessary for the species. An individual can survive without it (as for example, in the case of celibate monks or priests) but not a species. An example of a natural but not necessary desire is the desire for luxury food. Although food is necessary for survival, luxury food is not—one can survive on a diet of basic food alone. If, however, one can obtain luxury food without too much effort, the desire-satisfaction that results from the consumption of luxury food poses no problem. It is only when the desire for luxury food or other natural but not necessary desires are difficult to pursue and satisfy that they should be eliminated; for the desire-frustration that can result from one's dependence on such desires and inability to satisfy them could cause one the pain of anxiety and frustration. Thus, paring one's desires back to the desires that are both natural and necessary is thus a good strategy for desire-satisfaction and happiness; for the elimination of the pain of unfulfilled desires in the present and the anxiety that such desires may not be fulfilled in the future leads to tranquility or *ataraxia*: freedom from anxiety and freedom from fear about the present and the future.

Vain desires include desires for power, wealth, and fame among others. They are difficult to satisfy because there is no natural limit to them. They are characterized by a tendency to escalate without limit to further desires for more power, wealth, or fame. Unlike the natural and necessary desires that are naturally limited and easy to satisfy, vain desires have the feature that the more one has, the more one wants. Such desires are not natural to human beings but are artificially inculcated by society's false or at least misguided beliefs about what we need to be happy. Unlike the natural but unnecessary desires that are optional so long as they can be satisfied easily, vain desires should, according to Epicurus, be eliminated, as they are the ones to cause us anxiety, pain, and unhappiness; a good reason for avoiding them.

The Virtues

In addition to the pursuit of pleasure and avoidance of pain for the attainment of happiness, Epicurus also counsels the cultivation and inculcation of virtues such as prudence, courage, justice, and moderation (the cardinal virtues) as the best and most effective means for enabling us to pursue the right pleasures and avoid the ones likely to cause us pain and unhappiness. In contrast to Plato, Aristotle, and the Stoics, who thought that the virtues were inherently valuable, Epicurus thinks that they are valuable only insofar as enabling us to identify and pursue the pleasures that are conducive to happiness and avoid those conducive to greater pain and unhappiness. This is in keeping with the correct idea that what is desired is not always equivalent with what is desirable. For some, desires that result in greater trouble, pain, and unhappiness than the pleasure they produce are not desirable for that very reason.

Summary of an Epicurean Good Life

In summary, we can say that Epicurean *eudaimonia* is primarily but not exclusively the pursuit and satisfaction of simple and natural pleasures that are easy to fulfill and whose enjoyment does not require too much effort; in addition, it includes the inculcation of the cardinal virtues as enabling dispositions and means for achieving that aim, for the ultimate purpose of attaining happiness through ataraxia or tranquility: freedom from anxiety and fear, both present and future. Thus Epicurean happiness, in contrast to purely hedonistic happiness that is generally amoral (neutral with regard to ethical conduct or character), is conceptually committed to both ethical conduct and an ethical character, for both are, according to Epicurus, the necessary means of enabling an individual to pursue the right pleasures and avoid those that are likely to cause them anxiety, thus potentially resulting in unhappiness. Insofar as the ultimate goal of each individual person is the attainment of happiness, the pursuit of pleasures likely to cause unhappiness is instrumentally self-defeating and thus undesirable for the individual persons themselves.

A Sustainable Good Life

Setting aside for the moment any fine-tuning and conceptual adjustment that an Epicurean notion of desire and desire-satisfaction may require, does an Epicurean philosophy overall as expounded in this chapter meet the three essential conditions for a sustainable good life, and, in so doing, does it offer a practical approach to address the problem of a sustainable good life as defined earlier? I will argue that it does: First, it meets the hedonic condition. For by basing eudaimonia on pleasure, it directly addresses the source of the problem of a sustainable good life, namely our individual human desires. Moreover, Epicurean philosophy provides a rational and systematic method for *prudentially* evaluating the comparative eudaimonic merits of different types of desires within a structured, hierarchical taxonomy of desires that is sensitive to empirical evidence.

With the passage of time, for example, some desires once considered unnecessary because they related to what was previously considered as luxury goods, such as the desire to possess a personal computer, for example, might now be considered a necessary desire and thus indispensable for even an Epicurean lifestyle. For example, the problem of the digital divide suggests that as a practical necessity for accessing and disseminating information, the possession of a computer with access to the Internet is a practical necessity for everyone, especially in a time period described as the information age. Moreover, insofar as accessing and disseminating information is also vital to our well-being, Internet access via a computer as well as via other technological artifacts, such as mobile telephones, for example, that afford us the means for effectively acquiring and using information, is also crucial to our individual and collective eudaimonia. Hence, the access and dissemination of information through information communication technologies (ICTs) is important and necessary for the satisfaction of the eudemonic condition for a sustainable good life.

One possible objection to the claim that the proposed Neo-Epicurean solution to the problem of a sustainable good life meets the hedonic condition is that it cannot address everyone's particular hedonic desires, and hence some people reluctant to give up their particular cherished hedonic desires would not comply with it. The Neo-Epicurean solution proposed is an internalist solution for a sustainable good life. To the extent that certain desires are no longer deemed sustainable on the basis of empirical evidence, then the proposed solution offers individuals adequate rational motivation to curb their unsustainable desires. The whole point of a Neo-Epicurean solution to the problem of a sustainable good life is that it systematically offers a rational and practical method of objectively evaluating the sustainability of hedonic desires for the attainment of a good life. Being a materialist theory of desires (Epicurean philosophy is a materialist philosophy), it is not prescriptive in the choice of particular desires. It simply provides a rational methodology for evaluating desires in terms of (a) their capacity for the attainment of a good life and (b) their capacity for being sustainable. These two together, therefore, provide a method for finding a solution to a sustainable good life.

Second, in counseling and recommending the prudential inculcation of the virtues of character, including that of justice as essential means for allowing an agent to choose wisely one's desires and pursuit of pleasures for the attainment of *eudaimonia*, Epicurean philosophy also addresses the ethical condition, if not intrinsically like Stoicism, at least instrumentally and prudentially. Finally, it also addresses the eudaimonic condition in advocating the preference for desires that minimize anxiety and maximize well-being in terms of *ataraxia* or tranquility that result in a peaceful and untroubled mind. However, it does so both realistically and moderately. For if people can pursue the pleasures of a luxurious lifestyle or those of extreme wealth, power, and fame (those of the vain desires) without causing themselves or others any undue anxiety or harm with the potential diminution of their own *eudaimonia* and/or that of others, this would be consistent with Epicurean philosophy. So long as such a lifestyle was capable of resulting in the attainment of *eudaimonia* consistent with the satisfaction of one's prudentially chosen desires and moreover, was also ethical with regard to oneself and other people.

However, an important and crucial proviso is that such natural but unnecessary pleasures for luxury and the pleasures of fame and power must be sustainable in keeping with the conditions for a sustainable good life as previously discussed. Insofar as such pleasures are not sustainable, they should be avoided, as they are likely to fall foul of the ethical condition for a sustainable good life, even though they may meet the hedonic condition. Recall that any theory for a sustainable good life must meet the three conditions, of the hedonic, the ethical, and the eudemonic.

A SUSTAINABLE GOOD LIFE AND TECHNOLOGY

Due to constraints of space, I won't be able to offer a detailed analysis of how Epicurean philosophy can be applied to particular types of technology to offer an approach to a solution to the problem of a sustainable good life. In the remainder of the chapter, however, I will briefly explore in outline how a Neo-Epicurean philosophy as described here can in principle be mapped onto technological approaches that offer solutions to the problem of sustainability. Take, for example, the dual-flush toilet and shower heads that are designed for efficient use of water; recycling of water; recycling of consumables, such as plastic, glass, and paper; installation of electrical household appliances; as well as the design and production of transport vehicles including cars, buses, trains, and airplanes that are more energy efficient (the new Boeing aircraft, *Dreamliner*, is made mostly out of composite materials, such as carbon fiber, for more energy efficiency is a case in point).

In all these examples, what seems crucial, however, is the role that people's desires play in motivating *compliance* with these energy saving technologies. People must want to comply and moreover be persuaded that there are justified reasons for complying with such technological energy-saving policies and practices for the policies and practices to work. The installation of an energy-consumption meter inside one's home, for example, that allows the monitoring of the overall energy consumption of a household would be a practical technological device for encouraging compliance with efficient energy use. Crucially, however, individuals must first be motivated to want to install such a device within their homes. Such initial motivation can best be primed by rationally targeting the consumers' desires, as suggested by the Epicurean desire-satisfaction and eudaimonic model proposed in this chapter. I think it is clear from the analysis offered so far that what would be required for practical solutions to the problem of a sustainable good life is a mixture of internalist approaches, such as the one advocated here, and externalist approaches based mainly on emerging technologies. Briefly, I state the following interim theses, which are subject to further empirical investigation:

1. All things being equal, in any proposed solutions to the problem of a sustainable good life (a) necessary and natural desires (A-desires) should be given prudential priority over the other two types of desires, namely, (b) the natural but unnecessary desires (B-desires), and (c) the open-ended vain desires, such as those for power, excessive wealth, and fame (C-desires). This is so for at least three reasons:

1.1 A-desires being necessary cannot be avoided and must to some minimal degree at least be satisfied for the preservation and sustainment of life. So any savings in terms of resources or energy in proposed solutions to the problem of a sustainable good life must first target B-desires and C-desires before targeting A-desires. For example, the controversial production of biofuels so as to relieve the demand for oil, especially in the form of ethanol to be used as an alternative energy source for transport vehicles, in particular cars, has reduced the supply of certain staple foods, such as grains and rice, which has caused food prices to rise sharply, resulting in food shortages that have sparked food riots in some developing counties. The Epicurean model of a sustainable good life would counsel that the production of biofuels must not hinder nor diminish the production of staple foods necessary for sustaining a basic healthy lifestyle for individuals and communities whose life-survival diets are dependent on staple foods such as rice and wheat.

1.2 Pending further empirical investigation, it is likely that B-desires and C-desires are more energy and resource consumptive and, therefore, these being the least necessary desires, should be targeted first in any proposed solutions to the problem of a sustainable good life; that is, first C-desires, then B-desires, and then only if absolutely necessary, A-desires. For example, it has been reported that "the rising demand for flat-screen television sets may have a greater impact on climate change through global warming than the world's largest coal-fired power stations" (Sample 2008). This is because the manufacturers of flat TV screens use a greenhouse gas called nitrogen trifluoride, which is 17,000 times more potent than carbon dioxide. To make things worse, unlike other greenhouse gases such as carbon dioxide, sulfur hexafluoride (SF_6), and perfluorocarbons (PFCs) emissions of nitrogen trifluoride are not restricted by the Kyoto Protocol. According to an article published in the journal *Geophysical Research Letters*, Michael Prather and his colleague Juno Hsu, claim that the production of nitrogen trifluoride in 2008 was equivalent to 67 million tons of carbon dioxide, with "'a potential greenhouse impact larger than that of the industrialized nations' emissions of PFCs and SF_6, or even of the world's largest coal-fired power plants" (2008).

Insofar as this report is accurate, and insofar as the production and supply of nitrogen trifluoride for use in the manufacturing of flat-screen TVs that significantly contributes to the pollution of the biosphere relates to a consumer demand of the type B-desire specification, namely, natural but unnecessary desires, those desires should give way to A-desires, namely, a desire for a reduction in greenhouse gases that pollute the biosphere and hence reduce the potential for a sustainable good life. Although one could reasonably argue that flat-screen TVs can contribute to a good life outside the constrains of sustainability, one must also acknowledge that flat-screen TVs may not be conducive to a *sustainable good life* and, as such, their production is problematic and should therefore be reduced if not eliminated or at least supplied at a price that includes an appropriate tax penalty that reflects the negative cost to a sustainable environment.

1.3 However, even if B-desires and C-desires prove not to consume more resources or energy than A-desires, A-desires being necessary for life-preservation should be given prudential priority in any minimization of consumption of resources and energy schemes proposed in solutions to the problem of a sustainable goof life. Some B-desires, for example, such as flying business or first class, might not consume more energy than flying economy class, but maintaining and running several sports cars in one's car pool might.

1.4 Marketing and advertising create and manufacture desires for people, especially the B-desires type that contributes to the inflation of unsustainable consumer desires. This in turn contributes to an increase in the magnitude of the problem of a sustainable good life (Spence 2005). Insofar as this is the case, the marketing and advertising industries should be encouraged to reform either internally or externally through government sanctions so as to comply with the conditions for a sustainable good life in keeping with the Neo-Epicurean model proposed here.

THE TRAGEDY OF TECHNOLOGY

In conclusion, I wish to briefly signal an interesting paradox about technology and a sustainable good life.[5] The problem, in short, is this: insofar as technology offers an external solution to the problem of a sustainable good life, it might also contribute to that problem by creating its own B-desires (for example, an ever-increasing plethora of technological devices that consume rather than preserve scarce resources), which as we saw, are potentially energy-hungry and thus nonsustainable in a technological world of ever-diminishing resources. If so, it seems that technology itself should be made compliant to the conditions for a sustainable good life and become sustainable itself. This could be accomplished by giving preference to designing technologies that comply with a sustainable good life along the Neo-Epicurean approach argued for in this chapter.

NOTES

1. An earlier version of this chapter, "Groundwork for a Neo-Epicurean Approach to a Sustainable Good Life in a Technological World" first appeared in the *Australian Journal of Professional and Applied Ethics* in 2008.
2. I owe the suggestion for the formulation of this argument to Philip Brey.
3. Several articles reported Australia's Emission Trading Scheme in the weekend issue of the *Sydney Morning Herald*, July 5–6, 2008.
4. See Aristotle's *Nicomachean Ethics*.
5. I owe this insight to Adam Briggle.

REFERENCES

Garnaut Climate Change Review. (2011). Retrieved from http://www.garnaut-review.org.au/update-2011/garnaut-review-2011.html

O'Keefe, Tim. (2001). "Epicurus." The Internet Encyclopaedia of Philosophy. Retrieved November 12, 2004, from http://www.iep.utm.edu/epicur/

Prather, Michael J., and Juno Hsu. (2008). "NF3, the Greenhouse Gas Missing From Kyoto," *Geophysical Research Letters*, vol. 35, L12810, doi:10.1029/2008GL034542.

Sample, Ian. (2008, July 4). "TV Screen Gas 'Worse Than Coal,'" *The Sydney Morning Herald*.

Spence, H. Edward. (2005). "The Commodification of Values: What Plato and Epicurus Can Teach us about Advertising Ethics—Ancient Wisdom for a New World." In *Proceedings of the 10th International Conference on Corporate and Marketing Communications*, edited by Ioanna Papasolomou (Nicosia, Cyprus: Intercollege) pp. 14–29.

Spence, H. Edward. (2008). "Groundwork for a Neo-Epicurean Approach to a Sustainable Good Life in a Technological World," *Australian Journal of Professional and Applied Ethics*, vol. 10, nos. 1 and 2, pp. 73–81.

Zupko, Jack. (2010) "John Buridan." *Stanford Encyclopedia of Philosophy*. Retrieved July 14, 2011, from http://plato.stanford.edu/entries/buridan/

11 Cell Phones, iPods, and Subjective Well-Being

Valerie Tiberius

> All of the questions we used to ask about cyberspace are now more rewardingly asked about 'non-mediated experience'—those increasingly rare moments when we disconnect from all media. The digital is now the rule, while the non-mediated becomes ever more the exception. (Howell 2007)[1]

Mediated experience is becoming very common. It is increasingly unusual to have an experience that isn't interrupted by a cell phone call, accompanied by a podcast coming through your ear buds, or filtered through the lens of a digital camera. Digital technology has changed social interaction in more direct ways. For example, there are now "mobile clubbing" events, during which people dance "together" (in the same space) to their own music coming through their own music device. One might think that (at least in some cases) having our experiences interrupted or mediated by new technology is bad for us. By constantly attending to how to take the best shot of the mountain, we may miss its awesome beauty. Having a conversation with a friend that is interrupted by ringtones and brief conversations with other people affects the character of the conversation. And a constant podcast soundtrack to our walks distracts our attention from where we're walking and the people we meet along the way.

Of course, we are getting used to this way of experiencing life. And younger people are either getting used to it more quickly or are too young to have known anything different. Moreover, we (many of us, often) like our technological distractions. Nevertheless, I suspect I am not alone in thinking that something important has been lost now that we are constantly accessible and always multitasking. One might think that it would be good for us to have some unmediated experiences, even if our preferences adapt to ubiquitous technologies in such a way that we don't want to be without our cell phones and iPods. But how can we make sense of this thought without making controversial assumptions about objective values? How can we make sense of this thought and still give sufficient authority to the subjective point of view in matters of well-being? It might seem that the thought that mediated experiences are sometimes bad for us challenges subjective authority in a way that is incompatible with subjective theories. I don't think this is the case.

To answer these questions, we need a theory of well-being that both takes the subjective point of view seriously and makes room for claims about well-being that go against people's actual preferences and attitudes. We can see how simple desire-satisfaction theorists will have difficulty with this task because they privilege actual preferences (see chapter 7 in this volume). In the first two sections of this chapter, I introduce a more promising theory, the value-based life-satisfaction (VBLS) theory of well-being. Briefly, according to VBLS, to achieve well-being is to be disposed to endorse how your life is going overall on the basis of your values. In the last section before the conclusion, I apply the theory to the case of mediated experiences.

LIFE-SATISFACTION AND THE PSYCHOLOGICAL RESEARCH ON CONTEXT EFFECTS

I call the account I favor the VBLS account (for more on the standard life-satisfaction theory, see chapter 5 in this volume). According to this account, put colloquially, your life goes well for you if you have a good sense of what matters in life and you feel good about your life because you are achieving it. More precisely, and very briefly, to achieve well-being (or at least a crucial component of it) is to endorse how your life is going overall on the basis of your values, where those values suit your affective make-up and withstand your own reflective scrutiny. Endorsement, on this view, is a pro-attitude that includes a cognitive and an affective component.[2] Values are complex patterns of attitudes toward activities, relationships, goals, aims, ideals, principles, and so on. I'll say more about values in a minute, but first let me explain why it is important to add values to the life-satisfaction account of well-being.

The VBLS account takes a life-satisfaction account of well-being, represented in philosophy by L. W. Sumner (1996), and adds "value-basing," that is, the idea previously mentioned that what is relevant to well-being is how our lives are going with respect to what matters to us (our values). The reason for modifying the account in this way comes from a body of work in the psychological literature on life-satisfaction that demonstrates the context sensitivity of judgments of life-satisfaction. According to this research, judgments of life-satisfaction are sensitive to irrelevant changes in context, such as changes in weather (for an overview of the research, see Schwarz and Strack 1999).

For example, some studies have shown that item order on questionnaires influences subjects' reports of life-satisfaction. Strack, Martin, and Schwarz (1988) asked students two questions: "how happy are you?" and "how many dates did you have last month?" When the general happiness question was presented first, there was no correlation between the two responses, but when the dating question was presented first, the correlation rose to 0.66. Other studies indicate that judgments of overall life-satisfaction vary depending on the contrast that the subject considers at the time.

This phenomenon is illustrated by a study that demonstrated the effect that having a handicapped person in the room has on people's reports about their well-being (Strack, Schwarz, Chassein, Kern, and Wagner 1990). People tend to evaluate their lives as going better when there is a handicapped person in the room than they do when the handicapped person is not there, presumably because the handicapped person changes the evaluative standard that the person uses in her judgment.

The fact that life-satisfaction judgments are relative to context in these ways has suggested to critics that either there is no stable underlying psychological state of life-satisfaction (which makes life-satisfaction a poor candidate for an essential component of well-being) or self-reports of global life-satisfaction are a poor measure of it (Schwarz and Strack 1999). In psychology, these concerns have been addressed to some extent by recent studies that show that there is sufficient stability in life-satisfaction judgments. According to these studies, life-satisfaction judgments do correlate well with judgments of domain satisfaction. People who are satisfied in the domains they think are most important report high overall life-satisfaction consistently, and changes in domain-satisfaction are a good predictor of changes in overall life-satisfaction (Pavot and Diener 1993; Schimmack, Diener, and Oishi 2002).

This life-satisfaction research in psychology may be enough to show that we are warranted in taking life-satisfaction judgments to be evidence of some relatively stable underlying psychological state. However, this research doesn't solve a different problem—a philosophical problem—that the context effects highlight. This problem is the problem of the normative significance of life-satisfaction, and this is the problem that value-basing is meant to solve. Context effects draw our attention to the possibility that how satisfied we are with our lives may be sensitive to the wrong things. If it is, then life-satisfaction would lack intuitive normative significance.

To illustrate this point, consider an account that identified well-being with 'satisfaction with how your life is going with respect to trash pick-up.' Such an account may solve the problem of instability—how satisfied people are when they attend to the trash may be quite stable over time—but such satisfaction just seems irrelevant to well-being insofar as well-being is something we have reason to promote for others and something worth promoting for ourselves. However much stability would be achieved by specifying the relevant type of satisfaction in this way, it does nothing to ensure an account that makes life-satisfaction normatively significant. If the degree of satisfaction we experience with our lives overall depends on what we base our assessment on, or what we attend to when we make this assessment, we need to ensure that this basis is normatively significant.

The VBLS account takes the relevant satisfactions to be satisfactions with how your life is going with respect to what matters to you, your values. Values, as I said previously, are complex patterns of attitudes toward activities, relationships, goals, aims, ideals, principles, and so on. This isn't the place to defend a complete conception of a value, but there are two main features of values that are important in the current context (see Tiberius 2000, 2008 for more details). First, values are normative for the person who has them: We take our values to be justified by the facts and our other commitments and hence to provide good reasons for doing things. Second, value commitments include a motivating, affective component: Part of what it is to care about something in the way distinctive of valuing is to have some positive affective or emotional response toward it. Because values have these features, values can be more or less ideal. Our values can be more or less justified (by our own reasons) and more or less well-suited to our motives or affective dispositions.

We can now make sense of the possibility of error in our judgments of life-satisfaction. First, because our values can be more or less appropriate, our judgments of life-satisfaction can be criticized on the grounds that the values on which these judgments are based are not the right ones. Second, self-reported life-satisfaction can fail to track well-being when it is not based on what matters to us (as would happen when life-satisfaction is heavily influenced by irrelevant circumstances, such as a change in the weather, or when life-satisfaction is based on an inaccurate perception of how well we are achieving what we value).

For example, consider a person who values always being accessible by cell phone or email. This commitment figures into her planning, she has reasons for it, and it suits some of her affective dispositions (being accessible makes her feel important, which she enjoys, for example). Yet, there are two ways in which this value commitment might not be ideal. First, it might be that she has other affective dispositions that are frustrated by it (for example, a desire for quiet time to herself), so that the commitment does not fit her affective nature overall.

Second, it may be that this value does not meet her own standards of reflection. For example, perhaps she thinks her values ought to be informed and she has a false perception of how much her accessibility contributes to people thinking she's important. Furthermore, her assessment of how well her life is going may be in error if it is not really based on her values. For example, if the constant chattering and texting actually offends her friends and degrades the relationships that she values, her report of high satisfaction would not reflect this. So, there are several ways to make sense of the idea that constantly mediated experience does not contribute to a person's well-being, even though it is what the person wants or chooses.

THE VALUE BASED LIFE-SATISFACTION ACCOUNT AS A NORMATIVE ACCOUNT OF WELL-BEING

Now that we have a rough idea of what the VBLS account is, I can outline what I take to be the main argument for it. The main argument for this account is that it makes better sense of well-being as a normative notion than competing views. To count as normative, an account of well-being has to do (at least) three things: It has to allow for some diachronic stability in well-being attributions, it has to make room for the possibility of error, and it has to be appropriately related to people's motivations in a way that matches up with what people take to be normative for them.[3] The most popular alternatives to the life-satisfaction view fail the third criterion. Hedonism fails because, although we are motivated by pleasures, we're not motivated only by pleasure and we sometimes regard pleasures as inappropriate. Eudaimonist or objective list theories fail insofar as they claim that something can be good for a person even if that person has no relevant motivation to promote or pursue it. As we have seen, there are some reasons to think that the standard life-satisfaction account fails the first and second criteria. As we have also seen, however, the VBLS account does give well-being diachronic stability and does make sense of the possibility of error.

Much more would need to be said to fully articulate and defend VBLS (see Tiberius and Plakias 2010 for a more thorough exploration), but I hope I have said enough to pave the way for a discussion of the topic with which I opened. The purpose of introducing the VBLS account was to provide an example of a subjective theory of well-being that will allow us to think about how subjective theories can respond to worries about the authority of subjective experience.

VALUE-BASED LIFE-SATISFACTION AND ADAPTIVE PREFERENCES

Before we turn to a discussion of the particular case of technologically modified experiences, it will be helpful to notice that this is just one case of the more general phenomenon of adaptive preferences. The problem of adaptive preferences is usually discussed in the context of oppression (Sen 1987). Because people adapt to their circumstances, a person may be quite-satisfied with her very limited, objectively impoverished life under oppression. This presents a problem for subjective theories of well-being because it seems that subjective theories such as preference satisfaction and life-satisfaction theories are saddled with the unattractive view that the oppressed person who has adapted to her lot is faring just as well as the person with many more options and capabilities who is equally satisfied with them.

In the case of technologically mediated experiences, insofar as we have adapted to our cell phones, iPods, and digital camcorders, our subjective satisfaction may not be affected by these technological changes. Applying VBLS, a person who has adapted to the technology might have suitable values, think about them when she assesses how well her life is going, and be perfectly well satisfied even though she is missing something that someone whose experiences are not technologically mediated enjoys. In other words, it might seem that on VBLS there is no room for the claim that the proliferation of technologically mediated experiences is bad for us, as long as we adapt to it.

At this point, I should say something to address skepticism about the intuition that there is something in all this mediation by technology that is bad for us. Not everyone shares this intuition, and even those of us who do have it might be concerned that it is just the result of being old and stuck in our ways and that it should not carry much weight. Fair enough—I don't have a knock-down argument against these sources of doubt. (Although, for support of my worries about mediated experience, see Rosen 2004). Nevertheless, I do think there are things that the VBLS view can say to accommodate intuitions about adaptive preferences like these and that it is worth exploring these resources even if we ultimately think that this particular intuition does not need accommodating. The case of technologically mediated experience might, in fact, illuminate more serious cases because it does not involve the complicating factors of oppression and injustice. Moreover, future advancements may make the worries about digitally mediated experiences more severe and compelling to more people. For example, consider the possibility of contact lenses that provide constant information to the visual field about nearby options for consumption. Would we adapt to such an intrusion? It seems likely that we would. Would it be good for us? Not obviously, according to VBLS.

The VBLS account makes room for the possibility that your well-being is affected by changes to which we adapt in two ways. (Now I am just applying the points about error made earlier.) First, a person's values may adapt in a way that makes them less successful in terms of justification and emotional suitability. For example, if we adapt to technological mediation of experience by valuing that mediation and placing less value on unmediated experiences, and if our changed values are actually less suited to our affective character, we have grounds according to VBLS for claiming that our well-being has been affected.

For example, consider again the person who values having and using the latest technology and who takes no notice of the way in which her cell phone interrupts conversations with friends and makes the friends feel angry and insulted. If friendships are, in fact, more important to the positive emotional outlook of this person than cellular availability, then we can say—even if the person is satisfied with how her life is going—that her well-being is decreased. Her life is going well according to her values, but she would be better off with different (more ideal) values, even from her own point of view. Or, consider the person whose values have adapted to his iPhone such that he places a very high value on playing Tetris. Such a person might report high life-satisfaction as long as he has access to his phone, but we are not thereby forced to say that he has achieved a high level of well-being. Instead, VBLS allows us to ask whether these values are a good fit. If, for example, our game player is turning off sought-after romantic partners, or if he is foregoing experiences that would be better than Tetris at reducing undesirable stress, then there are grounds from within his own subjective point of view for saying that his values are problematic.

This way of limiting subjective authority is similar to what other subjective theories of well-being allow. Philosophers such as Sumner have answered the problem of adaptive preferences by adding idealizing constraints to the subjective states that count as constitutive of a person's well-being. For Sumner, we have reason to doubt a person's own assessment of how her life is going when we have reason to believe that her assessment is uninformed or non-autonomous. These constraints allow for judgments about a person's well-being that contradict that person's own assessment. Any view that includes such constraints must answer questions about the grounds and demandingness of these standards.

On the first question, Sumner argues that a lack of information (or autonomy) will only undermine a person's authority about her well-being if she herself would assess things differently if she were informed and autonomous. I agree with Sumner on this point: Claims about a person's well-being that contradict that person's assessment on grounds of justification or emotional suitability are only legitimate insofar as the person herself endorses these standards for improvement of her values. This preserves the connection to the subject that befits a theory of well-being. The second question is about how ideal a person's subjective states must be to count as having authority. Some opt for hyper-idealization in the form of full information and complete rationality. I think this is a mistake and that we ought to think in terms of meeting the standards to various degrees, depending on the context (Tiberius and Plakias 2010).

The VBLS account allows for a second way in which subjective authority about well-being might be challenged in cases of adaptation. VBLS allows for the possibility that a person is less successfully pursuing (or living up to) her values even though she feels just as satisfied as someone who is more successful. In such cases, we have grounds for saying that the person is worse off than she thinks she is. This can help with the problem of adaptive preferences when the pursuit of some values is hampered by the change. For example, if we assume that people value deeply felt experiences, and if it turns out that technologically mediated experiences (say, of nature) are less deeply felt than nonmediated experiences, then we can say that adaptation to technology makes us worse off, even if we don't notice the ways in which it does so. Similarly, if we assume that people value close personal relationships with others, and if being "plugged in" causes a kind of disengagement that negatively affects such relationships, then once again we have grounds for saying that we are worse off, even if our level of satisfaction has adapted to the change.

One might think that this is really just a point about the measurement of well-being: Asking people how satisfied they are with their lives doesn't tell us how satisfied they really are with their lives, one might argue, because people ignore certain dissatisfactions when they make such reports. While this could be what's going on, there is another possibility. It may be that we really are accurately measuring felt satisfaction, but, according to VBLS, felt satisfaction isn't all there is unless that's all you value. As long as a person values something other than her own satisfaction, her assessment of how well her life is going in accordance with her values can be in error. A person may be genuinely satisfied with how her life is going, on the basis of her values, but nevertheless be incorrect about how well she is achieving these values.

With respect to this second way in which subjective authority is limited, VBLS is similar to desire-satisfaction theories of well-being. Such theories can make sense of this second kind of error because whether or not we actually achieve the ends we desire is independent of whether we think that we have. The VBLS account is similar to desire satisfaction theories, because values as I understand them are a subset of preferences or desires, broadly construed. (Values are those desires that we take to be normative.) But VBLS differs from desire satisfaction theories in that it takes as central to well-being the subjective sense of satisfaction with how one's life is going. VBLS is a hybrid theory, then, that preserves important aspects of each of its theoretical parents. I have tried to argue (albeit briefly) that this hybrid offers valuable resources for accommodating intuitions about preferences that have adapted to ubiquitous technology.

To summarize this argument, let's take the example of the consumer contact lenses I mentioned earlier.[4] Imagine a contact lens that knows as much as Amazon.com about your consumer preferences and displays personalized information about opportunities in your environment to purchase goods and services. These lenses would be like the iPhone apps that give you information about restaurants, bars, shops, and theaters near your current location, only it would be constantly available in your visual field. Let's also imagine that such technology is available and most people want it. How could a subjective theory such as VBLS make sense of the idea that these lenses would not be good for us, despite the fact that we want them? Here's what VBLS can say: First, if we want the lenses but the benefits they provide (more opportunities to spend money) are not valued by us, they are not (necessarily) good for us.[5] Second, if we do value the benefits, but this value is inappropriate (that is, if it doesn't suit our emotional nature or wouldn't survive our own reflection) or is less appropriate than other values with which it competes, then the lenses aren't good for us. Third, if using these contact lenses renders us less successful at pursuing the values we have or less satisfied with how our lives are going overall, then they are not good for us. Given the weak link between satisfying consumer preferences and long-lasting satisfaction, it seems likely that this third possibility will turn out to be true. I won't argue further for this claim, though; my point here is just to show that there are a number of ways that a subjective theory such as VBLS can provide grounds for thinking that certain technologically mediated experiences are not good for us, even if we want them, like them, or "couldn't live without them."

CONCLUSION

Mediated experience might be bad for us even if we like it. It might be bad for us if our valuing it does not suit our affective natures or stand up to our own reflective scrutiny; or it might be bad for us if liking it prevents us from pursuing our other values in a satisfying way. The VBLS account of well-being makes room for both of these thoughts. Still, VBLS is a subjective theory of well-being according to which what is good for a person depends on her attitudes; in this theory, nothing is good for a person that wouldn't seem good from the person's own point of view, after correcting her point of view along dimensions that the person herself regards as improvements. What the VBLS account of well-being encourages us to see is that not every subjective attitude is equally deserving of attention.

NOTES

1. Quote by William Gibson, author of *The Neuromancer* and coiner of the term "cyberspace."
2. Here I follow L. W. Sumner (1996), the main philosophical advocate of the life-satisfaction view.
3. The idea here, found also in Korsgaard (1996) and Smith (1994), is that to explain normativity, an account of a normative concept must capture motivation and justification.
4. The idea is inspired by various science fiction movies and by an interview with Michio Kaku in the *Los Angeles Times* (Khan 2011). Kaku remarks that in the future, "The Internet will be everywhere, including in your contact lenses—when you blink, you'll go online."
5. They may still be good as a means to something else we value.

REFERENCES

Howell, Nic. (2007, August 2). "NMA Futures—William Gibson," *New Media Age*. Retrieved from http://www.nma.co.uk/features/nma-futures-william-gibson/34418.article

Khan, Amina. (2011, June 3). "Online with the Blink of an Eye and Other Marvels in our Future." *Los Angeles Times*. Retrieved from http://articles.latimes.com/2011/jun/03/science/la-sci-michio-kaku-20110604. Last accessed June 15, 2011.

Kosgaard, Christine. (1996). *The Sources of Normativity*. Cambridge: Cambridge University Press.

Pavot, William, and Ed Diener. (1993). "Review of the Satisfaction with Life Scale," *Psychological Assessment*, vol. 5, pp. 164–172.

Rosen, Christine. (2004). "Our Cell Phones, Ourselves," *The New Atlantis*, vol. 6, pp. 26–45.

Schimmack, Ullrich, Ed Diener, and Shigehiro Oishi. (2002). "Life-Satisfaction Is a Momentary Judgment and a Stable Personality Characteristic: The Use of Chronically Accessible and Stable Sources," *Journal of Personality*, vol. 70, no. 3, 345–384.

Schwarz, Norbert, and Fritz Strack. (1999). "Reports of Subjective Well-Being: Judgmental Processes and Their Methodological Implications." In *Well-Being: The*

Foundations of Hedonic Psychology, edited by Kahneman, Daniel, Ed Diener, and Norbert Schwarz (New York: Russell Sage Foundation Press), pp. 61–84.
Sen, Amartya. (1987). *On Ethics and Economics*. Oxford: Oxford University Press.
Smith, Michael. (1994). *The Moral Problem*. Oxford: Blackwell.
Strack, Fritz, Leonard Martin, and Norbert Schwarz. (1988). "Priming and Communication: Social Determinants of Information Use in Judgments of Life Satisfaction," *European Journal of Social Psychology*, vol. 18, pp. 429–442.
Strack, Fritz, Norbert Schwarz, Brigitte Chassein, Dieter Kern, and Dirk Wagner. (1990). "The Salience of Comparison Standards and the Activation of Social Norms: Consequences for Judgments of Happiness and their Communication," *British Journal of Social Psychology*, vol. 29, pp. 303–314.
Sumner, Leonard W. (1996). *Welfare, Happiness, and Ethics*. New York: Oxford University Press.
Tiberius, Valerie. (2000). "Humean Heroism: Value Commitments and the Source of Normativity," *Pacific Philosophical Quarterly*, vol. 81, no. 4, pp. 426–446.
Tiberius, Valerie. (2008). *The Reflective Life: Living Wisely with our Limits*. Oxford: Oxford University Press.
Tiberius, Valerie, and Alexandra Plakias. (2010). "Well-Being." In *The Moral Psychology Handbook*, edited by Doris, J. and the Moral Psychology Research Group (Oxford: Oxford University Press) pp. 401–431.

Part IV

Information Technology and Well-Being

12 New Social Media and the Virtues[1]

Shannon Vallor

'The good life' is a universal and perennial human concern. Most thinkers hold that the good life either requires or is greatly enhanced by close personal relationships, and any careful inquiry into the impact of modern technology on the good life must address the social dimension of our technological age. This requires a philosophical appraisal of recent technology-driven changes in the way that many humans build and sustain intimate relationships, and more specifically, how such technologies may impact the development of certain social *virtues* essential to the good life. Unfortunately, the influence of technology on the virtues is largely ignored in the otherwise burgeoning research on the ethical impact of technology. In a preliminary and partial attempt to address this lacuna, I explore here one set of technology-mediated shifts in human relationships, namely, those driven by new social media, such as Facebook, MySpace, and Twitter.

THE VIRTUES OF A CHARACTER-BASED ACCOUNT

It seems highly likely that Web 2.0 technologies and their subsequent iterations will shape the communicative habits of future generations to some significant degree. If this assumption is correct, then the need for reflection on the ethical impact of such changes is equally clear. The rise of new social media has already provoked considerable debate among scholars over their likely effect on the well-being of future generations. The outcome of the debate remains equivocal: some conclude that new media promote the well-being of young people by helping them develop essential social and technical skills (Ito et al. 2009) or by increasing social capital and self-expression (Ko and Kuo 2009), whereas others warn of dangers ranging from neurological deficits (Derbyshire 2009) to impaired immune function and cardiovascular disease (Sigman 2009). Largely overlooked, however, is a question of tremendous social and ethical importance: What impact will habitual use of new social media have on the development of users' *character*, and in particular, on their development of various social *virtues* essential to the good life?

The framing of this question relies on one of the oldest theoretical frameworks for exploring the nature of the good life and the effective means of its realization, namely Aristotelian virtue ethics. This framework, laid out in Aristotle's *Nicomachean Ethics*, assumes the existence in many human beings of certain stable character traits, including virtues. A virtue is a habituated tendency to express one's capacities in action *excellently*, that is, in a manner that allows one to flourish in one's physical and social environment.[2] Virtues, then, are traits that dispose their possessors toward activity that promotes human flourishing, whereas vices are traits disposing their possessors toward activity that inhibits human flourishing. Virtues and vices are themselves formed by the repetition of specific kinds of activity, and, once formed, are reliably and habitually expressed (Aristotle 1984, book 2). One's moral character, then, is dependent for its formation on repeated engagement in activities leading to settled habits of excellence. On this view, virtues are neither innate (although natural tendencies can enhance or inhibit their development), nor acquired simply by cognizing certain norms, values, or rules for right action. This is important for several reasons.

First, if Aristotle is correct, then moral development of individuals cannot be predicted simply by examining what they think, feel, or believe—we also have to know what kinds of actions they will get in the habit of doing. As Aristotle notes, once we realize that the differences in our moral character arise largely from differences in the actions to which we become habituated, it becomes apparent that "It makes no small difference, then, whether we form habits of one kind or of another from our very youth; it makes a very great difference, or rather *all* the difference" (1984, 1103b25). Thus, to the extent that a technology transforms activities that young people regularly perform, it potentially reshapes their moral development. This gradual effect is distinct from a technology's immediate social consequences, although no less important. Yet most current studies of new media focus on their relatively immediate psychosocial impact. The most common empirical measure of the influence of social media on quality of life, namely *social capital*, tells us much about how these technologies can expand, deepen, narrow, or weaken the social connections and support available to users (Ko and Kuo 2009; Ellison, Steinfeld, and Lampe 2007; Franzen 2003; Wellman, Haase, Witte, and Hampton 2001). However, it reveals nothing about what is happening to the moral character of the persons enjoying these gains or suffering these losses.

For a user might truthfully report gains in social capital as a result of heavy engagement in new social media, despite having suffered a lasting injury to her character through the development of certain vices, or the inhibition of specific virtues that would have developed were it not for her use. The reverse could also be true, that is, losses in social capital correlated with new social media practices cannot exclude the possibility that users may cultivate important virtues through those practices. An examination of the impact of new social media on the virtues is therefore a critical component of any thorough assessment of the relationship of those media to the good life. We may find that new media practices are complementary, inhibitory, or neutral to the development of specific virtues, or we may find that they promote the development of new but compensatory virtues. Whatever the answer may be, it is of great practical importance that we ask the question.

There is one further point worth making. Because virtues and vices are states of character developed over time and through habit, they are very resistant to change. A true virtue is not easily uprooted, but neither is it easy for us to rid ourselves of our vices. For this reason, we would do well to consider the relationship between new social media and the virtues now, when the phenomenon is fairly new and still open to social critique, than later, when opportunities for institutional and personal reform may be more difficult to come by.

THE GOOD LIFE AND COMMUNICATIVE VIRTUES

For Aristotle, the good or 'blessed life' of *eudaimonia* entails a rich social context of friendship (1984, 1097b14, 1155a5).[3] Ideally, friends personally embody in their character, and reflect for one another, concrete aspects of the good to be loved and admired and, in turn, support one another in the appreciation and active pursuit of virtue (Aristotle 1984, book 9). Which character traits are strongly associated with such friendships? Book 4 of the *Nicomachean Ethics* describes a number of social virtues. This is not the place to debate, as many have, the correctness of Aristotle's taxonomy of the virtues. I will focus on three virtues, two of which have close counterparts in Aristotle's account, and one that does not. These are *patience, honesty,* and *empathy*. They belong to a class I will call *communicative* virtues, for they develop primarily (although not exclusively) through communicative practices.[4] A thorough justification of my selection of these virtues is beyond the scope of a short chapter such as this, as is a full account of the relationship of new social media to the virtues. At best, my analysis can serve as a provisional sketch of this territory, serving as a guidepost for the thorough, rigorous, and empirically grounded analyses that must follow.

Patience

While the Aristotelian virtue often translated as patience (*praotes*) is defined as slowness to anger (1984, 1125b25), as a communicative virtue patience demands a far richer description. Patience is critical for sustaining close relationships. It becomes an ingrained habit through the mastery of communicative practices such as listening. The patient person can allow a friend to tell a lengthy story without jumping in and finishing the story herself. She can listen to a deeply personal revelation without interrupting with 'hey, that reminds me of this thing that just happened to me!' Once patience becomes not just a momentary indulgence of the other but an enduring and habitual element of one's character (that is, a virtue), it facilitates trust by communicating to others that your interest in them does not end with their ability to keep you constantly pleased or fascinated.

Patience helps us to 'ride out' moments of irritation, boredom, or incomprehension rather than terminate dialogue or attempt to force it into a more satisfactory state by changing the subject. It carries us through periods of conflict or misunderstanding, allowing us to reconnect with our partner on the other side of the communicative breach. The richest joys of communication often come from being patient enough to grasp what is being said, to finally get the joke, or to hear a challenging truth. Patience demonstrates to another the depth of one's commitment to the relationship and builds trust and confidence in its future by showing that the relationship's continued existence is not wholly dependent on its momentary rewards.

The immediate physicality of face-to-face communication often forces us to be patient, even when we would rather 'tune out' or 'switch off', to use telling metaphors. Yet this is precisely what builds patience as a virtue, not just a momentary resignation to the absence of an escape route. Many might recall the experience of being a sullen teenager stuck for a long afternoon with an elderly relative, surrounded by painful silence or the oppressive ticking of a grandfather clock, fighting the urge to flee and eventually finding a point of engagement that one realizes only in hindsight was the first seed of a mature and rewarding connection with another generation. The point of the example is that virtues are characteristically *difficult* to acquire, especially in the beginning when our existing motivations often pull us in the opposite direction (Aristotle 1984, 1109b). One therefore requires, in addition to our existing motives, situational 'gradients' that exert pressure on us to move in the virtuous direction. The social strains and burdens of face-to-face conversation have historically, and across cultures, often been rich sources of such pressure. We must ask whether and how new social media can also support such communicative breakthroughs, and the virtues they help to engender.

Today's technologies provide us with an ever-widening horizon of escape routes from any interaction that has lost its momentary appeal and are widely celebrated by users precisely for their capacity to liberate us from the uncomfortable strains and burdens of conventional communication. I can terminate a Facebook chat that has become contentious or tiresome with a hurried 'bye,' or click away from a friend's blog, without the price that must be paid for walking away from a face-to-face conversation.

In the latter practice, the gaze of the morally significant other holds me respectfully in place and solicits my ongoing patience. This solicitation is a critical element in my moral development; although I might ignore its pull, it creates an important situational gradient in the direction of patience. Can new social media create *technological gradients* that bring a power comparable to the gaze to an online exchange, and if so, what form will they take? The brevity of the formats most privileged by social networking sites obviates the need for patience: the tweet, the 'poke,' the status update. Email is on a steep decline among teens, in part because it is perceived as a slow and ponderous means of communication by comparison with new social media (Lenhart, Madden, Smith, and Macgill 2007, 20). The instant digestibility of these transactions is among the features driving the technology's appeal, as the success of Twitter indicates; but do such 'social' media thereby defeat their own aims to enrich human sociability? Or will developers of social media find ways to help new generations of digital natives ford across communicative breaches and experience the rewards of patience?

Honesty

Honesty is widely characterized in the media as threatened by the Internet, chiefly by the ability in many Web contexts to misrepresent one's age, gender, or other personal attributes. For social networking sites where the only tie between members is a common hobby or interest, this remains a danger, although peer auditing makes this more difficult on community-based networks such as Facebook. But honesty as a virtue goes well beyond being truthful about one's social identity and warrants a broader view. While it includes truth-telling, honesty as a virtue encompasses words *and* actions, hence it represents something akin to personal integrity (Aristotle 1984, 1127b). Honesty as a virtue is the willingness in communication to put one's authentic self in play, and it involves the assumption of certain risks, including social judgment and rejection. It can be argued that new social media may actually promote greater risk taking than face-to-face communications, which may be perceived as 'higher stakes' encounters requiring more caution and restraint. But one must also ask how users put their authentic selves in play when they create profiles offering a carefully edited, tightly controlled image of themselves, one perhaps aimed more at drawing in as many 'friends' as possible than at exposing one's authentic personality.[5]

We should also recall that virtues typically represent a 'mean' between two extremes of vice, one a deficiency, the other excess (Aristotle 1984, p. 1109b25). Hence, while deceitfulness represents a deficiency of candor, and honesty and openness characterize the virtuous mean, a tactless or vulgar lack of reserve is viewed as a vice, as openness taken to excess. The first wave of media coverage of social networking sites highlighted employers who surf MySpace and Facebook to screen out candidates whose online posts or pictures display a dangerous lack of personal reserve. While a close, trusting relationship cannot be built or sustained without honesty, reckless candor can bring a relationship to a premature end.

Like all virtues, then, honesty demands the perceptiveness and cognitive agility of the prudentially wise person, the ability to read the right course of action from the situational variants at hand. This involves much more than the belief that honesty is the ethical choice, or even the will to be truthful or act with integrity, but something far more difficult to acquire: the capacity to be truthful "at the right times, in the right way, to the right persons and in the right degree" (Aristotle 1984, 1109a25). And, like all virtues, acquiring that capacity depends in large part on the situational gradients we encounter in our moral development, whether they are the truth-telling practices of those we come to imitate or the technological opportunities to represent ourselves in various authentic or inauthentic ways.

Arguably, modern societies have fallen well short of the virtuous mark in this regard, even prior to the information age. This invites the moral creativity of IT developers to change the landscape by creating more technological opportunities for honesty. Despite the concerns noted earlier, one can argue that this has already begun: consider the many online networks that provide support for gay, lesbian, bisexual, or transgendered youth whose authentic selves are locally suppressed, or the rise of blog culture that offers virtually unlimited opportunities for self-expression and personal reflection. Indeed, it makes little sense to ask whether new social media have either inhibitory or salutary effects on honesty in *general*; rather, we ought to ask the following questions: first, what are the salient technological and contextual differences between those social media that tend to promote honesty and those that do not; and, second, how can we use this information to ethically inform and shape the development, promotion, and use of new social media?

Empathy

Unlike patience and honesty, empathy has no close relative in Aristotle's account of the moral virtues. Empathy is the capacity for feeling with and for others, and as a virtue, it is a reliable disposition to share in their joys and sufferings. Aristotle *does* claim that empathy is an essential mark of genuine friendship, rooted in the way in which a true friend becomes 'another self' whose joys and sorrows are experienced as one's own (1984, 1166a8). However, Aristotle seems to believe that as an affective response, empathy irrupts naturally and involuntarily in such relationships, rather than abiding as a disposition that can and must be cultivated, as is the case with virtues (1984, 1157b35).

Yet many of us know that empathy does indeed often require cultivation; it is often a daunting challenge to take on the emotional load of another's situation, requiring significant investment and risk on our part. The habit of doing so is not guaranteed by intimacy itself; I may put up 'walls' to insulate myself from the pain of even someone I genuinely love. The habit of being empathetic is in fact one of the most difficult to cultivate, for it requires maintaining a delicate balancing act between openness to the affective life of others and preservation of adequate emotional resources for oneself. Given that empathy *is* indeed a cultivated disposition, and that according to Aristotle (as well as our ordinary moral intuitions) it satisfies the other element of the defining criterion for a social virtue, namely, that it tends to promote intimate friendship and thus the good life, it seems reasonable to conclude that empathy is in fact one of the virtues, and that Aristotle was simply mistaken in concluding otherwise.[6]

An important precondition for empathy in its traditional expression is access to the embodied presence of another's emotion. The ability to feel *with* others entails the capacity to discern and identify bodily with another's suffering, even suffering not caused by a physical injury or illness. New social media, however, raise legitimate concerns with regard to the cultivation of this capacity. Consider that as Danah Boyd notes, online communication can eliminate "visceral reactions that might have seeped out in everyday communication" (2008, p. 129). The editing out of some visceral reactions, such as anger or disgust, might aid communication by encouraging greater trust and openness. But what if empathy is built on visceral responses of its own?

Many of us know that the most immediate comfort to a grieving friend is often not words but a physical embrace—often prompted by a visceral reaction to the bodily display of her suffering. When a friend literally *feels* us grieving with them in such an embrace, one communicates a kind of empathy that no mediated condolence could facilitate. In many circumstances we would, on selfish grounds, *rather* have a physical buffer between ourselves and our friend's pain. But by relinquishing that buffer, one gains an experience that, if repeated, eventually deepens one's character as a friend. To what extent can this virtue also be encouraged by new social media? Or does an absence of embodied presence present an insuperable obstacle to the richest expressions of empathy? If so, will heavy use of such sites tend to impoverish empathy by providing safer and less demanding ways to engage the suffering of others?

Yet in posing these questions, we must also recognize the value of social networking sites that link people with serious and rare illnesses, victims of violent crime, or those suffering and alienated in other ways. Such sites can cultivate empathy by allowing otherwise isolated people to share their sufferings and joys with others in a unique position to understand them, a phenomenon that undeniably stands among the noblest social uses of new social media.

We must still inquire, however, to what extent new social media can create technological gradients toward empathy for those who do not find themselves in such circumstances, and who may as yet have little experience with the deepest sufferings (and joys) of humankind.

CONCLUDING REFLECTIONS

One theme has begun to emerge from these analyses. Most new social media are designed with the aim of making human communication easier, less risky, and more immediately satisfying. The motivations behind such design choices are no mystery. Yet virtues are typically developed as a consequence of performing actions that are, at least initially, *difficult*. As Aristotle noted, virtues result from actions that pull against tendencies that would otherwise lead us away from excellence (1984, 1109a23). This eventually becomes easier as right action becomes habitual, and as its repetition gradually engenders a reliable faculty of moral perception, an acute and intuitive sense of where the right action lies among the many alternatives. But these mature advantages are gained only through initial effort, effort that necessarily exposes us to risk, discomfort, and/or delay of gratification. Can new social media support our choices to take on such risks and challenges, or will they instead provide only an ever-widening array of easier alternatives?

Certainly, the virtues I have mentioned are promoted only by traditional forms of communication at their *best*. Even then, these traditional forms are not without their own inherent defects, and new social media clearly have the potential to address some of those limitations and move human communication forward, making social ties more rewarding, flexible, and enduring. They may even facilitate new communicative virtues not previously recognized or given their due. Yet such advances may be sporadic, transient, or outweighed by contravening effects if new social media are designed and driven by market pressures alone, with indifference to communicative virtues and their essential role in developing and sustaining human connections. Given that how we learn to communicate is ultimately how we learn to enjoy the good life with others, it is time for the designers, marketers, and users of new social media to engage in serious reflection on the importance of these virtues and to invest in the challenge of building on them.

ACKNOWLEDGMENTS

I wish to thank the participants in the 2008 Good Life in a Technological Age workshop at the University of Twente, as well as the editors of this volume for their many insightful comments and critiques. I owe particular gratitude to Charles Ess, whose guidance and encouragement has been invaluable. This project was supported by the Markkula Center for Applied Ethics at Santa Clara University through the generosity of a Hackworth Grant. Finally, I must thank Erin Berkenmeier for her laudable efforts as my research assistant on this project.

NOTES

1. An earlier, abbreviated version of this chapter appeared in 2010 as "Social Networking Technology and the Virtues," *Ethics and Information Technology*, vol. 12, no. 2, pp. 157–170.
2. 'Flourishing' in this sense presupposes moral success, excluding the possibility of, say, a thief who 'flourishes' in her environment by being an 'excellent'

thief. One who flourishes, lives in a manner that would invite the general admiration and moral approval of other members of the community.

3. *Eudaimonia* entails a far more active, holistic, and externalized kind of well-being than is implied by the subjective-emotive orientation that informs most contemporary uses of near-equivalent terms, such as 'happiness,' 'fulfillment,' 'well-being,' and 'satisfaction.'
4. There are, of course, moral virtues not enacted primarily in communicative action (e.g., temperance). There are also nonmoral virtues ('skills') that contribute significantly to individual flourishing. Notably, Ito and colleagues (2009) stress the value of social media for helping youth acquire technological competence and cultivate creativity through content-generation.
5. This invites consideration of Turkle's (1995) reflections on identity play and the self. Here, however, I presuppose the notion of an authentic self, multifaceted and evolving but rooted in one's personal history and actual dispositions/capacities.
6. This judgment is reinforced by the recognition of empathy as a virtue by comparably august figures in the virtue tradition, including David Hume (although Hume uses the term 'sympathy,' Michael Slote [2007] notes that 'empathy' appears to be closer to his meaning) and Confucius (whose term '*shu*' can best be translated as 'empathy').

REFERENCES

Aristotle. (1984). "Nicomachean Ethics." In *The Complete Works of Aristotle*, edited by Jonathan Barnes. Princeton, NJ: Princeton University Press.

Boyd, Danah. (2008). "Why Youth (heart) Social Network Sites: The Role of Networked Publics in Teenage Social Life." In *Youth, Identity and Digital Media*, edited by David Buckingham (Cambridge, MA: MIT Press) pp. 119–142.

Derbyshire, David. (2009, February 24.). "Social Websites Harm Children's Brains: Chilling Warning to Parents from Top Neuroscientist," *Daily Mail* (UK).

Ellison, Nicole B., Charles Steinfeld, and Cliff Lampe. (2007). "The Benefits of Facebook 'Friends': Social Capital and College Students' Use of Online Social Network Sites," *Journal of Computer-Mediated Communication*, vol. 12, no. 4, article 1. Retrieved from http://jcmc.indiana.edu/vol12/issue4/ellison.html

Franzen, Axel. (2003). "Social Capital and the Internet: Evidence from Swiss Panel Data," *Kyklos*, vol. 56, no. 3, pp. 341–360.

Ito, Mizuko, Heather A. Horst, Bittanti Matteo, Danah Boyd, Becky Herr-Stephenson, Patricia G. Lange, C. J. Pascoe, and Laura Robinson. (2009). *Living and Learning with New Media: Summary of Findings from the Digital Youth Project*. Cambridge, MA: MIT Press.

Ko, Hsiu-Chia, and Feng-Yang Kuo. (2009). "Can Blogging Enhance Subjective Well-being Through Self-disclosure?" *CyberPsychology and Behavior*, vol. 12, no. 1, pp. 75–79.

Lenhart, Amanda, Mary Madden, Aaron Smith, and Alexandra Macgill. (2007). *Teens and Social Media*. Pew Internet and American Life Project. Retrieved from http://www.pewinternet.org/Reports/2007/Teens-and-Social-Media.aspx

Sigman, Aric. (2009). "Well Connected? The Biological Implications of 'Social Networking,'" *Biologist*, vol. 56, no. 1, pp. 14–21.

Slote, Michael. (2007). "Famine, Affluence and Virtue." In *Working Virtue*, edited by Walker, Rebecca L. and P. J. Ivanhoe (Oxford: Clarendon Press), pp. 279–296.

Turkle, Sherry. (1995). *Life on the Screen: Identity in the Age of the Internet.* New York: Simon and Schuster.

Wellman, Barry, Anabel Quan Haase, James Witte, and Keith Hampton. (2001). "Does the Internet Increase, Decrease or Supplement Social Capital? Social Networks, Participation, and Community Commitment," *American Behavioral Scientist*, vol. 45, no. 3, pp. 436–455

13 Web 2.0

Community as Commodity?

Diane P. Michelfelder

The emergence of the so-called "Web 2.0" or "social Web," on which users generate and share content through activities such as blogging, media sharing, social networking, tagging, tweeting, virtual world gaming, wiki developing, and the like, has spawned a number of philosophical questions in its wake. The one I will take up in this chapter has a broad scope: How does everyday experience with the social Web help to shape individual well-being and the quality of life? According to the founder of Facebook, as of July 2010, over 500 million people were active users of that social networking site, a figure that leapt to 750 million in 2011 (Zuckerberg 2010; Protalinski 2011). Around the world, people sent nearly 50 million tweets per day in January 2010 as compared to just 5,000 a day in 2007 (Weil 2010). In 2007, there were half a million photos uploaded to Flickr (Kullin 2010); by September 2010, the number had risen to 5 billion (Sheppard 2010). Based on statistics such as these, social websites can be said to be thriving. To what extent do they also contribute to human flourishing—as Aristotle would put it, to living well and doing well?

Given the broad scope of this question, a refinement is immediately in order. In what follows, I will look at this question through the lens of the distinction between community and commodity. I have deliberately selected this distinction for one particular reason. At a time when the social Web was barely in its infancy, in two separate articles published in the same year, communications researcher Darin Barney and philosopher Albert Borgmann used it as a frame for taking a critical look at the relationship between Internet use and human flourishing (Barney 2004b; Borgmann 2004).[1] Drawing on his analysis of the pioneering online community The Well, Borgmann argued that the moral and social benefits to individual flourishing made possible by communities embedded in physical space, benefits such as mutual support, shared affection, and assistance in identity-formation, could not be realized by communities whose presence was exclusively online.[2] Similarly, although taking "surfing the Net" to be the primary way individuals interacted with online communication technologies, Barney concluded that by virtue of their very "make-up" such technologies inherently "conspire more with commodity than community"

(Barney 2004b, p. 32), an observation that Borgmann would have had little difficulty endorsing.[3]

Such views, however, rest on the assumption that a bright line can be drawn between geographically bounded and Web-based communities, an assumption that is itself in turn rooted in the idea that a firm distinction exists between the virtual and the real. In 2004, these assumptions may have been plausible. At this point in time, they are no longer sustainable.[4] What Barney said of the Web, that it "undermines our inhabitation in a common world of things," a world that is the precondition for life lived in community with one another, has lost applicability less than a decade later (Barney 2004b, p. 48). Far from being undermined, with social media websites increasingly depending on visual images of real things, people, and places for their content, for example, with Facebook expanding its features to include Skype, one might wonder if such inhabitation is made much more meaningful rather than less.

Should we then take the social Web as having the opposite influence from what Barney took it to have? Does it in fact "conspire" more with community then commodity and, in so doing, make a positive contribution to human flourishing? My response to this question: yes under some conditions and no under others, may seem paradoxical and counter-intuitive. Suppose we look at a social utility such as Facebook. On the surface, it might seem to be a model case of how the social Web conspires more with community than commodity. I propose it does the reverse, and so can be pointed to in support of Barney's original claim. On the other hand, at least some online communities who owe their very existence to particular commodities with which they are connected, such as MMOGs (massively multiplayer online games), can arguably be seen in the opposite light. In conspiring more with community than commodity, they serve to build social virtues such as trust and to make a positive contribution to identity formation and individual well-being.

In turning now to defend these two claims in the next two sections of this chapter, my intent is not to settle debate over the community versus commodity question. It is simply to provoke discussion and to show how any answer to this question may be more complex than one might initially think, a result which in turn has implications for future research into questions about the ethical impacts of the Web.

Before going on, though, a couple of definitions are needed. In talking about 'commodity,' I will be keeping Borgmann's well-known definition in mind: a commodity is a good that is "instantaneous" in its availability as well as "ubiquitous, easy, and safe" (Borgmann 2000, p. 421) and obtainable in a context involving marketing and consumer relations. And, I will take it that individuals form a "community" when, based on shared interests or values, they connect with one another to further develop these interests and values and, consequently, develop their own self-identity, through participating in practices specific to the group.

Community in this sense could be said to be somewhat analogous to Aristotle's notion of friendship, which he even identifies at one point in the *Nicomachean Ethics* as being a kind of community (Aristotle, 1171b32). As just noted, it could be formed among those who spend time conversing with one another in the pursuit of a mutual interest or interests. They would see this time as pleasantly spent and contributing to the formation of self-identity. And, they could, as a result of spending time together, grow and develop as individuals by becoming more accomplished in the pursuit of the particular shared interest or interests around which the community is gathered. By using 'community' in this way, I am deliberately trying to steer a course between setting the bar for what counts as a community so high that social Web communities would *ipso facto* have difficulty being included, and setting it so low that nearly any association of individuals would be ruled in.[5]

SELF-IDENTITY AND SELF-COMMODIFICATION

Borgmann's argument that online communities work to the detriment of the goods genuine communities can provide rests, as previously mentioned, on a sharp split between the "online" and "off-line" worlds. Participation in the former, distanced from the tangible realities of place, other people, and things, requires that individuals live "on screen" (Borgmann 2004, p. 63). To live well on screen demands the creation of a persona or an artificial "virtual self." A virtual self shares in the attributes of a commodity: it is a self that is immediately, constantly, and readily available to others. Much like any other commodity, a virtual self can be readily "taken off the market" at will, with little loss to one's actual self. Hence, Borgmann concludes that participation in online communities works to the advantage of commodity rather than community, as participation in these communities depends on a process of self-commodification.

I agree with Borgmann that when self-commodification plays a role in online interactions with others it serves to undermine community, but I take a different view of what it means to commodify oneself in the context of the social Web. Rather than seeing this activity as involving the construction and presentation of a "virtual" self that can be abandoned at any point in time, I take it to be the all-too-factual presentation of oneself to particular online communities, to advertisers and data collectors, and often even to the public at large. In willingly and knowingly abandoning their privacy, understood as control over personal data, individual members of online communities and social networking sites act in ways that make a significant contribution to the commodification of the self. Although I take the voluntary giving up of control over personal information to play a relevant role in the process of self-commodification, I am not claiming here that abandoning privacy in this sense is a necessary condition for community.

Rather, through the voluntary sharing of personal information that leads to self-commodification within the setting of a social network, what is put at risk is the ability to obtain one of the key affordances of community: namely, the possibility of increasing the understanding one might have of who one is and of developing one's sense of self-identity.

Let me now turn to Facebook to illustrate this point.[6] Someone might well question this choice on the basis that Facebook *prima facie* falls short of being a community, given the variety of reasons why people choose to join this site. Facebook is, though, commonly and informally spoken of as being a world-wide community. It also serves as a platform for a host of individual Facebook groups, which are explicitly represented as communities and whose aims are to create community, for example, Facebook pages created by universities for incoming classes of students. Rather than saying that Facebook *prima facie* falls short of being a community, it seems more reasonable to say that it *prima facie* fits under the "weak" definition of community described above.

In thinking about ways in which Facebook is linked to commodification, what for many comes most immediately to mind is the way this site (much like many social networking websites) matches its members with marketers, so that advertising targeted to a particular member's presumed preferences and interests will appear on that individual's Facebook page. On occasion, these efforts backfire, as witnessed by Facebook's Beacon project, an initiative in which Facebook members' purchasing information on their advertising partners Websites was mined and passed on to the Facebook friends of those who made the purchases without their knowledge or permission.[7] In general, however, Facebook has been able to develop its strategies for connecting users with advertisers without much controversy, defending, on the official Facebook blog, investing in these relations as a way of making the user experience more personalized and meaningful:

> We believe that more personalized social advertising complements the ways that people use Facebook every day—to discover, share and connect with the people and the world around them. Whether it's a new car, a local dentist, or a t-shirt, many of the things you discover on Facebook come from your friends. Just as advertising has changed the experience of searching on the web, we think advertisers are a meaningful part of the social graph and the online experience as a whole. (Sandberg, 2010)

Although fostering the consumptive habits of its members is perhaps the most obvious way in which commodification is reflected in Facebook, it is not by itself a philosophically significant reason for thinking that social networking sites conspire more with commodity than community. In developing "personalized social advertising," Facebook necessarily relies on demographic information provided by its members: the more such information it has, the more it can provide to the creators of targeted advertising with whom it has partnered. This need provides the framework for how Facebook pages are designed, as it is by incorporating such features as "status updates" and "liking" into this design that its members are encouraged to provide a steady stream of information about the very latest concerning whatever they happen to be doing, thinking about, or feeling.[8] In this way, Facebook promotes the self-commodification of its members, as it is through "liking" and providing "status updates" that Facebook members shape their availability to others in ways that reflect the characteristic features of a commodity.

In other words, to join Facebook is arguably to enter into a process of self-commodification distinguished by features such as the continuous availability of information about oneself, as well as ease of access to that information, aided by another one of Facebook's features, News Feed. Other ways of connecting with others on the social web that do not have the character of social utilities also have elements that connect to self-commodification. Google's Gmail, for example, has an "availability for chat" feature, which lets others know whether or not one is online. As anecdotal evidence suggests that those who leave on this feature feel some sense of moral obligation to let others know when, for whatever reason, they are unavailable to chat, availability becomes the default mode. But Facebook combines a number of features that systematically work in favor of getting people to commodify themselves. If in the further development of Facebook "passive applications" are also incorporated into this combination (Protalinski 2011) that would take up the "burden" of an individual's needing to do something to share information, it would work in the direction of strengthening how Facebook "conspires" more with commodity than community.

Readers reaching this point might easily wonder whether it is possible to argue that Facebook encourages self-commodification without going so far as to draw the conclusion that it "conspires" more with commodity than community. In tracing back the success of Facebook's status update feature to its philosophical roots in David Hume's discussion of natural sympathy, Abrol Fairweather and Jodi Halpern have proposed that status updates help individuals to extend their sympathy to a more diverse group of individuals than they would in the "offline" world (Fairweather and Halpern 2009, p. 197). A 2011 report by the Pew Research Center's Internet and American Life project indicated that Facebook users who visit the site more than once a day are significantly more likely to believe that other people could be trusted than those who use the Internet but not Facebook and those who do not use the Internet at all (Rainie, Purcell, Goulet, and Hampton 2011). More generally, sharing information with others and so creating and strengthening relational bonds helps to sustain individual well-being. Because the formation of communities depends on people who are sympathetic toward one another, trustful, and willing to share, so this argument might run, Facebook makes an important contribution toward building community despite having a structure that encourages self-commodification of its users.

The question here, though, is not whether Facebook makes no or little meaningful contribution to the formation of community and so to the positive goods that community can offer. Even while encouraging community in ways such as the ones just mentioned, Facebook does more by way of self-commodification. Tipping the scales in the direction of conspiring more with commodity than community is the fact that by supporting self-commodification, Facebook works to undermine ways that it could serve as a context within which individuals could actively engage in shaping their own identities. The discovery and shaping of one's identity requires more than sharing information with others. It demands both an openness to change and a willingness to be challenged, as well as an ability to look forward to the future and to take risks that initially could be quite disorienting. The self-commodification that is involved in being a friend to others on Facebook, however, favors a presentation of the self that emphasizes the immediacy of the present and the status quo of one's interests and values over their development and the uncertainties to which such development can lead.

A status update, for instance, offers a small piece of information about someone's present experiences that can be immediately taken in or consumed (through clicking on "like," for example) without the effort of reflection. To put this point a little differently, the abundance of weight given to the present moment in its character of immediacy on Facebook works to the detriment of gaining new perspectives and new understandings of who one is, and likewise to the detriment of thoughtfully considering rather than simply reacting to the postings of others. In response to Fairweather's and Halpern's conclusion that status updates, although not necessarily having moral significance in themselves, can open "new windows for the moral imagination to wander through" (2009, p. 198), one might well wonder whether the opposite is more likely the case. Where these new windows do open up may be in unexpected locations, as the following section will attempt to show.

SELF-IDENTITY, COMMUNITY, AND ONLINE GAMING

In addition to sites such as Facebook that are primarily designed to facilitate social interaction, the social Web also includes sites that, although not primarily designed to be social networks themselves, include and are dependent on elements of social networking. Among these websites are massively multiplayer online games, or MMOGs, such as World of Warcraft (WoW). In WoW, individual players can join forces with one another within the context of guilds to better their opportunities for advancing in the game, and can participate in multiple discussion forums devoted to the game. In this section, I want to turn to a discussion of MMOGs and WoW in particular for the purpose of making a case for the second half of the argument described at the beginning of this chapter: namely, that some sites on the social Web, which on the surface appear to be deeply rooted in practices associated with commodification and consumption, can be said to conspire more with community than commodity. In this discussion, attention to the place in these sites afforded to identity-formation, through the development of a player's moral imagination and moral-reasoning abilities, will play a key role.

At first glance, the idea that an online game such as WoW could be said to conspire more with community than commodity would not seem to be particularly viable. WoW is not only the most popular online game, it also grosses the largest amount of revenue, primarily through subscriptions. The design of WoW (for example, moving from level to level) invites its subscribers to invest a great deal of time in making virtual money (gold) themselves, as it is necessary for purchasing equipment such as weapons, as well as the consumables that are part of the game itself.[9] More important, because WoW is fundamentally a virtual world (Azeroth) in which the Alliance is pitted against the Horde, any player new to the game needs to begin by choosing an avatar belonging to one side or the other. The fact that one needs an avatar or online persona in order to play WoW would seem to support Borgmann's view of the relationship between virtual worlds, virtual selves, and commodification. One's avatar identity can even extend beyond the boundaries of the game itself, as it is also possible to keep this identity when participating in WoW discussion forums affiliated with Blizzard Entertainment, the company responsible for WoW's production. After the company moved in 2010 to make forum participants post messages using their real names, the pushback—focused on concern over exposing one's actual self to the world at large—was such that it forced Blizzard to withdraw the requirement. And, adding to the notion that players of WoW, much like participants in Facebook, might be engaged in a process of self-commodification is the fact that through the use of special software, it is possible to have one's avatar play WoW while one's "real self" is elsewhere engaged.

At the same time, although WoW certainly reflects features supportive of self-commodification in its character as a game, it likewise reflects the character of a community in its multiplayer dimensions. Adam Ruch observes that:

> While Blizzard unquestionably provide a great deal of 'content' in the form of a world to explore and monsters to battle, this is not the sole asset of the game. That is the sole value of a typical single-player game, but an MMO provides a new value: social interactions. WoW is marketed as an MMO, it *is* an MMO, and the average consumer (even those who don't play themselves) have a vague understanding that this means it is a multiplayer game. . . . The fact that there are players

> playing WoW makes it a more attractive game to play. The way the game is devised to be spatial enables players to join each other 'there.' That is, in a space, to meet up, to explore together, to fight together, to fight against each other, to run away from each other, etc. The persistent nature of the world allows the kind of long-lasting relationships needed to succeed in the end-game adventures, to form. (2009)

It is possible to extend the concept of "exploring together" to exploring what to do when needing to make at least three types of moral decisions arising in the course of playing WoW. One type of decision would be what to do when confronted with activity on the part of other players that is generally considered within the culture of the game to be cheating or otherwise offensive behavior, such as griefing (see Warner and Raiter 2005, p. 48). Another would be whether or not to engage in such activity oneself. A third would be to engage in behavior that is explicitly contrary to the rules of the game as defined by Blizzard but condoned by many other players.

The need to make these types of decisions within the context of WoW presents moral challenges to a player, especially to a newcomer to the game. In facing such moral challenges, players are put in positions where they must take the sorts of risks and look at matters from new perspectives that contribute to the development of their self-identity as moral agents. Writing specifically about the negative reaction to a design feature within WoW introduced by Blizzard, which many players perceived as encouraging players to engage in unethical action,[10] Miguel Sicart notes that the example shows that:

> . . . players act as moral beings, that they reflect upon those values that are contained in the system of the game, and that they evaluate them keeping in mind the values of the game world, of the player's community, and ultimately cultural values. (2005, p. 17)

Sicart's point can be interpreted as suggesting that because players are interested in having their gameplay be "good" in the sense of being pleasant and enjoyable, they will press to have values explicitly represented in the game that will uphold the game's "well-being" (2005, p. 17). We could add to this that when players view particular rules of WoW (or its terms of use, see Ruch 2009) as being inessential to the game's "well-being" and to the pleasantness of their playing experience, they may very well decide to ignore them without necessarily framing their behavior as immoral. In either event, a player will reflect on what counts as responsible gameplay, and so develop their own sense of moral responsibility, and hence self-identity, accordingly.

One might react to the conclusion just drawn by questioning whether a player who acts responsibly in the sense just mentioned within the context of WoW can really be said to be exploring and enhancing his or her more general sense of moral responsibility. After all, one might say (particularly someone who wholeheartedly accepts Johan Huizinga's belief that games formed "magic circles" within but separate from the larger context of life) that it is not really you that is playing WoW, but rather your virtual character. As a result, no implications follow regarding the development of your own moral imagination. And, consequently, no implications follow regarding your own sense of identity, as a result of your addressing a moral dilemma confronting your virtual character who finds itself in a situation of great uncertainty, in the presence of many other virtual characters, in the context of WoW.

Such a view, predicated on there being a distance between fictional and nonfictional worlds, suffers from oversimplification. Going back once more to Ruch, we find the following observation:

> So, these are *real* people in a mediated environment, interacting with each other, through their utterances, actions, thoughts and emotions. Their project of self-creation (either a deliberate project in a role-playing sense, or a subconscious project of simply trying to maintain a stable presence) in this generally unfamiliar but exciting venue occurs largely through their interaction with each other, which is in turn created by utterances between characters, actions performed in relation to other characters, and thoughts and emotions arising from these interactions. (2009)

If we accept that what goes on in WoW is not simply a matter of avatars or virtual characters interacting with one another but rather human beings interacting with one another through the mediation of virtual characters, it might then be reasonable to draw the conclusion that in making decisions of an ethical nature within WoW, we are not simply making decisions that impact the moral dimensions of our virtual characters, but our own character *simpliciter* (see also Spence 2008). In offering affordances that relate positively to the development of one's own character, WoW and its affiliated discussion forums can be seen as lending themselves more to the formation of community than commodity.

CONCLUSION

In the course of this chapter, I have been concerned with making and briefly defending three points. First, the ethical question as to whether the Web in its current stage of development as the so-called Web 2.0 or "social" Web does or does not further the growth of community over commodity is a complex one to which there is no single, yes or no answer. Second, at least some components of the social Web that at first glance might be thought of as "conspiring" more with the formation of community than commodity (or more to the formation of commodity than community) may actually be doing the opposite. Third, a key element in determining the direction of this "conspiracy" has to do with the capacity for developing one's sense of self-identity. Overall, as mentioned earlier, my intent in this chapter is not to lay any matters to rest—save perhaps the idea that it is good to exercise caution in taking "the social Web" to be an entity about which it is possible to make broad-sweeping philosophical claims—but rather to generate fruitful ongoing discussion. To that end, I hope this is a useful contribution.

NOTES

1. Of the three websites mentioned in the first paragraph, only Facebook existed in 2004, with participants restricted to members of the Harvard community.
2. Borgmann based his argument on two contentions: first, that creators behind online communities inevitably need to turn to the corporate world (and so to the world of marketing and consumption) for financial support; and second, that individuals who participate in these communities, free from the constraints of time and space, turn themselves into commodities by presenting themselves not as who they are but as artificial selves or personae (Borgmann 2004, p. 64).
3. Following Borgmann's distinction between things and devices, Barney claims online communication technologies are devices, as they lighten the efforts required of us to obtain information or to communicate with one another to the point of "disburdenment": a salient characteristic of devices, and in particular of the commodity aspect of devices, as Borgmann understands them to be (Borgmann 1992, 1984).
4. Jonny Hartz Søraker makes a similar point in chapter 15 of this volume: Virtually Good: Disclosing the Presuppositions Behind the Claimed Inferiority of Virtual Worlds.
5. Borgmann, I believe, is among those who set an ambitious bar for what counts as a community. He does not set it nearly as high as say, Alasdair MacIntyre sets it for political communities, which as he sees them cannot thrive without "the shared possession of a rationally justifiable conception of human good" (MacIntyre 1994, p. 8). Still, Borgmann nevertheless maintains that genuine communities contribute to one's sense of identity, to what makes life pleasant to live, and (most important for this discussion) serve as settings in which "one finds or works out one's reason for living" (Borgmann 2004b, p. 63). Although Barney intentionally avoids proposing a definition of "community" in order not to divert his reader's attention to the debate over community between liberal thinkers and communitarians (Barney 2004b, p. 49, footnote 4), the overall thrust of his argument suggests he would be in accord with Borgmann's point of view. The sense of community I am proposing in this chapter could be construed as a "weak" sense of community as compared to the "strong" sense found in Borgmann: as nothing rides on whether its members share a common geographical location, relate to one another in person in the context of a setting of tangible things, or find a community to be the convivium within which life is worth living.
6. Other social networking websites could be substituted for Facebook in this analysis, such as MySpace and Twitter. Google's new social media site, Google+, appeared too late for consideration for this chapter.
7. It took the combined efforts of bloggers, thousands of online protesters, plus an article in the *New York Times* to get Facebook to alter its policies to change participation in Beacon to be opt-in.

8. The sheer amount of information shared on Facebook is impressive: four billion "things" a day, according to Mark Zuckerberg in an announcement in 2011. Just as critical to the analysis here is his observation that the average Facebook user shares twice as much information in 2011 as in the year before, and that such sharing would continue to rise at an exponential rate (Protalinski 2011).
9. See the "Guide to Making Money" on WoWWiki; http://www.wowwiki.com/Guide_to_making_money
10. The design feature, Honor, rewarded players for killing other players in one-on-one situations, but did not offer any disincentives for killing in dishonorable ways.

REFERENCES

Aristotle. (2002). *Nicomachean Ethics* (Christopher Rowe, trans., Introduction and Commentary by Sarah Broadie). Oxford: Oxford University Press.

Barney, Darin. (2000). *Prometheus Wired: The Hope for Democracy in the Age of Network Technology*. Sydney: University of New South Wales Press.

Barney, Darin. (2004a). *The Network Society*. Cambridge: Polity Press.

Barney, Darin. (2004b). "The Vanishing Table: Or Community in a World That Is No World." In *Community in the Digital Age: Philosophy and Practice,* edited by Feenberg, Andrew and Darin Barney (Lanham, MD: Roman and Littlefield) pp. 31–52.

Benhabib, Seyla. (1992). *Situating the Self: Gender, Community, and Postmodernism in Contemporary Ethics*. New York: Routledge.

Bloor, Chris. (2009). "What Are Friends For?" In *Facebook and Philosophy,* edited by D. E. Wittkower (Chicago: Open Court Press) pp. 147–157.

Borgmann, Albert. (1984). *Technology and the Character of Contemporary Life*. Chicago: University of Chicago Press.

Borgmann, Albert. (1992). *Crossing the Postmodern Divide*. Chicago: University of Chicago Press.

Borgmann, Albert. (2000). "The Moral Complexion of Consumption," *The Journal of Consumer Research*, vol. 26, pp. 418–422.

Borgmann, Albert. (2004). "Is the Internet the Solution to the Problem of Community?" In *Community in the Digital Age: Philosophy and Practice,* edited by Feenberg, Andrew and Darin Barney (Lanham, MD: Roman and Littlefield) pp. 53–67.

Dreyfus, Hubert. (2009). *On the Internet*. New York: Routledge.

Fairweather, Abrol, and Jodi Halpern. (2009). "Do Status Updates Have Any Value?' In *Facebook and Philosophy,* edited by D. E. Wittkower (Chicago: Open Court Press) pp. 191–199.

Kullin, Hans. (2010, September 18). "Flickr Reaches Five Billion Photos," Mea Culpa. Retrieved from http://www.kullin.net/2010/09/flickr-5-billion-photos/

MacIntyre, Alasdair. (1994). "The Privatization of Good: An Inaugural Lecture." In *The Liberalism-Communitarianism Debate,* edited by C. F. Delaney (Lanham, MD: Roman and Littlefield) pp. 1–17.

Newman, Jarad. (2010 December). Posts on World of Warcraft from the Technologizer, July 7, 9 and 15. Retrieved from http://technologizer.com/tag/world-of-warcraft/

Protalinski, Emil. (2011, July 11). "Zuckerberg: 4 Billion "Things" Are Shared on Facebook Every Day," Friending Facebook blog. Retrieved from http://www.zdnet.com/blog/facebook/zuckerberg-4-billion-8220things-8221-are-shared-on-facebook-every-day/2020

Ranie, Lee, Kristen Purcell, Lauren Sessions Goulet, and Keith N. Hampton. (2011, June 16). "Social Networking Sites and our Lives," *Pew Research Center Publications*. Retrieved from http://pewresearch.org/pubs/2025/social-impact-social-networking-sites-technology-facebook-twitter-linkedin-myspace

Ruch, Adam. (2009, November). "World of Warcraft: Service or Space?" *Game Studies*, vol. 9, no. 2. Retrieved from http://gamestudies.org/0902/articles/ruch

Sandberg, Sheryl. (2010, July 6). "The Role of Advertising on Facebook," The Facebook blog. Retrieved from http://blog.facebook.com/blog.php?q=the+role+of+advertising+on+facebook

Sandel, Michael. (1998). *Liberalism and the Limits of Justice*. Cambridge: Cambridge University Press.

Sheppard, Zack. (2010, September 19). "5,000,000,000," Flickr Blog. Retrieved from http://blog.flickr.net/en/2010/09/19/5000000000/

Sicart, Miguel. (December 2005). "Game, Player, Ethics: A Virtue Ethics Approach to Computer Games," *International Review of Information Ethics*, vol. 4, pp. 13–18.

Spence, Edward. (2008). "Metaethics for the Metaverse: The Ethics of Virtual Worlds," *Current Issues in Computing and Philosophy*, edited by Adam Briggle, Philip Brey, and Katinka Waelbers. Amsterdam: IOS Press.

Warner, Dorothy E., and Mike Raiter. (December 2005). "Social Context in Massively-Multiplayer Online Games (MMOGs): Ethical Questions in Shared Space," *International Review of Information Ethics*, vol. 4, pp. 46–52.

Weil, Kevin. (2010, February 22). "Measuring Tweets," Twitter blog. Retrieved from http://blog.twitter.com/2010/02/measuring-tweets.html

Zuckerberg, Mark. (2010, July 21). "500 Stories," The Facebook blog. Retrieved from http://blog.facebook.com/blog.php?post_start=10

14 Types of Internet Use, Well-Being, and the Good Life

Ethical Views from Prudential Psychology

Omar Rosas

Over the last decade, psychological research has made it evident that Internet-related activities have considerable effects on people's mental and social lives. By considering "Internet use" as a one-dimensional construct, early studies posited that Internet use has paradoxical consequences for people's well-being. Although Internet use had been thought to enhance communication, information seeking, and computer-based competence and thus was able to foster social involvement and support, it turned out to be a way of leaving people isolated, stressed, and depressed (Kraut et al. 1998). Although follow-up studies showed that the initial effects dissipated over time (Kraut et al. 2002), further research corroborated the recurrent presence of some of those problematic consequences (Ybarra, Alexander, and Mitchell 2005). This negative impact of Internet use on people's psychological well-being was contrasted, though, with more positive appraisals, particularly those showing that Internet-based activities also provide personal gratifications or positive outcomes in relation to fun, entertainment, social interaction, and relaxing activities pursued by most Internet users (LaRose, Mastro, and Eastin 2001).

Today, several researchers are both questioning the theoretical validity and empirical testability of "Internet use" as a one-dimensional construct and are endeavoring to isolate different types of Internet use to examine psychological consequences of specific Internet-based practices (Morgan and Cotton 2003; Valkenburg, Peter, and Schouten 2006). The rationale behind this theoretical and empirical move is that the Internet, considered as an information and communication tool, is neither entirely good nor entirely bad with regard to the effects it is able to produce. This implies a shift in focus away from taking Internet use as a unified category and reconsidering the Internet as a multidimensional space providing different kinds of e-services and e-contents that may have discrete and measurable effects on users' well-being.

Yet the state of the art is still Janus-faced. On the one hand, several scholars—including developmental, social, and life span psychologists—argue that multiple benefits can be derived from Internet experiences, including a sense of self-efficacy (Whitty and McLaughlin 2007), improvements in learning (Jackson et al. 2006), increasing social ties (Karavidas, Lim, and Katsikas 2005; Zhao 2006), and a feeling of mastering one's environment (Amichai-Hamburger and Furnham 2007). On the other hand, social psychologists and psychiatrists claim that Internet-related practices may foster a number of negative effects, including loneliness (Sum et al. 2008), problematic (e.g., "addictive") use (Caplan 2003, 2007), asking and sharing information related to self-injurious behavior (Whitlock, Powers, and Eckenrode 2006), impoverished personal and family relationships (Mesch 2006), and lower psychological and physical well-being (Yellowlees and Marks 2007), among others. Given the variety and complexity of Internet-related activities, it seems manifest that different modes of Internet use have the potential to yield both psychological harms and benefits to users.

In light of these psychological studies, one basic question can be raised: What are the implications of psychological research on Internet use for philosophical reflection on the good life in a technological age? To answer this question, this chapter will be structured as follows. The next section will introduce what Haybron (2008a and 2008b) has recently labeled *prudential psychology*, that is, psychological research concerned with matters of well-being. The trends belonging to prudential psychology will be briefly introduced to identify conceptual interfaces between them and their philosophical counterparts. The following section will briefly report some case studies in which a particular trend in prudential psychology has been applied to different types of Internet use. This section succinctly highlights the nature of well-being measures and the scope of their empirical results. The final, section will draw on the previous section to underline some implications of prudential psychology analyses of Internet use and well-being for philosophical reflection on the good life.

PRUDENTIAL PSYCHOLOGY, WELL-BEING, AND THE GOOD LIFE

After more or less 50 years of psychological work concerned with several aspects of mental illness, the new field of *positive psychology* emerged as a philosophically inspired research program into the positive aspects of people's affective lives, personality traits, and social institutions. As Seligman (2003, p. xi) has put it, "The time has finally arrived for a science that seeks to understand positive emotion, build strength and virtue, and provide guideposts for finding what Aristotle called the 'good life.'" This "positive turn" poses the challenge of suitably articulating contemporary psychological research on positive emotion, strength, well-being, and virtue in terms of philosophical reflection on the good life.

To take up the challenge, contemporary philosophers and psychologists have argued for the need to drop their rigid disciplinary blinders and engage in interdisciplinary work to elucidate complementary dimensions of well-being. While recognizing that each discipline has its own methodological premises and explanatory aims, these scholars have claimed that ethical and psychological research on well-being can benefit from each other in at least two ways: (1) psychological research can raise problems or lend support and credibility to ethical accounts of well-being insofar as the latter's formulation and/or application implies empirical claims (Haybron 2007; Tiberius 2006, 2008), and (2) ethical analysis can inform psychological approaches to well-being and even shape the kind of empirical research it makes sense to do (Haybron 2008a).

In this spirit, Haybron (2008a, 2008b) has coined the term *prudential psychology* to categorize the field of positive psychology that both parallels the related field of moral psychology and, following the philosophical usage of "prudential," denotes matters of well-being. Yet a note on terminology is needed here. Although the concept of "well-being" is sometimes used as a synonym for "the good life," the latter usually implies a broader notion, including not just a morally good but a choice-worthy life, a life that is good, all things considered. I will here follow Haybron (2008b, p. 36) when he argues that the good life "functions as an umbrella concept encompassing the domain of values that matter in a person's life, and can be employed within any ethical framework." This implies that whatever the philosophical approach implemented, we can reasonably argue that well-being belongs to the aforementioned domain of values and, consequently, any improvement in a person's well-being can be seen as a suitable component of that person's good life.

Prudential psychology can be analyzed in terms of subjective and objective evaluative stances. Subjective approaches are principally represented by two trends. The first is *subjective well-being*, which is often called "hedonic" well-being and concerns the scientific analysis of how people evaluate their lives on the following dimensions: overall life satisfaction, satisfaction with important life domains, positive affect, and low levels of negative affect (Diener, Oishi, and Lucas 2003). The second is *psychosocial well-being*, which consists of a cluster of views according to which well-being refers to assessments of intrasubjective dimensions, such as depression, loneliness, overall life satisfaction, self-anchoring, self-esteem, hardiness, interpersonal relations, and perceived control (O'Hare et al. 2003; Brummett et al. 2007).

Objective approaches, which draw on an Aristotelian view of *eudaimonia* defined as leading a well-lived and flourishing life, principally comprise three trends: (1) *psychological well-being,* which is the view that well-being, understood as human self-fulfillment, concerns an individual's striving to realize his or her own true potential. To evaluate such self-fulfillment, researchers have posited six basic measurable dimensions: self-acceptance, purpose in life, environmental mastery, positive relationships, personal growth, and autonomy (Ryff and Singer 1998, 2008); (2)

social well-being is based on a five-dimensional model comprising: social integration, social contribution, social coherence, social actualization, and social acceptance as core components of well-being. The idea behind this approach is that human beings are essentially social creatures whose quality of social interactions has a significant bearing on both how an individual perceives herself as a social being and how she actualizes her inner potentials in a social environment (Keyes 1998). Finally, (3) *self-determination theory* advances a view of well-being that focuses on the process of living well rather than on the outcomes of a happy (i.e., hedonic) life. It posits that fulfillment of the psychological needs for competence, autonomy, and relatedness is of paramount importance for factors associated with well-being, such as psychological growth, integrity, and the experiences of vitality and self-congruence (Ryan and Deci 2001; Ryan, Huta, and Deci 2008).

Trends in prudential psychology share common interfaces with philosophical theories of the good life. Such interfaces include the evaluative stance being adopted (subjective/objective) and the theoretical tenets underlying specific perspectives (e.g., hedonism/positive emotion, eudaimonia/flourishing, perfectionism/realizing human potential, capabilities/satisfying basic psychological needs). Given its scientific aims, prudential psychology goes a step further by operationalizing and measuring constructs that are designed to tap into the cognitive, affective, and behavioral dimensions implied in well-being and the good life. Although such measures and the outcomes so obtained may seem controversial in the eyes of some theoretical researchers, one should recognize that they are helpful to (1) examine what people feel and think about their own lives, (2) critically assess the scope and limits of ongoing prudential psychology research, and (3) suggest alternative directions and methodologies for future interdisciplinary investigation into the nature of well-being and the good life.

USES OF THE INTERNET AND PRUDENTIAL PSYCHOLOGY

To examine how prudential psychology has been applied to issues in Internet use and well-being, a short selection of psychological studies implementing a particular trend will be briefly covered here. It should be noted, though, that given the huge amount of research published every year on this topic, this selection is just a snapshot of the current theoretical approaches, methodologies, and domain-specific results.

Internet Use and Subjective Well-Being

In an online survey study, Patti M. Valkenburg, Jochen Peter, and Alexander P. Schouten (2006) examined the consequences of friend networking sites for Dutch adolescents' well-being and self-esteem. Measures included use of friend networking sites, frequency of reactions to profiles, tone of reactions to profiles, relationships established online, social self-esteem, and the satisfaction with life scale (Diener, Emmons, Larsen, and Griffin 1985). The authors claimed that (1) Internet use has been treated in many studies as a one-dimensional construct, which has yielded inconsistent findings, and (2) at least a distinction between social and nonsocial Internet use is required to adequately investigate Internet effects on self-esteem and well-being. In their view, social self-esteem and well-being are more likely to be affected if the Internet is used for communication than for information seeking. The study demonstrated that adolescents' self-esteem was solely affected by the tone of the feedback that they received on their profiles. Positive feedback enhanced adolescents' well-being and negative feedback decreased their well-being.

Internet Use and Psychosocial Well-Being

Scott E. Caplan (2003) examined the relationships between *problematic internet use* and loneliness and depression in young American undergraduates. Problematic internet use was conceptualized as comprising those maladaptive cognitions and behaviors involving Internet use that result in negative academic, professional, and social consequences. The author measured and analyzed mood alteration, perceived social benefits, perceived social control, withdrawal compulsivity, excessive Internet use, and negative outcomes. Psychosocial well-being was assessed through the Beck depression inventory (Beck, Steer, and Brown 1996) and the UCLA loneliness scale (Russell, Peplau, and Cutrona 1980) to account for depression and loneliness respectively. Caplan's theoretical perspective suggests that problematic psychosocial predispositions lead individuals to excessive and compulsive computer-mediated social interaction, which, in turn, worsens their problems. The results of the study supported the claim that preference for online socialization is a key contributor to the development of problematic Internet use, which in turn has negative effects on undergraduates' well-being.

Internet Use and Psychological Well-Being

Yiwei Chen and Anna Persson (2002) examined (1) the impact of Internet use on young and older American adults' well-being, and (2) group differences between older Internet users and nonusers in terms of psychological well-being and personality traits. Psychological well-being and personality traits were assessed through Carol D. Ryff's psychological well-being scale (Ryff and Singer 1998) and a mini-marker scale adapted from the five-factor model (McCrae and John 1992) respectively. As far as Internet use and well-being are concerned, the authors found no significant correlations between Internet use (defined as time spent online) and psychological well-being in both young and older adults. Older Internet users scored higher than nonusers on two dimensions of psychological well-being, namely "personal growth" and "purpose in life." Although the study did not find any correlations between average time spent online and psychological well-being in older users, the authors contended that being able to learn and use the Internet seemed to make a difference between users' and nonusers' well-being.

Internet Use and Self-Determination Theory

Richard M. Ryan, C. Scott Rigby, and Andrew Przybylski (2006, study 4) examined the effects of Massively Multiplayer Online (MMO) games on young adult players' need for competence, autonomy, and relatedness. The study demonstrated that autonomy and competence provide significant explanations of players' motivation and enjoyment. In addition, because of the interactive structure of MMOs, the need for relatedness emerged as an important satisfaction that promoted a sense of presence, game enjoyment, and an intention for future play. Game enjoyment and preference for future play were significantly accounted for by psychological need satisfactions. The authors contend, though, that there were many limitations to their studies: (1) their measures are both new and in need of more extensive construct validation, (2) their laboratory experiments artificially induced people to engage in games, thereby ruling out factors such as self-selection and voluntary involvement in game playing, and (3) different genders, game contents, and interfaces may bring about different effects on, and relate differently to, motivational and need satisfaction variables.

Internet Use and Social Well-Being

Alberta Contarello and Mauro Sarrica (2007) examined the effects of Internet representations (e.g., information, communication, and emotional texture) on psychology undergraduates' social well-being through Keyes's (1998) scale. The study revealed that undergraduates' perception of any change that has occurred since the Internet entered their lives is on the whole one of improvement. Positive results include participants' feelings of higher closeness and contribution both to society in general and to their own community, and a growing sense of coherence and actualization. However, the authors also found that despite the aforementioned positive effects of Internet technology, participants overall have less trust in people they meet online and tend to look for certainty and comfort within their own communities. Based on these mixed results, the authors contend that greater access to rich and varied online information may carry with it greater levels of uncertainty and fuzziness, thereby affecting key dimensions of Internet users' social well-being.

IMPLICATIONS OF PRUDENTIAL PSYCHOLOGY FOR PHILOSOPHICAL REFLECTION ON INTERNET USE, WELL-BEING, AND THE GOOD LIFE

Let's now turn to the potential implications of prudential psychology for philosophical reflection on Internet use, well-being, and the good life. It should be noted that such implications will depend on how one conceives of the interaction between empirical research and philosophical reflection. I will outline here two possibilities of interaction and draw some of their implications for the issue in question. First, one can implement a *top-down* way in which prudential psychology conceptions of well-being are formally tested for descriptive and normative adequacy. Such a test would include, for example, a substantive analysis of psychological conceptions of well-being the outcomes of which will, in turn, determine the adequacy and normative implications of empirical results for Internet practices, well-being policies, and conceptions of the good life. Second, one can also implement a *bottom-up* way in which empirical results are (a) taken as the object of ethical reflection, (b) heuristically implemented to unveil normative features implied in different types of Internet use, and (c) operationalized in terms of the supporting or question-raising role they play in assessing the application of available theories of well-being and the good life to Internet-related practices. Although both methodological ways can be argued to be complementary to each other, each one provides us with distinguishable epistemic attitudes and procedural patterns.

The top-down way will start from established philosophical views of well-being and look for formal compatibilities between them and their psychological counterparts. Once a particular view or a suitable combination of views is favored and its explanatory and normative authority is settled, empirical results will be evaluated in terms of their relevant support to the endorsed view. In this sense, we could, for example, argue for the explanatory and normative authority of eudaimonic views of well-being and evaluate the extent to which psychological well-being, social well-being, and self-determination theory research lends support to the normative adequacy of capabilities, flourishing, or virtue views. This evaluation would provide normative guides and public policy suggestions concerned with the consequences of Internet-based services and contents for users' individual and social functioning, their satisfaction of basic psychological needs, or their living a virtuous life. However, despite its explanatory and normative strength, this way of articulating both philosophical and psychological domains may unilaterally favor a particular theoretical trend to the detriment of other suitable and/or equally compelling views.

The bottom-up way, by contrast, will recognize from the outset a plurality of philosophical and psychological views about well-being. But instead of engaging in a quest for the explanatory authority of a particular theoretical view, it will take psychological data as a means to assess the empirical validity of the cognitive, affective, motivational, behavioral, and social background assumptions of available theories of well-being. In this sense, one could, for instance, assess the extent to which problematic Internet use thwarts users' autonomy, positive emotion, social relatedness, self-fulfillment, and self-control, which are considered by various philosophical and psychological views of well-being as components of healthy human functioning and a healthy life. By focusing on empirically observable indicators of problematic Internet use (e.g., predictors of compulsive use, personality profiles of compulsive users, users' assessments of life satisfaction before and after using Internet-based applications, discrete effects of particular Internet applications, etc.), psychological research can help unveil implicit notions of good, right, suitable, bad, wrong, and inappropriate use of Internet applications. In addition, those implicit notions can be integrated into philosophical, normative frameworks to examine whether such frameworks are compelling enough to judge maladaptive Internet-related practices or whether they need some empirically inspired fine-tuning.

CONCLUSION

The aforementioned implications make it evident that the time has arrived for prudential psychology and philosophical reflection to engage in an interdisciplinary work aimed at normatively assessing the effects of Internet technology on people's well-being and the good life. Consequently, it is now up to researchers working at the interface between these two domains both to build their interdisciplinary frameworks and to reap the fruits of their psycho-philosophical collaboration.

REFERENCES

Amichai-Hamburger, Yair, and Adrian Furnham. (2007). "The Positive Net," *Computers in Human Behavior*, vol. 23, pp. 1033–1045.

Beck, Aaron T., Robert A. Steer, and Gregory K. Brown. (1996). *Beck Depression Inventory-II (BDI-II) Manual*. San Antonio, TX: Psychological Corporation.

Brummett, Bradley B., Jay C. C. Wade, Joseph G. Ponterotto, Brett Thombs, and Charles Lewis. (2007). "Psychosocial Well-Being and a Multicultural Personality Disposition," *Journal of Counseling and Development*, vol. 85, pp. 73–81.

Caplan, Scott E. (2003). "Preference for Online Social Interaction: A Theory of Problematic Internet Use and Psychosocial Well-Being," *Communication Research*, vol. 30, no. 6, pp. 625–648.

Caplan, Scott E. (2007). "Relations Among Loneliness, Social Anxiety, and Problematic Internet Use," *CyberPsychology and Behavior*, vol. 10, no. 2, pp. 234–242.

Chen, Yiwei, and Anna Persson. (2002). "Internet Use among Young and Older Adults: Relation to Psychological Well-Being," *Educational Gerontology*, vol. 28, pp. 731–744.

Contarello, Alberta, and Mauro Sarrica. (2007). "ICTs, Social Thinking, and Subjective Well-Being: The Internet and Its Representations in Everyday Life," *Computers in Human Behavior,* vol. 23, pp. 1016–1032.

Diener, Ed, Robert A. Emmons, Randy J. Larsen, and Sharon Griffin. (1985). "The Satisfaction with Life Scale," *Journal of Personality Assessment,* vol. 49, no. 1, pp. 71–75.

Diener, Ed, Shigehiro Oishi, and Richard E. Lucas. (2003). "Personality, Culture and Subjective Well-Being: Emotional and Cognitive Evaluations of Life," *Annual Review of Psychology,* vol. 54, pp. 403–425.

Haybron, Daniel M. (2007). "Life Satisfaction, Ethical Reflection, and the Science of Happiness," *Journal of Happiness Studies,* vol. 8, no. 1, pp. 99–138.

Haybron, Daniel M. (2008a). "Philosophy and the Science of Subjective Well-Being." In *The Science of Subjective Well-Being,* edited by Michael Eid and Randy Larsen, pp. 17–43. New York: Guilford Press.

Haybron, Daniel M. (2008b). *The Pursuit of Unhappiness. The Elusive Psychology of Well-Being.* New York: Oxford University Press.

Jackson, Linda A., Alexander von Eye, Frank Biocca, Gretchen Barbatsis, Yong Zhao, and Hiram E. Fitzgerald. (2006). "Does Home Internet Use Influence the Academic Performance of Low-Income Children?" *Developmental Psychology,* vol. 42, no. 3, pp. 429–435.

Karavidas, Maria, Nicholas K. Lim, and Steve L. Katsikas. (2005). "The Effects of Computers on Older Adults," *Computers in Human Behavior,* vol. 21, pp. 697–711.

Keyes, Corey Lee M. (1998). "Social Well-Being," *Social Psychology Quarterly,* vol. 61, no. 2, 121–140.

Kraut, Robert, Sara Kiesler, Bonka Boneva, Jonathon Cummings, Vicki Helgeson, and Anne Crawford. (2002). "Internet Paradox Revisited," *Journal of Social Issues,* vol. 58, no. 1, pp. 49–74.

Kraut, Robert, Michael Patterson, Vicki Lundmark, Sara Kiesler, Tridas Mukopadhyay, and William Scherlis. (1998). "Internet Paradox: A Social Technology that Reduces Social Involvement and Psychological Well-Being?" *American Psychologist,* vol. 53, no. 9, pp. 1017–1031.

LaRose, Robert, Dana Mastro, and Matthew S. Eastin. (2001). "Understanding Internet Usage: A Social-Cognitive Approach to Uses and Gratifications," *Social Science Computer Review,* vol. 19, no. 4, pp. 395–413.

McCrae, Robert R., and Oliver P. John. (1992). "An Introduction to the Five-Factor Model and Its Applications," *Journal of Personality,* vol. 60, pp. 175–215.

McKenna, Katelyn Y. A., and John A. Bargh. (2000). "Plan 9 From Cyberspace: The Implications of the Internet for Personality and Social Psychology," *Personality and Social Psychology Review,* vol. 4, no. 1, pp. 57–75.

Meerkerk, Gert-Jan, Regina J. van den Eijnden, Ad A. Vermulst, and Henk Garretsen. (2009). "The Compulsive Internet Use Scale (CIUS): Some Psychometric Properties," *CyberPsychology and Behavior,* vol. 12, no. 1, pp. 1–6.

Mesch, Gustavo S. (2006). "Family Relations and the Internet: Exploring a Family Boundaries Approach," *Journal of Family Communication,* vol. 6, no. 2, pp. 119–138.

Morgan, Charlie, and Sheila R. Cotton. (2003). "The Relationship Between Internet Activities and Depressive Symptoms in a Sample of College Freshmen," *CyberPsychology and Behavior,* vol. 6, no. 2, pp. 133–141.

O'Hare, Thomas, Margaret V. Sherrer, Hilary S. Connery, Jay Thornton, Annamaria LaButti, and Kelly Emrick. (2003). "Further Validation of the Psycho-Social Well-Being Scale (PSWS) with Community Clients," *Community Mental Health Journal,* vol. 39, no. 2, pp. 115–129.

Russell, Dan, Letitia A. Peplau, and Carolyn E. Cutrona. (1980). "The Revised UCLA Loneliness Scale: Concurrent and Discriminant Validity Reference," *Journal of Personality and Social Psychology*, vol. 39, no. 3, pp. 472–480.

Ryan, Richard M., and Edward L. Deci. (2001). "On Happiness and Human Potentials: A Review of Research on Hedonic and Eudaimonic Well-Being," *Annual Review of Psychology*, vol. 52, pp. 141–166.

Ryan, Richard M., Veronika Huta, and Edward L. Deci. (2008). "Living Well: A Self-Determination Theory Perspective on Eudaimonia," *Journal of Happiness Studies*, vol. 9, no. 1, pp. 139–170.

Ryan, Richard M., C. Scott Rigby, and Andrew Przybylski. (2006). "The Motivational Pull of Video Games: A Self-Determination Theory Approach," *Motivation and Emotion*, vol. 30, no. 4, pp. 347–363.

Ryff, Carol D., and Burton H. Singer. (1998). "The Contours of Positive Human Health," *Psychological Inquiry*, vol. 9, no. 1, pp. 1–28.

Ryff, Carol D., and Burton H. Singer. (2008). "Know Thyself and Become What You Are: A Eudaimonic Approach to Psychological Well-Being," *Journal of Happiness Studies*, vol. 9, no. 1, pp. 13–39.

Seligman, Martin E. P. (2003). *Authentic Happiness. Using the New Positive Psychology to realize Your Potential for Lasting Fulfillment*. London: Nicholas Brealey.

Sum, Shima, R. Mark Mathews, Ian Hughes, and Andrew Campbell. (2008). "Internet Use and Loneliness in Older Adults," *CyberPsychology and Behavior*, vol. 11, no. 2, pp. 208–211.

Tiberius, Valerie. (2006). "Well-Being: Psychological Research for Philosophers," *Philosophy Compass*, vol. 1, no. 5, pp. 493–505.

Tiberius, Valerie. (2008). *The Reflective Life: Living Wisely with Our Limits*. New York: Oxford University Press.

Valkenburg, Patti M., Jochen Peter, and Alexander P. Schouten. (2006). "Friend Networking Sites and Their Relationships to Adolescents' Well-Being and Social Self-Esteem," *CyberPsychology and Behavior*, vol. 9, no. 5, pp. 584–590.

Whitlock, Janis L., Jane L. Powers, and John Eckenrode. (2006). "The Virtual Cutting Edge: The Internet and Adolescent Self-Injury," *Developmental Psychology*, vol. 42, no. 3, pp. 407–417.

Whitty, Monica T., and Deborah McLaughlin. (2007). "Online Recreation: The Relationship between Loneliness, Internet Self-Efficacy, and the use of Internet for Entertainment Purposes," *Computers in Human Behavior*, vol. 23, pp. 1435–1446.

Ybarra, Michele L., Cheryl Alexander, and Kimberly J. Mitchell. (2005). "Depressive Symptomatology, Youth Internet Use, and Online Interactions: A National Survey," *Journal of Adolescent Health*, vol. 36, no. 1, pp. 9–18.

Yellowlees, Peter M., and Shayna Marks. (2007). "Problematic Internet Use or Internet Addiction?" *Computers in Human Behavior*, vol. 23, pp. 1447–1453.

Zhao, S. (2006). "Do Internet Users Have More Social Ties? A Call for Differentiated Analyses of Internet Use," *Journal of Computer-Mediated Communication*, vol. 11, pp. 844–862.

15 Virtually Good?

Disclosing the Presuppositions Behind the Claimed Inferiority of Virtual Worlds

Johnny Hartz Søraker

Whether my life fares well or not seems inextricably related to whether the states of affairs that contribute to my well-being are real or not. Thus, we are constantly exposed to and troubled by claims of unreality: Is this real love? Get a real life! What is my real identity? Is she a real friend? Am I escaping my real obligations? Is there really a God? We are either relieved or disappointed when a particularly lifelike dream turned out not to be real, or by realizing that a particular state of affairs turned out not to be true after all. When so much of our lives and well-being is tied up with concerns and claims about reality, it should come as no surprise that the impact of *virtual* events and experiences on our lives is a controversial topic. Indeed, philosophers, policymakers, researchers, and journalists often make claims to the effect that wasting one's life on virtual surrogates for the real thing amounts to being bereaved of what *real* life has to offer; that virtual worlds, entities, and experiences might give immediate gratification, but not deliver the kind of authentic happiness that a life engaged with reality can offer.

The problem with these kinds of claims is that they often rest on a large set of implicit presuppositions—presuppositions that are far from uncontroversial. The purpose of this chapter is to disclose some of the most common presuppositions and show how contentious they are—and to illustrate the importance of making these presuppositions explicit when we make claims about how particular technologies may affect our well-being. In this chapter, I will focus on graphical, three-dimensional, multiuser environments, such as Second Life or World of Warcraft—which I will refer to as 'virtual worlds.' I will use virtual relationships and communities as the main examples throughout, because their supposed inferiority to actual relationships and communities underpins some of the most commonly made criticisms of virtual worlds.

The claim that we, in effect, ought to choose real events and experiences over virtual surrogates has been repeatedly put forward both in and outside academia. Most commonly, these claims single out some feature thought to be essential to well-being, such as social relationships, education, experiencing nature, and so forth, and then argue that any virtual simulation or 'surrogate' of these will be of less value and that we therefore should pursue their real counterparts instead. Some of the most frequent claims, which will be the main examples throughout this chapter, relate to the supposed inferiority of virtual relationships and communities.

Hubert Dreyfus is one of many philosophers who have criticized virtual communities as inferior to actual ones because we cannot have the same range of movements and expressions of the body in virtual communities, nor a sense of context, commitment, or shared risk taking. Furthermore, the lack of physical context fosters what he characterizes as the nihilist, irresponsible, and often uninformed nature of virtual communities (Dreyfus 2004). Albert Borgmann, on the basis of a distinction between instrumental, commodified, and final communities, argues that virtual communities and relationships can at best be instrumental or commodified, because they do not contain "the fullness of reality, the bodily presence of persons and the commanding presence of things" found in final communities (Borgmann 2004, p. 63).

In a similar fashion, Darin Barney (2004) sees virtual communities as inferior due to their lack of physical *practices;* and Howard Rheingold argues that the lack of spontaneous bodily gestures and facial expression is the reason for the "ontological untrustworthiness" of virtual acts of communication (Rheingold 2000, p. 177). A related argument has also been made by Langdon Winner, who argues that virtual communities ought not to be regarded as communities at all, because this ignores the importance of "obligations, responsibilities, constraints, and mounds of sheer work that real communities involve" (Winner 1997, p. 17).

What all of these and many other criticisms have in common is that the virtual "surrogate" lacks something found in the actual counterpart, and that this lack translates to reduced value—and because of this we *ought* to choose the real over the virtual. However, many other premises must be presupposed for this claim to hold. Indeed, it can be reconstructed as building on a generic line of reasoning as follows: *Some x is good (i.e., conducive to your well-being), this type/token x can be carried out (e.g., sports), constructed (e.g., communities), and/or experienced (e.g., nature) in both the actual and virtual worlds, but the actual is more conducive to well-being and insofar as the actual x becomes replaced by the virtual x, we settle for lesser well-being if we choose the virtual x.* This is the line of reasoning I will analyze in this chapter, with the main purpose being to show how contentious many of the underlying presuppositions are.

In what follows, I will try to keep my analysis relevant to as many different kinds of claims about virtual inferiority as possible so I will primarily talk about virtual 'x' compared to actual 'x.' That is, when I speak of a general act, event, or experience, regardless of it being virtual or actual, I will simply use 'x' (e.g., relationships in general, be they actual or virtual). When I discuss their particular realization, I will refer to them as 'virtual x' (sometimes shortened to 'Vx') or 'actual x' (sometimes shortened to 'Ax'). As mentioned, I will primarily use actual/virtual communities and relationships as examples throughout. With those clarifications in place, I proceed by looking at six particularly important and contentious presuppositions, many or all of which are presupposed when arguing that we ought to choose actual x over virtual x because the latter is less valuable (i.e., less conducive to well-being).

PRESUPPOSITION 1: X IS GOOD

The first and most fundamental presupposition implicit in any claim about inferiority is a certain conception of the Good. Whether or not a virtual x is less valuable than an actual x requires that x, actual and/or virtual, is 'Good.' Good, in this context, does not refer to its *ethical* goodness but to its conduciveness to well-being (what is often referred to as *prudential* value). For instance, in his *Nicomachean Ethics* (book 8, p. 1), Aristotle famously claims that friendship is necessary for having a good life, which gives rise to the question whether certain kinds of friendships are more or less valuable than others.

Judgments about some x being valuable in this sense are often made on the basis of an implicit *theory of well-being*. Theories of well-being can broadly be characterized as belonging to one of three approaches: *hedonism, desire-fulfillment*, and *objective list theories* (Parfit 1986). In short, *hedonist theories* of well-being propose that the ultimate measure of one's well-being is the individual's experience of pain and pleasure. *Desire-fulfillment theories* focus on the satisfaction of more complex desires, which can include long-term goals and even posthumous events. Some desire-fulfillment theories require the desires to be made on the basis of well-informed deliberation, arguing that well-being consists in satisfying the desires you would have had if you were sufficiently informed. Finally, *objective list theories* differ from both of the aforementioned because they prescribe certain substantive goods that are necessary, sufficient, and/or essential for well-being, goods that can contradict the individual's experience. For instance, only an objective list theory can claim that a particular "ingredient," such as friendship, is *necessary* to have a good life, and that anyone who fails to live her life accordingly would be destined not to live the best of lives.

Depending on which theory of the good life we (implicitly) hold, when and why some x is of value will differ tremendously—even for relatively uncontroversial values. For instance, a hedonist account would consider relationships as conducive to well-being only insofar as they generate pleasure. A desire-fulfillment theory would consider relationships valuable only insofar as they fulfill a desire. Most objective list theories hold relationships as one of the essential ingredients to a good life, but may hold other values higher. When it comes to more controversial "values," such as materialist goods, spirituality, meaningfulness, and so forth, there is very little consensus concerning which x's are Good. For instance, saying that we ought to choose actual spirituality over virtual spirituality will have no relevance to someone who does not regard spirituality as Good at all.

What makes for a good life has been the subject of heated debate since ancient Greece, but it is rarely discussed in claims about the inferior value of virtual worlds. When claiming that virtual relationships are less valuable than actual relationships, it is usually taken for granted that relationships are conducive to well-being. Although most theories of well-being, as well as empirical research (Diener and Seligman 2002, p. 83), suggest that relationships are conducive to well-being, it is far from uncontroversial and—more important—not necessarily a universal value. It is not my purpose in this chapter to argue for or against the value of relationships, but only to point out that such values are often presupposed without proper attention to whether they are necessary, essential, important, or constitutive of well-being. In any sufficiently robust evaluation of the value of virtual worlds, such presuppositions must be made explicit and justified—for instance, by way of a robust philosophical theory of well-being and/or by drawing on empirical research.

PRESUPPOSITION 2: THE VIRTUAL X (Vx) IS EASIER TO ATTAIN THAN THE ACTUAL X (Ax)

The second presupposition is perhaps better described as the root of the "problem." If x is good, then it would not be a widespread problem if people could choose the virtual x over the actual x if the actual was just as easily attainable. To draw an analogy: if someone discovers that the planet Venus is made of gold, this would not be a threat to the value of gold on earth as long as the gold on Venus is significantly more difficult to get hold of. That said, virtual surrogates *are* often easier to attain, because there are fewer physical obstacles related to physical appearance (age, gender, looks, etc.), physical status indicators (cultural indicators, social status, etc.), and the absence of other physical limitations, such as geographical distance, physical disabilities, and so forth. This has rightly raised concern for the substitution of the actual for the virtual.

If we carry the previous presupposition (x is good) and add the presupposition that the virtual x tends to be easier to attain than the actual x, this leads to one of the first branching points: If x is good and the virtual x is easier to attain than the actual x, we could just as well conclude that this is good news because we now have more opportunities for attaining x. If friendship is necessary for a good life and virtual worlds allow you to have vastly more friends, this would seem to imply that virtual worlds could vastly improve your well-being. Thus, claiming that we ought to choose the real over the virtual x requires another important step, which is to argue that the virtual x is less valuable than the actual x.

PRESUPPOSITION 3: Ax IS MORE VALUABLE THAN Vx

If people choose the more easily attainable virtual x over the actual x, this only becomes potentially detrimental to well-being if the virtual x is less valuable than the actual x—if it is less conducive to well-being. Indeed, the question whether the actual x is better than the virtual x is the main presupposition behind claims regarding the impact of virtual worlds on our well-being. The reason I single it out as one among other presuppositions is that it is not problematic in itself if a particular Vx may be less valuable than Ax—not unless the other presuppositions hold true (as I hope will be clear by the end of this chapter).

There are many ways in which this presupposition can be examined because it partly envelops all the other ones, but one approach is to draw on the so-called *principle of formal equality*.[1] The principle of formal equality states that a difference in treatment or value between two kinds of entities can only be justified on the basis of a relevant and significant difference between the two. For instance, many issues in animal ethics can be approached by first discussing the theoretical differences between humans and other animals (e.g., that some animals have a significantly less developed central nervous system) and subsequently discuss to what degree these differences are relevant to their value (e.g., that the less developed CNS indicates little or no ability to experience pain). More generally, for something to be of more or less value than something else, the two must be significantly different in one way or another, and we must be able to justify why this difference entails a difference in value.

Furthermore, to avoid question begging, this difference should be grounded in empirical and/or theoretical properties that are open to debate. The questions become, then, (1) are there theoretically significant differences between the virtual and actual x, and if so, (2) is this difference relevant to their value? Using relationships as an example: (1) are there theoretically significant differences between virtual relationships and actual relationships, and if so, (2) is this difference relevant to their value.

To illustrate what such a deliberation may consist in, I will (too briefly) discuss these questions in some more detail with regard to relationships. One apt starting point is Dean Cocking and Steve Matthews's (2000) provocative paper on the unreality (or illusion) of close friendships in virtual worlds (or, text-based communication in general), particularly because their reasoning lies close to the principle of formal equality. In a nutshell, Cocking and Matthews claim that virtual friendships currently do not allow for nonvoluntary self-disclosure (theoretically significant difference) and that genuine friendships can only be established on the basis of nonvoluntary self-disclosure (hence relevant to their value). According to this account, the only way to genuinely know someone is to spend considerable amounts of time in their physical presence—a conclusion that typically follows from this kind of "cues filtered-out" approach (Joinson 2003, pp. 25–37).

It is important to note that Cocking and Matthews see self-disclosure as necessary for genuine friendship, and that they regard *nonvoluntary* self-disclosure as more revealing of our true nature. Even if we agree with the former, as most accounts of friendship would, there are reasons to disagree with the latter. Could virtual worlds afford other means of self-disclosure that may be as revealing as nonvoluntary self-disclosure in physical proximity? Adam Briggle argues that the lack of nonvoluntary cues in virtual worlds may actually be an advantage when it comes to self-disclosure. This is not a critique of the criterion as a condition for friendship, but rather an observation to the effect that we may be less likely to disclose ourselves in the actual world. In a sense, Cocking and Matthews may be right that we have more *nonvoluntary* self-disclosure in the actual world, but one of the reasons we need to resort to such cues is that we often "wear masks, play roles, and fit molds" (Briggle 2008, p. 75)—and it is often easier to drop those masks in virtual worlds, thereby eliminating the *need* for involuntary self-disclosure.

This has also been suggested by empirical research, such as Nick Yee's (2006) comprehensive survey of 30,000 virtual world users, where approximately 30% reported having told personal issues or secrets to their virtual friends that they have never told their real-life friends. There are also numerous other studies that suggest the same (see e.g., Joinson 2003, pp. 130–133; and Bargh, McKenna, and Fitzsimons 2002, p. 33). This short example is meant to illustrate how complex such a deliberation can be.[2] The question of whether a virtual x is more or less valuable than an actual x will in most cases be a multifaceted issue, where some differences (e.g., less nonvoluntary self-disclosure) are counterbalanced by other differences (e.g., more voluntary self-disclosure).

If we (hypothetically) agree with all the presuppositions reviewed so far, we get the following line of reasoning: some x is good (i.e., conducive to well-being), but the actual x is better than the virtual x, so that even if the virtual x is easier to attain, we ought to pursue the actual x. The conclusion seems intuitive, but there are (at least) three other presuppositions that need to be addressed—all of which are immensely important (and often neglected) caveats to any claim about the inferior value of virtual experiences.

PRESUPPOSITION 4: MAXIMAX VERSUS MAXIMIN DECISION STRATEGY

In decision theory, one of the most fundamental distinctions regarding choice under uncertainty is that between Maximax and Maximin decision strategies. Maximin means that you choose the option whose worst outcome is least harmful (MAXImizing the MINimum gain), whereas Maximax means that you choose the option whose best outcome is the most rewarding (MAXimizing the MAXimum gain). To put it bluntly, Maximin entails that you do not aim for optimal results but settle for safer, less rewarding results. Maximax entails that you shoot for the optimal results, even if this could mean that you end up empty-handed or worse. Thus, even if you agree that the actual x is vastly superior to the virtual, one could argue (in line with presupposition 2) that there is a higher chance of ending up empty-handed if all one's efforts are targeted at the actual x.

For instance, whether or not to spend your time and energy on acquiring actual relationships depends on your stance on all the previous presuppositions, as well as your general strategy in life. On a Maximin strategy, it might be just as rational to settle for easier-to-attain virtual relationships even if these are less valuable. What is important is that two individuals might be equally rational and well-informed and have the exact same risks and possibilities, yet rationally choose differently because of a difference in decision strategy—and the choice of strategy will often be closely related to one's general conception of what constitutes a good life.

This presupposition rarely forms an explicit part of any theory of well-being. Even objective list theories rarely address how far we should go to acquire the ingredients to a good life. This becomes especially problematic when there are potentially inferior surrogates available. I take this issue to be crucial to any claim about the value of virtual worlds and entities, and it is an important caveat to any claim about how something may affect our well-being. Being of less value does not necessarily entail that we should pursue that which is of higher value—the risks involved must be taken into consideration.

PRESUPPOSITION 5: Vx TENDS TO REPLACE Ax

Whether or not Vx tends to replace Ax is a collection of interrelated presuppositions. A trivial but nonetheless plausible objection to the critics would be to argue that most of us are perfectly capable of balancing our actual and virtual activities. Thus, the critics' arguments seem to be aimed at people who lack this ability and become enslaved by technology despite their better judgment. Although this is probably wrong for the majority of users, this does not mean that we should discount the criticism on this ground. The human ability to become addicted to various kinds of stimuli or to show poor judgment calls when faced with temptation should be relatively uncontroversial.

Another objection could be that the very distinction between actual and virtual—the so-called 'magic circle' (Huizinga 2003)—does not make sense. Although the border between actual and virtual is often diffused, I believe that a distinction between the two is important in many respects. I think Edward Castronova gets it right when comparing the relation between the actual and the virtual with a membrane, one that is "quite porous. Indeed, it cannot be sealed completely; people are crossing it all the time in both directions, carrying their behavioural assumptions and attitudes with them" (Castronova 2005, p. 147). In other words, there is an important distinction between virtual and actual, but that does not mean that whatever happens in virtual worlds stays in virtual worlds—or vice versa. As Mia Consalvo argues, we do integrate virtual worlds into our daily lives, but the rules and rewards that apply to each virtual world still form an important boundary (2007, p. 190). This is also in line with Jaron Lanier's description of virtual entities as being "indistinguishable [from the actual entity] in some practical context, while it remains distinguishable in another" (Lanier 1999).

Although Borgmann ultimately concludes that virtual worlds only lead to triviality, he is probably correct that they are most rewarding when the veil of ambiguity (i.e., the detachment from the actual world) is "dense and thick" (Borgmann 1999, p. 189). Thus, for many virtual acts, events, and experiences, we can legitimately and usefully draw a distinction between actual and virtual. This does not have to be more controversial than the fact that certain acts, events, and experiences are different when taking place in a virtual world than they would be if they took place in the actual world.

To better illustrate how important this presupposition is, it is instructive to combine the presuppositions previously outlined, and see whether any conclusion follows unless we also assume that Vx replaces Ax. If we, for illustrative purposes, grant that x is good, Vx is easier to attain than Ax, Ax is better than Vx, and that we ought to follow a Maximax decision strategy, does this yield the conclusion that we ought to choose Ax? This only follows if you cannot have both Vx and Ax at the same time—if, as it were, the balance between virtual and actual experiences is a zero-sum game. If it is the case that Vx does not actually replace Ax, then there is no problem choosing Vx because it will not affect the attainment of Ax.

If we look to empirical research, the consensus seems to be that virtual relationships usually do not significantly replace actual relationships. Wendy Griswold and Nathan Wright state that "research exploring how the internet fits into pre-existing activities generally conclude that . . . the internet complements and supports offline practices rather than displacing, undermining, or competing with them" (Griswold and Wright 2004, p. 206). Moreover, a recent report on usage of screen-based technologies (TV, computers, portable devices, and so forth) concluded that although there are big differences between what type of screen-based activity we spend our time on, "total screen time was strikingly similar" (Council for Research Excellence 2009, p. 50). In a recent survey, the conclusion was that the extent of social isolation has hardly changed at all since 1985, and Internet activities appear to have little or a positive relationship to engagement in local communities (Hampton, Sessions, Her, and Rainie 2009, pp. 4–5). These findings suggest that virtual worlds primarily replace *other* screen

Table 15.1 Virtual as Replacement or Encouragement of Good and Bad

	Actual x is good	Actual x is bad
Virtual x replaces actual x	Virtual relationships replace actual relationships	Virtual acts of violence replace actual acts of violence
Virtual x encourages actual x	Virtual relationships encourage actual relationships	Virtual acts of violence encourage actual acts of violence

activities, in particular watching TV. Thus, although there are certainly individual extremes, there does not seem to be a general tendency for virtual activities to replace actual ones to a significant degree.

Whether a given virtual x typically replaces (partially or fully) its potentially more valuable actual counterpart is ultimately an empirical question, but it is interesting to note that the relation between Vx and Ax often seems to be chosen, intentionally or not, with a certain conclusion in mind. That is, engaging in virtual activities has been criticized on grounds of its replacement of the actual thing when the actual thing is considered to be good, *but* it has also been criticized on the grounds of its *encouragement* of the actual thing when the actual thing is considered to be bad. This can perhaps best be illustrated by table 15.1.

Criticism of virtual activities presuppose either that the virtual replaces something good or encourages something bad. However, the opposite line of reasoning is far from improbable—at least not *a priori*. If it is the case that engaging in virtual acts of violence can encourage actual acts of violence, why is it not the case that engaging in virtual communities *encourages* engagement in actual communities? Or, vice versa, if it is the case that engaging in virtual communities replaces engagement in actual communities, why is it not the case that engaging in virtual acts of violence *replaces* engagement in *actual* acts of violence? Although empirical research suggests that such form of catharsis is not a very efficient form of treatment (Seligman 1994, pp. 238–239), the important point here is that this is not something that should be presupposed *a priori*. These are difficult empirical and probably inherently subjective questions, and it is unlikely that one kind of relation holds for all cases, so we cannot simply assume that certain experiences will replace or encourage their actual counterparts.[3]

Let us pause for a moment and review the presuppositions so far. The generic line of reasoning is as follows: x is good, but despite the fact that Vx is easier to attain than Ax, we ought to follow a Maximax decision strategy according to which we ought to choose Ax because Ax is more valuable than Vx, and Vx tends to replace Ax. If we agree with all of this, then surely the conclusion follows? Not necessarily. There is at least one more presupposition that must be considered.

PRESUPPOSITION 6: Vx COMES WITH THE OPPORTUNITY COST OF Ax

The notion of 'opportunity cost' is primarily a term from economics and relates to the choice between two or more desirable but mutually exclusive actions. To use a simple example, if you decide to go to the movies, this comes with the opportunity cost of not being able to go to the theatre; the value of going to the movies is reduced because of the disvalue ("cost") of not going to the theatre. Thus, 'opportunity cost' is an apt term for describing the relation between two activities that are both desirable, but where one is less so than the other. What is most interesting for present purposes is that the relevant opportunity costs to take into consideration are only those that are sufficiently plausible. That is, there would be no reduced value—no opportunity cost—of going to the movies if there were no other desirable option. In particular, there would be no reduced value if you did not forego any realistically achievable and desirable opportunities—if there was no chance for you to go to the theatre anyway. As long as we do not regard Vx as inherently bad, but only bad insofar as it replaces a more valuable Ax, the same consideration is relevant to claims about inferiority as well. That is, even if the reconstructed previous argument holds in general, in some cases, the ability to attain Ax is so severely limited that it would be irrational to forego Vx. In other words, there is no opportunity cost involved in choosing Vx because Ax is so unrealistic. This is similar to presupposition 2, but in these cases, Vx is not merely *easier* to attain than Ax, relatively speaking, but Ax is so unrealistic that there is nothing to lose by choosing Vx.

The difference between presuppositions 2 and 6 can be illustrated by way of Derek Parfit's argument that some pleasures are infinitely better than others (Parfit 2004, pp. 19–20). For instance, the experience of Mozart may be infinitely better than the experience of Muzak. If so, as Thomas Hurka interprets it, "its loss could not be made good by any increase in the quantity or quality of muzak" (Hurka 1993, p. 82). This would be of relevance to presupposition 2, because the infinitely higher value of certain goods makes it irrelevant that it is *more* difficult to achieve. However, if there is *no* opportunity cost, the infinitely higher value of whatever it is that is unattainable does not matter anymore. If it were impossible to listen to Mozart, then the proposition that it is infinitely more valuable than Muzak would not be relevant. This illustrates how lack of opportunity costs drastically changes our deliberations.

Less abstractly, if a person has severely limited opportunities to attain an actual friendship, it would be irrational for this person to forego the opportunity to have virtual friendships instead (unless they are seen as *inherently* bad). At the other end of the extreme, a person with no problems attaining actual friendships would be irrational in choosing virtual over actual friendships, if all the presuppositions previously outlined are true. The important point is this: Whether to choose Vx over Ax is *still* contingent and a matter of degree, not a matter of universal principle—*even if* we were to agree with all the controversial presuppositions previously disclosed. This is particularly important when it comes to the question of whether we should allow technological surrogates for people who, for reasons out of their control, have no social network.

CONCLUSION

To sum up, we can now combine all of these presuppositions into a line of reasoning that would, if *all* the presuppositions were true, entail a clear recommendation to choose Ax over Vx:

> *(1) X is Good, but despite the fact that (2) Vx is easier to attain than Ax, we ought to (3) follow a Maximax decision strategy according to which we ought to choose Ax whenever (4) Ax is more valuable than Vx, and (5) Vx tends to replace Ax, and (6) when we have the opportunity to choose Ax (i.e., there are opportunity no costs).*

In terms of the recurring example in this chapter:

> *Friendship is Good, but despite the fact that virtual friendships are easier to attain than actual ones, we ought to follow a Maximax decision strategy according to which we ought to choose actual friendships whenever they are more valuable than virtual ones, when virtual friendships tend to replace actual ones, and when you have the opportunity to choose actual friendship.*

To use another example, to illustrate the generalizability of this line of reasoning:

> *Education (or gaining knowledge) is Good, but despite the fact that education is easier to obtain in virtual worlds, we ought to follow a Maximax decision strategy according to which we ought to obtain education in the actual world instead, when such education is more valuable, when virtual education tends to replace actual education, and when you have the opportunity to obtain actual education.*

The reasoning is straightforward enough, but *only* if all the presuppositions hold true. What I have tried to indicate in this chapter is not that they are *false,* but that they are all contentious and will often differ from case to case and from individual to individual. The purpose of disclosing some of these underlying presuppositions has been to show that claims about the effect of virtual worlds on well-being should only be made on the basis of a thorough evaluation of these presuppositions, ideally spelled out in detail, so as to enable the reader to more easily decide whether the conclusion of the aforementioned line of reasoning holds true in her case.

NOTES

1. For more on this principle and its use in practice, see Gosepath (2008), Søraker (2007) and Wetlesen (1999).
2. I have discussed many more of these similarities and differences elsewhere (Søraker 2010b), which more clearly shows how complex this issue is, but I have left much of this discussion out for the purpose of brevity.
3. To complicate matters further, whether or not virtual activities replace or encourage real activities ultimately rests on whether we presuppose an emotivist or rationalist account of the impact of virtual experiences—that is, whether virtual experiences (primarily) affect our deliberation process or our emotional dispositions. I have argued elsewhere (Søraker 2010a) that we ought to take rationalist accounts more seriously.

REFERENCES

Aristotle. (2009). *The Nicomachean Ethics* (D. Ross, trans.). Oxford: Oxford University Press.

Bargh, John A., Katelyn Y. A. McKenna, and Grainne M. Fitzsimons. (2002). "Can You See the Real Me? Activation and Expression of the "True Self" on the Internet," *Journal of Social Issues*, vol. 58, no. 1, pp. 33–48.

Barney, Darin. (2004). "The Vanishing Table, or Community in a World That Is No World." In *Community in the Digital Age: Philosophy and Practice*, edited by Feenberg, A. and D. Barney. Oxford: Rowman and Littlefield.

Bentham, Jeremy. (1823). *An Introduction to the Principles of Morals and Legislation*. London: W. Pickering.

Borgmann, Albert. (1999). *Holding On to Reality: The Nature of Information at the Turn of the Millennium*. Chicago: University Of Chicago Press.

Borgmann, Albert. (2004). "Is the Internet the Solution to the Problem of Community?" In *Community in the Digital Age: Philosophy and Practice*, edited by Feenberg, Adam and Darin Barney. Oxford: Rowman and Littlefield.

Briggle, Adam (2008). "Real Friends: How the Internet Can Foster Friendship," *Ethics and Information Technology*, vol. 10, no. 1, pp. 71–79.

Castronova, Edward. (2005). *Synthetic Worlds: The Business and Culture of Online Games*. Chicago: University of Chicago Press.

Cocking, Dean, and Steve Matthews. (2000). "Unreal Friends," *Ethics and Information Technology*, vol. 2, no. 4, pp. 223–231.

Consalvo, Mia. (2007). *Cheating: Gaining Advantage in Videogames*. Cambridge, MA: MIT Press.

Council for Research Excellence. (2009). *Video Consumer Mapping Study*. Retrieved December 2, 2009, from http://www.researchexcellence.com/VCMFINALREPORT_4_28_09.pdf

Diener, Ed, and Martin Seligman. (2002). "Very Happy People," *Psychological Science*, vol. 13, no. 1, pp. 81–84.

Dreyfus, Hubert L. (2004). "Nihilism on the Information highway: Anonymity versus Commitment in the Present Age." In *Community in the Digital Age: Philosophy and Practice*, edited by Feenberg, Adam and Darin Barney. Oxford: Rowman and Littlefield.

Gosepath, Stefan. (2008). "Equality." In *The Stanford Encyclopedia of Philosophy*, edited by Edward N. Zalta (Fall 2008 ed.). Stanford: Stanford University.

Grinde, Bjørn. (2002). "Happiness in the Perspective of Evolutionary Psychology," *Journal of Happiness Studies*, vol. 3, no. 4, pp. 331–354.

Griswold, Wendy, and Nathan Wright. (2004). "Wired and Well Read." In *Society Online: The Internet in Context*, edited by Howard, Philip N. and Steve Jones. London: Sage.

Hampton, Keith N., Lauren F. Sessions, Eun Ja Her, and Lee Rainie. (2009). "Social Isolation and New Technology: How the Internet and Mobile Phones Impact Americans' Social Networks." Pew Internet and American Life Project. Retrieved December 10, 2009, from http://www.pewinternet.org/~/media//Files/Reports/2009/PIP_Tech_and_Social_Isolation.pdf

Huizinga, Johan. (2003). *Homo Ludens: A Study of the Play-Element in Culture*. London: Routledge.

Hurka, Thomas. (1993). *Perfectionism*. Oxford: Oxford University Press.

Joinson, Adam N. (2003). *Understanding the Psychology of Internet behaviour: Virtual Worlds, Real Lives*. New York: Palgrave MacMillan.

Lanier, Jaron. (1999, Fall). *Virtual Reality*. Retrieved January 6, 2008, from http://findarticles.com/p/articles/mi_m0GER/is_1999_Fall/ai_56457593

Parfit, Derek. (1986). *Reasons and Persons*. Oxford: Oxford University Press.

Parfit, Derek. (2004). "Overpopulation and the Quality of Life." In *The Repugnant Conclusion: Essays on Population Ethics*, edited by Ryberg, Jesper and Torbjörn Tännsjö. Dordrecht: Kluwer.

Pascal, Blaise. (1670/2003). *Pensées* (W. F. Trotter, trans.). Mineola, NY: Dover.

Peterson, Christopher. (2006). *A Primer in Positive Psychology*. Oxford: Oxford University Press.

Rheingold, Howard. (2000). *The Virtual Community: Homesteading on the Electronic Frontier*. Cambridge, MA: MIT Press.

Schwartz, Barry. (2004). *The Paradox of Choice: Why Less Is More*. New York: HarperCollins.

Seligman, Martin E. P. (1994). *What You Can Change and What You Can't: The Complete Guide to Successful Self-Improvement*. New York: Knopf.

Søraker, Johnny Hartz. (2007). "The Moral Status of Information and Information Technologies: A Relational Theory of Moral Status. In *Information Technology Ethics: Cultural Perspectives*, edited by Hongladarom, Soraj and Charles Ess. Hershey, PA: Idea Group.

Søraker, Johnny Hartz. (2010a). "The Neglect of Reason: A Plea for Rationalist Accounts of the Effects of Virtual Violence." In *Emerging Ethical Issues of Life in Virtual Worlds*, edited by Wankel, Charles and Shaun K. Malleck. Charlotte, NC: Information Age.

Søraker, Johnny Hartz. (2010b). *The Value of Virtual Worlds and Entities: A Philosophical Analysis of Virtual Worlds and Their Potential Impact on Well-Being*. Doctoral dissertation, University of Twente, Enschede, Netherlands.

Wetlesen, Jon. (1999). "The Moral Status of Beings Who Are Not Persons: A Casuistic Argument," *Environmental Values*, vol. 8, pp. 287–323.

Winner, Langdon. (1997). "Cyberlibertarian Myths and the Prospects for Community," *SIGCAS Computers and Society*, vol. 27, no. 3, pp. 14–19.

Yee, Nick. (2006). "The Demographics, Motivations and Derived Experiences of Users of Massively-Multiuser Online Graphical Environments," *PRESENCE: Teleoperators and Virtual Environments*, vol. 15, no. 3, pp. 309–329.

Part V

Medical and Agricultural Technology and Well-Being

16 What's Wrong with Techno Food?

David M. Kaplan

Everything we eat has been grown, raised, or processed in some way. All of our food undergoes some transformation from its original state. Even the most ecologically attuned organic farming and ranching use technologies to alter plants or animals into food. We use simple technologies for slicing, cooking, and storing; complex ones for pasteurizing, freezing, and flavoring. On a scale from least to most processed, first comes raw food (especially organically grown), then whole food (sometimes cooked), then natural food (no artificial ingredients), and finally conventional food (often with artificial ingredients). What we eat and drink has always been technologically transformed to some degree. But does that mean that the very idea of a "natural food" is dubious if it always involves some human intervention? Can we speak meaningfully of natural and artificial foods if both involve at least some processing? Even if we grant the inevitability of technological mediation, what it is about high-tech processing like irradiation that makes it more controversial than something wholesome (but no less technological) like plowing? In short, what's wrong with techno food, and is it really worse than natural food?

I believe a case can be made for the superiority of natural and minimally processed food but the reasons are subtle and the differences more social than scientific or metaphysical. Although scientists can tell us quite a lot about, for example, the nutritional content of food or the environmental impacts of agriculture, few of us are in any position to validate their claims. We simply have to trust them. And although metaphysics is more accessible to laypeople, that discourse only tells us what things are, not what we ought to do. Essences have no moral authority; only we do. We have to decide for ourselves what foods we should make and eat. There is no need to rely on science or metaphysics for guidance. We can figure out what to do about our food without knowing precisely what it is—just like we can make an omelet without knowing exactly what eggs are or where they come from. We can decide whether or not to make and eat techno foods using only concepts that are in principle available to everyone, such as common sense and public debates about what is right and what is good. No expertise is needed to raise and contest claims about justice and the good life.

So what, if anything, is wrong with techno food? Their production, distribution, and consumption can sometimes violate rights and fail to foster the good. They can be unjust and in poor taste—and this is an argument anyone can make.

NATURAL FOOD

It is far from clear what food even is, much less when processed. It is obviously something edible (although so are water and drugs), something that contains naturally occurring ingredients (although entirely synthetic food would be considered food), something that provides nutrients (although often minimal), and something that is intended to be eaten (although garnishes and edible centerpiece displays are difficult border cases). Processed food is any raw ingredient of plant or animal origins transformed into food (although fungi are also considered food, as are inorganic components in food, such as salt and baking powder). The methods of food processing are just as important in the controversies surrounding it as the ingredients. Simple food processing, where naturally occurring ingredients undergo change, is rarely controversial. But more intensive processing and the addition of synthetic ingredients often raise concerns. This class of techno foods includes genetically modified plants and animals, functional foods, and junk food. Members of this class of food items stretch the very definition of what food is and how we ought to produce it.

It is tempting to criticize processed foods as unnatural by-products of science and technology that should be avoided in favor of a more natural, wholesome diet. Nature is said to be a more reliable guide to diet than changing human preferences. Evolution shaped our dietary needs and preferences over thousands of years: techno foods defy the wisdom of nature. Science and technology supposedly make food worse by corrupting its nature. The more we tamper and tinker with our food, the worse we make it. Often science and technology are unfavorably compared to custom and habit. Our dietary practices developed over centuries: techno foods defy the wisdom of tradition. Food should be seen as more than just nutrients; eating as more than mere nutrition. The more we understand food, diet, and health in terms of scientific explanations and technical accomplishments, the more cut off we become from nature and our traditions.

Michael Pollan takes this kind of grim view of science and technology. His often repeated mantra is to "eat food, not too much, mostly plants." By "food" he means "real food"—something your grandparents might have grown up eating. We should avoid "fake food," the kind of junk we find in the center aisles of a typical supermarket. Pollan makes a strong distinction between natural food (tied to a traditional diet) and processed food (tied to industrial agriculture and science). All traditional diets are healthful—regardless of the tradition. So long as we avoid nontraditional diets we will be fine. The reason is that traditional diets understand that food is more than its chemical parts. Any particular food item should not be understood in isolation from the traditional diet. Food is always a part of a diet, which is itself part of an entire way of life. The key is to appreciate that food is culture and not just nutrition. We should think about it in terms of wholes not parts (Pollan 2009).

Pollan's critique is not new. In fact, it has a real pedigree. We find echoes of it in the American agrarian tradition, which stresses the role of farming and ranching in the formation of moral character and in preserving culture and traditions. By living a rural lifestyle connected to the climate and soil, we acquire a sense of identity and place that can only come about by direct contact with the land. Agrarian philosophy is critical of the social and environmental impacts of industrial agriculture. Wendell Berry argues that modern agriculture and exodus from farms to cities harms the environment, destroys communities, and eclipses the basic human dignity that comes from an agrarian lifestyle. "Eating," he famously says, "is an agricultural act" (Berry 1991, p. 145). We are all involved in agriculture because our daily food choices affect how the land is treated. People, food, and the environment are all connected in the act of eating—provided we eat the right foods in the right way. Artificial foods, produced by industrial agriculture and factories, disconnect us from the land and from each other. At the risk of oversimplifying: connection is good, disconnection is bad.

The worlds of food, technology, and tradition come together nicely in Albert Borgmann's, *Technology and the Character of Contemporary Life* (1984). One of his best examples of a focal activity is a traditional home-cooked meal enjoyed with friends and family. Fast food consumed on-the-go is food understood as a device. The problem, Borgmann says, is the more our lives are entangled with device-like things, the more cut off we are from what truly makes our lives meaningful. We need focal things to help us become more engaged, not devices that disengage. We need to be bound to nature and culture, not freed from them (Borgmann 1984). This critique of the dehumanizing character of science and technology is well-known. And by now, the critique of the critique is just as well-known. It is better to understand artifacts in terms of something other than just technical qualities because technologies are more than mere technical things. Technology cannot be value-neutral because people are not value-neutral; technology cannot determine history because it never is so independent from society to be in a position to cause it; and technology cannot be entirely good or bad because society is not entirely good or bad. Humans, things, and social and natural contexts all fit together like pieces of a puzzle.

Seen in light of the response to the putative dehumanizing nature of technology, the critique of techno food misses the point. The technical properties of food are not what makes it alienating because science and technology are not intrinsically alienating. Furthermore, food has neither been entirely natural nor entirely technological. It has always been a mélange.

The Romantic temptation is to view whatever is good or bad about food in terms of how much it connects us to nature and each other. Connection on this view is always good; disconnection is always bad. Nature and humanity are good; science and technology are bad.

And yet to view food as wholly natural is to ignore the role of science and technology in food production. Without processing, our food would be inedible or at least unavailable. And to treat science and technology as antithetical to food romanticizes nature and simpler ways of living that are supposedly more wholesome than technologically structured (possibly urban) lifestyles. Recourse to nature is futile; it tells us nothing until it is interpreted. And cultures and traditions—the supposed cures to dehumanization—are sometimes really awful. They can be repressive, patriarchal, racist, homophobic, cruel to animals, and so on. There's nothing morally special about the idea of a tradition, or a culture, or a connection. There are some things we really don't want to be connected to. And it is sometimes a good thing to be disconnected, for example, when attempting to be fair and impartial. Thankfully, the discourses of metaphysics and existential connection do not exhaust the analytic tools for understanding processed food. There is also the sober discourse of normative evaluation.

JUST FOOD

Let's briefly consider some examples of techno foods and then evaluate them in terms of the justice issues they might raise.

Genetically modified food. Genetically modified (GM) foods are plants and animals that have been altered using recombinant DNA technology, a technique that combines DNA molecules from different sources into a single molecule. The purpose of genetic modification is to produce new and useful traits otherwise unattainable through conventional techniques. Most often foods are genetically modified to contain their own pesticides or to be herbicide resistant, although a small percentage of crops are engineered to be nutritionally enhanced or drought resistant. GM food technology promises to increase yields, increase food security, and protect the environment. Critics of GM warn that they also pose health and environmental risks. They warn about the consequences of genetic pollution causing indestructible "super-pests" and "super-weeds" immune from pesticides and herbicides (Ellstrand 2002, pp. 325–330).

Animal biotechnology. Animal biotechnology applies recombinant DNA techniques to animals. The largest class of GE animals are designed to produce pharmaceuticals (also known as "agriceuticals"); another class is designed for industrial purposes; another for food. Livestock and fish are engineered to be disease resistant, have improved nutritional value, increased growth rates, decreased pollutants in their manure, or produce antimicrobials that target *E. coli* and *Salmonella.* GE animals promise increased resistance to disease, productivity, and hardiness; they yield more meat, eggs, and milk; and they provide more healthy food. Some critics are concerned that genetic manipulation is unnatural (Holland 1997); others warn that it simply makes the lives of animals worse (Comstock 2002).

Functional foods. A functional food, or "nutraceutical," is a food-based product that has added ingredients believed to provide health benefits. They are designed to assist in the prevention or treatment of disease, or to enhance and improve human capacities. They include products such as vitamin-fortified grains, energy bars, low-fat or low-sodium foods, and sports drinks. Functional foods eliminate properties from the food to make them more nutritious—even to replace medicine. Manufacturers can produce food items that make *general* health claims (to promote health) so long as they make no *specific* claims (to treat diseases). In the U.S., no proof of general health claims is required, nor is there a legal definition that distinguishes functional foods from ordinary food that includes dietary supplements. Critics warn that when food is designed and marketed to double as medicine, it should become subject to medical regulations rather than left to the whims of the market (Scrinis 2011).

Junk Food. Heavily processed foods—such as Twinkies, Chicken McNuggets, and Skittles—are the result of intensive technological intervention and possess little nutritional value. Junk food is a vague category. Typically, what it signifies are foods with long lists of chemical ingredients, sugars, fats, and milled grains. It lends itself to overeating and has been linked to the rise in obesity, type II diabetes, and heart disease. The reason junk foods promote overeating is because they are low in fiber, taste good, and offer a high number of calories in a small volume. They are typically high in sugar and fat, especially in liquid form. Critics of junk food warn of the health risks of a diet consisting of more than occasional junk food (Nestle 2007).

Let's assume there is no demonstrable evidence that processed food causes bad consequences. Assume genetically modified foods pose no risks of allergic reactions, antibiotic resistance, or carcinogens; no threat of genetic pollution, herbicide, or pesticide-resistant weeds, and no harm to monarch butterflies. Assume genetically engineered animals are safe to eat and do not cause the animals to suffer; functional foods are safe and effective; and junk food poses no health risks. Let's further assume that each of these processed foods have desirable consequences. They produce higher quality food, economically, and benefit the environment. Would there still be anything wrong with these processed foods?

Take GM food as an example. This class raises questions about intellectual property rights, the sovereignty of nations, and the right to food security. GM production, distribution, and consumption are (not surprisingly) driven by market imperatives. But unlike other consumer goods, they are the products of international agribusiness, supported by powerful Western governments and backed by international financial institutions. This powerful conjunction imperils human rights. Specifically, the trade policies enforced by the World Trade Organization (WTO), which require nations to purchase GM food, privatize public farms, and transform agricultural production from subsistence to export, violates the sovereignty of nations and threatens food security (i.e., the right not to be deprived of food). The WTO agreement on Trade Related Aspects of Intellectual Property (TRIPs) benefits agribusiness at the expense of developing nations by allowing for food and medicine that was once under the public domain to be privatized through global patent law. This lets food manufacturers modify traditionally bred seeds, patent them, and then sell them back to people who had always used them for free. It allows for the private sector to own the diversity of nature itself. Food security is then threatened when seeds become private property. At the very least, it increases food dependence on industrialized nations by developing nations (Kaplan 2004).

These rights-based arguments of course depend on some knowledge of states of affairs. But as moral arguments, they do not depend on scientific expertise or costly laboratories that deter public participation. GM food raises justice claims about intellectual property, the role of governments in ensuring fairness, and the right to food sovereignty and security. These are discourses anyone can participate in because they invoke principles, not facts.

Many of the same criticisms of genetically modified food apply to genetically engineered animals. The TRIPs agreement also covers microorganisms such as cell lines, genes, and animal varieties. Some critics contend that genetic manipulation violates an animal's intrinsic value (or its *telos*, its natural function) and that mixing the genes of different species tampers with the natural order. Others maintain the more moderate position that we should not engage in practices (using biotechnology or otherwise) that make food animals worse off than they are now. The relevant moral concerns are with privatization and patenting life, as well as basic issues in animal ethics. Unlike plant-based food, animal-based food raises additional moral concerns about the welfare, if not rights, of animals. Animals suffer in a way that plants do not. This ability to suffer means that we have to consider their rights, or at least our responsibilities, in addition to the rights of humans and duties of governments.

As for functional foods, the main moral problem is that they are consumer goods that double as medicine. If an individual chooses to support her health or treat disease by purchasing functional foods, and they are safe, effective, and consumed with knowledge, then there again is little reason to oppose them. But it is unwise to rely on market mechanisms to solve public health problems. That is not what markets are designed to do. When food and medicine are treated like any other consumer good, there is a real danger that our very dietary and medical practices ultimately serve the interests of others more than our own interests. On one hand, they are commodities like any other to be manufactured, sold, and consumed; on the other hand, they are uniquely situated at the nexus of diet and commerce, spanning the worlds of optional consumer goods and vital human needs. If the case can be made that the private medicalization of food threatens public health and the right to life, then governments must do more than protect the market for functional food. They must also regulate them in the public interest.

Finally, junk food, although inarguably unhealthy, does not obviously raise justice issues—except, of course, for the arguments usually made in their defense: consumer choice. The reason junk foods are so popular is because a lot of people like the way they taste. They are successful products in exactly the way that wheat grass juice and fish oil are not. Consumers have the right to choose whatever food they want to eat, and governments have a duty to protect this right. Protecting free choice and the private accumulation of capital is, however, only one among many government obligations. It also has the obligation to regulate industry; protect consumers, animals, and the environment; and enforce corporate liability.

The situation with junk food is not unlike functional foods: consumers make choices and governments have responsibilities. These consumer goods impact our lives as both consumers and citizens precisely because they are bound up with questions concerning individual liberty and the proper role of government. Arguably, citizens have a right to be protected from junk food—even the ones they willingly choose. Governments ought to regulate the advertizing and distribution of a known public health menace. They should not unfairly burden taxpayers with the health costs that should be borne either by junk food users, if the not manufacturers responsible for unsafe products.

GOOD FOOD

Just as often, the issue when we criticize techno food is a question about the good, not the right. We have a sense that there's something wrong with unnatural foods, and so we couch it in terms of something dire, such as health or human rights. But really what troubles us is something less certain and more interpretive, like the kind of lifestyle they reflect, how they taste, how they make our lives better or worse. These "soft impacts" are about the quality of life and how much we like or dislike the social and cultural influence of something (Boenink, Swierstra, and Stemerding 2011). By contrast, "hard impacts" consist of health, safety, cost, and human rights. Soft impacts can never be verified or tested in a laboratory. They are notoriously contentious and rarely enshrined in laws. The public is on its own to sort through lifestyle preferences and to condone or condemn uses of things. We rarely even make them explicit, but rather leave it up to custom or personal whim to figure out how to contend with the way they texture our lives.

So what exactly are the soft impacts of techno foods? What questions concerning the good life do they raise, and what reasons might be offered to oppose them? It is hard to say because there are an indefinite number of concerns that might be raised about processed food, depending on the situation in which they are encountered. They will be seen differently by different actors, in different contexts, in light of different questions. There is no single fact of the matter for something that can be so many different things to different people. This is not to say that all interpretations are equally valid, but simply that questions concerning the way processed foods affect us are interpretive, not empirical. They have less to do with science or rights than with judgment and taste—not in their narrowly aesthetic senses but in their broader meaning of wisdom and discernment. As Kant notes, there can be disagreement about taste but no dispute (because there are no universal criteria) (1987). Taste is a kind of judgment that cannot be reduced to rules or concepts, yet it claims validity: you either have good taste or no taste. When understood in its cognitive sense, taste is a powerful mode of social knowledge and one that is readily applicable to food choices. Such claims can be contested on rational grounds and might receive universal assent, but they are far less determinate than truth or rights claims. This is the kind of knowledge that governs most of our individual food choices as consumers.

Take, for example, genetically modified foods. In the U.S., most people surveyed respond negatively to them (Hallman, Hebden, Aquino, Cuite, and Lang 2003). Of those who have negative opinions, most believe they violate religious or ethical principles but have a hard time explaining which ones. When pressed, they say, "it is just wrong." In a recent poll, 87% said they want them labeled (CBS News 2008). In Europe, the situation is even more clear: The E.U. zero-tolerance policy states that any imported food or feed material cannot contain even trace amounts of GM substances that have not been approved by the E.U. Council. They are already labeled there, as well as in Japan and Australia. The reasons are similar to those offered in the U.S.: basically fear (Eurobarometer 2006). People are simply leery about them.

In addition to concerns about health and environmental consequences (i.e., hard impacts) people generally tend to be cautious about what they eat. GM foods seem foreign—more like chemicals than food. The very idea does not square with the traditional sense that food should not be too processed. Perhaps it shatters the illusion that agricultural is a time-honored practice tied to the rhythms of the seasons. Given the option, most people surveyed would opt for conventional over industrialized farming and natural over genetically modified food. The concern over GM food is quite real whether or not any ultimate justification can be offered for it. People seem to be more inclined to trust their gut than be persuaded by science when it comes to food.

The situation with GE animals is even more apparent. People surveyed in the U.S. have even stronger reservations about them than they do about GM food. Of those surveyed, 70% said they are "very uncomfortable" with the idea of engineering animals for food (Pew 2006). Those surveyed had a hard time explaining why. Perhaps the idea of slicing together animal genes seems creepy. Perhaps it evokes a "yuck factor," or the "wisdom of repugnance" that Leon Kass speaks about—a deeply seated negative response that goes beyond our ability to articulate it (Kass 1997). My gut tells me that there's something wrong with cloning, genetic engineering, and enhancement technologies. These prejudices are, of course, no less irrational that refusing to eat foods that are not Kosher or vegan. A claim that lacks evidence is not without reason.

The case against functional foods can also be framed as a matter of judgment (probably even better than as a justice issue). We can imagine any number of good reasons to avoid them and to judge them unfavorably. For example, you might argue that they are unnecessary for a healthy lifestyle; they seem like a gimmick; they probably taste worse (or at least no better) than regular food; they cost too much; they are sold in stores (or in the aisles) you typically don't go to; they blur food and medicine in a way that only dilutes a food heritage; people are skeptical about the food industry. These are all valid reasons to eschew functional foods: they make our foodscape worse, not better.

The same could be said of junk food. Perhaps they taste bad (in the ordinary sense); they are manufactured by large-scale industries that run counter to our expectations about proper food production; the large-scale manufacturers are part of a distribution network that undermines entrepreneurship; they are homogenizing and erode regionality and regional cuisine—one of the most important aspects of a culture. Junk food makes the foodscape worse, something to be opposed to on aesthetic and ethical grounds.

It can be treacherous even to venture into a discussion of techno food and the good without sounding like a moralizing wag—and equally difficult to praise without seemingly dwelling on the trivial. Nevertheless, philosophers can play a role in discussions about food and food policy by calling attention to the role that the right and the good play, both rhetorically and practically. We can call attention to the justice claims invisible to the empirical sciences; and the effects of poor judgment and taste on the quality of our lives. Once we sort through our arguments and distinguish claims about justice from those of appropriateness, we will have a more clear understanding of what can be done about it. Justice claims are better suited for litigation; quality of life claims are better suited for personal choice and public deliberation. Either way, we can decide what do make and what to eat in the absence of scientific evidence or metaphysical speculation into the ultimate nature of things. We should avoid Romantic critiques of techno foods that fetishize nature and are phobic toward science and technology. And we should avoid advocating for any food traditionalism and provincialism that stands in stark contrast to the universalist aspirations of philosophy.

REFERENCES

Berry, Wendell. (1991). "The Pleasures of Eating." In *What Are People For?* Berkeley: North Point.

Boenink, Marianne, Tsjalling Swierstra, and Dirk Stemerding. (2010). "Anticipating the Interaction Between Technology and Morality: A Scenario Study of Experimenting with Humans in Bionanotechnology," *Studies in Ethics, Law, and Technology*, vol. 4, no. 2, pp. 1–38.

Borgmann, Albert. (1984). *Technology and the Character of Contemporary Life*. Chicago: University of Chicago Press.

CBS News. (2008, May 11). Figuring Out What's In Your Food. CBS News/*New York Times* Poll. Retrieved from http://www.cbsnews.com/stories/2008/05/11/eveningnews/main4086518.shtml

Comstock, Gary. (2000). *Vexing Nature: On the Ethical Case Against Agricultural Biotechnology*. Norwell: Kluwer Academic.

Ellstrand, Norman C. (2002). "When Transgenes Wander, Should We Worry?" In *Genetically Modified Food: Debating Biotechnology*, edited by Ruse, Michael and David Castle. New York: Prometheus Books.

Eurobarometer. (2006). "Europeans and Biotechnology in 2005: Patterns and Trends," *Eurobarometer* 64.3.

Hallman, William K., Wilfred C. Hebden, Helen L. Aquino, Cara L. Cuite, and John T. Lang. (2003). *Public Perceptions of Genetically Modified Foods: A National Study of American Knowledge and Opinion* (Food Policy Institute Report No. RR-1003–004). New Brunswick, NJ: Rutgers University, Food Policy Institute.

Heasman, Michael, and Julian Mellentin. (2001). *The Function Foods Revolution: Healthy People, Healthy Profits?* London: Earthscan.

Holland, Allan. (1997). *Animal Biotechnology and Ethics*. New York: Springer.

Kant, Immanuel. (1987). *Critique of Judgment* (Werner Pluhar, trans.). Indianapolis: Hackett.

Kaplan, David M. (2004). "What's Wrong with Genetically Modified Foods?" In *Ethical Issues of the 21st Century*, edited by Frederick Adams. Charlottesville: Philosophy Documentation Center.

Kass, Leon. (1997, June 2). "The Wisdom of Repugnance," *The New Republic*, pp. 17–26.

Nestle, Marion. (2007). *What to Eat*. New York: North Point Press.

Pew Initiative on Food and Biotechnology. (2006). "Public Sentiment About Genetically Modified Food." Washington, DC: Mellman Group.

Pollan, Michael. (2009). *In Defense of Food: An Eater's Manifesto*. New York: Counterpoint.

Scrinis, Gyorgy. (2011). "Nutritionism and Functional Foods." In *Philosophy of Food*, edited by David M. Kaplan. Berkeley: University of California Press.

17 Human Enhancement and Well-Being

Bengt Brülde

New technologies might improve our well-being in many different ways, for example, by improving our health, our living and working conditions, and by making it possible to engage in interesting and enjoyable activities. It is worth noting, however, that there has been little or no increase in average happiness in industrially developed countries during the last 50 years (Argyle 2001; Layard 2005). This does not imply that technological developments are not beneficial, but it suggests that these benefits are often of a limited duration, for example, because we adapt rather quickly to external changes (Frederick and Loewenstein 1999).

Technological developments sometimes give rise to "ethical worries," for example, about the environment, social relations, or spiritual well-being. A central theme in the ethics of technology is *risks*, or possible harms. Technologies such as nuclear power may give rise to considerable harm, and even though the probability is rather low, this possibility has given rise to considerable reflection on how risks should be assessed, evaluated, and managed. For example, what makes a risk unacceptable? Risk awareness seems particularly high in the case of biotechnology. Some think that, for example, genetic engineering in agriculture or medicine is so risky that the whole enterprise should be stopped. In these cases, the focus is often on the physical harms that might result if the technology fails. The possible harms may also be social or economic, however, and social risks (e.g., increased inequality or oppression) are typically present when a technology succeeds rather than fails.

In this contribution, I focus on those biotechnologies that aim to improve human functioning or well-being by manipulating the human organism in some way or other. The manipulation involved can be of several kinds. Some techniques stimulate the brain in different ways and can be regarded as *neurotechnologies*, for example, deep brain stimulation (DBS), transcranial direct current stimulation (tDCS), or transcranial magnetic stimulation (TMS). In the case of *human genetic engineering*, the genotype of a person is manipulated to change her phenotype, or to determine the phenotype of a newborn. The organism can also be manipulated by pharmacological means (e.g., by drugs or growth hormones), by radiation, or by surgery.

Most of these biotechnologies have been developed for "therapeutic purposes"—for example, to cure or prevent disease or to eliminate defects—or for palliative purposes (to relieve suffering). These technologies sometimes give rise to undesirable side effects, but many of them seem to work rather well, including some emerging technologies, such as DBS. Other emerging technologies may prove helpful in the future, for example, gene therapy. These technologies need not be used therapeutically, however. They can also be used to enhance different properties of healthy people, for example, cognitive abilities, physical endurance, life spans, or appearances. Yet not all enhancements have developed out of therapeutic technologies. A good example of this is nootropics, that is, drugs, nutraceuticals, and functional foods that purport to improve human cognitive abilities.

THE NOTION OF ENHANCEMENT

Therapeutic interventions and enhancements both aim to improve our existence through manipulating our bodies in different ways. It is commonly suggested that enhancement and therapy are mutually exclusive categories, and that a biomedical intervention should count as enhancement if it is *not* curative. This view is problematic, however. First, it makes the notion of enhancement dependent on the notion of disorder, and all the obscurities that are associated with the latter notion are incorporated into the former. Second, if the view is combined with the normative idea that we should give priority to treatment, it has unacceptable consequences, because it entails that we should help a better functioning person if he suffers from an identifiable disorder. This suggests that the crucial question is where on the health scale (etc.) the improvement is intended to occur, for example, how well the person who receives the help is functioning on the whole (Brülde 2011). An intervention should count as enhancement if it aims at improving health above a certain level, for example, "normal" health.

Enhancement interventions can be classified on the basis of the type of technology involved (e.g., as neuroenhancement, genetic enhancement, or pharmacological enhancement), but they are more commonly classified on basis of what property or function they purport to improve. Some interventions aim to improve our cognitive abilities, for example, our intelligence, memory, or concentration. Other interventions aim to improve our physical performance or appearance, but physical enhancement can also aim at improving our capacity for extracting nutrients from current foods, help us adapt physiologically to climate change, improve our immune system and the body's ability to repair damaged tissues (Buchanan 2008a), or reduce our need for sleep. In the case of mood enhancement, the primary aim is to make us feel happier, more engaged, or more energized; and in the case of moral enhancement, the aim is to make us morally better, for example, by improving existing capacities for impulse control, sympathy, or altruism

(Buchanan 2008a). Biomedical technologies can also be used to improve our reproductive capacities or to extend life by intervening in the mechanisms of aging.

ENHANCEMENT AND ETHICS

Many of these possible enhancements would most probably, if successful, be beneficial, not just to those enhanced, for example, if people would become morally better and more productive, this would benefit society as a whole. The possibility of using emerging biotechnologies to enhance human capacities also gives rise to ethical concerns, however, which might give us reason to adopt a restrictive attitude toward these technologies. For example, it is sometimes argued that we should not "play God" or that it is always wrong to alter human nature (Buchanan 2008b), but the relevant worries are more commonly formulated in terms of possible harms. For example, several supposedly risk-free enhancements have later been associated with serious unanticipated side effects (Caplan and Elliott 2004), and genetic engineering might not just harm the person in the short run, it might also prove impossible to stop the process, as genetic changes may be transmitted to one's offspring.

However, most risk arguments purport to show that enhancement practices might give rise to harm *even if successful*, at least if widely used. These worries are often based on the assumption that "the most significant risks of enhancement are serious unintended social or collective harms" (Buchanan 2008a). As Elliott puts it (in Caplan and Elliott 2004), the "traditional worry about enhancement technologies is that users of the technologies are buying individual well-being at the expense of some larger social good." For example, a widespread use of these practices might reinforce perfectionist attitudes and make it harder to value human life in its imperfection. Enhancement practices might also reinforce "racist norms": "I may get cosmetic surgery for my 'Asian eyes' or use skin lighteners for my dark skin, but I reinforce the implicitly racist norms that say that Asian eyes or dark skin are traits to be ashamed of" (Caplan and Elliott 2004).

A more common concern is that a widespread use of enhancement might increase existing social inequalities. If the relevant technologies become available on the market, the competitive advantage of the privileged will be even bigger than today, both on the national and global levels, which is unfair.[1] This worry is based on the assumption that there will be a market in enhancements, and this holds for other worries too. For example, Alex Rajczi (2008) believes that the enhancement industry will advertise their products in a way that will make people overvalue these enhancements, which might lead them to devote time, energy, and resources to enhancements that could be better spent on pursuing happy or ethical lives. Elliott is also concerned by the possibility that the pharmaceutical industry will control the emerging enhancement technologies, and that "the self-improvement agenda will be set not by individuals, but by powerful corporate interests" (Caplan and Elliott 2004). All these possible developments may well have negative effects on individual well-being.

Now, the existence of such risks is not in itself an argument against developing these technologies, but they give rise to questions about how these risks should be managed, for example, how the enhancement market should be regulated. Another general question is when, if ever, enhancement is a legitimate goal of medicine (Brülde 2011). There are also worries related to specific technologies or specific areas of enhancement. For example, it might be argued that considerable life extensions can have all sorts of awkward consequences (Temkin 2008), or that enhancing cognitive capacities may enable truly evil individuals to produce weapons of mass destruction (Liao, Savulescu, and Wasserman 2008).

MOOD ENHANCEMENT AND LONG-TERM HAPPINESS: SOME PRELIMINARIES

In the rest of this chapter, my aim is to enrich the ethical discussion on enhancement by reflecting on a question of high ethical relevance, namely, what effects technologically induced mood enhancements are (if successful) likely to have on our long-term happiness. Suppose that effective and harmless mood-enhancing drugs can be developed, and that we will gain access to these drugs. It might seem obvious that this would make the world a happier place, but would this really be the case? For example, what indirect effects would a widespread use of mood enhancers have on the determinants of happiness?

"Mood enhancement" in the broadest sense can be defined as any intervention or method that aims at improving people's mood in some dimension or other, directly or indirectly (Brülde 2007a). Mood enhancement in the narrow sense only includes those biomedical techniques that aim at enhancing mood in a relatively direct way, for example, psychopharmaceuticals, recreational drugs, DBS, tDCS, and TMS. The relevant improvements can take place in the hedonic dimension (an intervention can make us feel better on the whole), but people sometimes have other mood variables in mind, such as alertness, energy, comfort, and relaxation. In this contribution, I restrict my attention to drugs that are designed to increase hedonic level directly.

If mood-enhancing drugs work, they do so by affecting the human brain. There are many neural mechanisms through which this can be achieved, and depending on which specific mechanism is activated or affected (e.g., dopamine or serotonin levels), the non-hedonic effects may vary significantly. Mood enhancement does not occur in a physiological vacuum, and it is hardly possible to affect mood without affecting something else at the same time (Brülde 2007a).[2] It is of course possible to *imagine* mood enhancers that have no cognitive or behavioral effects at all, or that make us feel exactly the way we feel when we are satisfied with our lives. These fantasies are extremely unrealistic in the pharmacological case, however, but for the sake of the argument, I will not pursue this issue any further.

The term "happiness" can be used in several ways, but if we restrict ourselves to the psychological uses of the term, we can distinguish three major approaches. On the *life satisfaction view*, happiness is regarded as a positive cognitive attitude toward one's life as a whole. To be happy is to evaluate one's own life in a positive manner, to regard it favorably. On the *affective view*, happiness is some kind of affective state. To be happy is to feel good. There are several versions of this view, and I have argued elsewhere that the most plausible version regards happiness as a high hedonic level, as a matter of how pleasant one's total mental state is (Brülde 2007a; 2007b). On the *hybrid view*, happiness is a combination of life satisfaction and subjective well-being. In my view, this is the most plausible conception of happiness, and it is also the most commonly adopted conception in the field (Brülde 2007b).

DOES MOOD ENHANCEMENT HAVE A DETRIMENTAL EFFECT ON HAPPINESS?

The arguments that purport to show that mood enhancements might have negative effects on our long-term happiness can be divided into three groups: (1) Mood-enhancing drugs may be detrimental to the happiness of the consumers of these drugs, by affecting some determinant of their long-term happiness; (2) these drugs may have negative effects on the happiness of the immediate environment of the consumers, for example, spouses or colleagues; and (3) a wide spread use of mood-enhancing drugs may have certain effects on society as a whole, which may be detrimental to the happiness of its citizens.

When assessing these arguments, I assume that mood-enhancing drugs work, that is, that they make us feel better in the short run, that there are no undesirable side effects, and that a regular intake makes us feel better over time.[3] Feel better than *what*? "Better than without the drug" begs the question. A more plausible answer is: "Better than without the drug, everything else (e.g., activities and relationships) being equal."

ARE MOOD-ENHANCING DRUGS BAD FOR THE INDIVIDUAL?

What the arguments of the first group purport to show is not so much that the long-term effects of mood-enhancing drugs are positively bad, but rather that such drugs would affect the person's future happiness negatively *in comparison to* some alternative method, or to no enhancement at all.

Let us first note that certain displeasures have direct positive effects on some factor that, in turn, has good effects on our long-term happiness. For example, some fears (like most pains) are functional. Negative moods may also be adaptive, for example, despondency may help us redefine our goals or take steps to develop new personal resources (Morris 1999). If a certain displeasure is an appropriate response to an untenable situation, the appropriate way to get rid of it is not to take drugs, but to deal with the underlying problem. It can also be argued that certain displeasures have positive effects on our personal development, which in turn has good effects on our long-term happiness, for example, by improving our relationships. Frustrations, disappointments, discomfort, and sadness may improve our self-knowledge and make us more resilient, less self-centered, more authentic, or more empathic. To use mood-enhancing drugs to eliminate all frustration would probably be detrimental to our personal development, which suggests that mood enhancement is not good under all conditions, and that the relevant drugs should be carefully administered.

Let us now shift our attention to the means or method used to improve mood, for example, on how a regular intake of mood-enhancing drugs would affect the person's long-term happiness as compared with other possible means.[4] There are many factors that have been shown to have a positive effect on happiness over time, for example, certain external living conditions, relationships, activities, and psychological factors. Is there any risk that mood-enhancing drugs will affect these important determinants of happiness in such a way that the net effect will be negative?

How happy someone is depends (to a small extent) on her immediate *living conditions*, for example, on her physical environment, income, social position, and level of education. How a frequent use of mood-enhancing drugs would affect these determinants depends on their non-hedonic effects, for example, how they affect activation levels. However, my guess is that the effect on the person's living conditions would be rather small, unless they prevent users from striving for goals and making efforts.

A person's happiness also depends on her health, activities (work and leisure), activity level, intimate relationships, and social network. These objective factors have a larger impact on her long-term happiness than her living conditions. Would a frequent use of mood-enhancing drugs have a detrimental effect on any of these determinants? Again, this depends on the relevant non-hedonic effects, for example, how the drug can be expected to affect sociability, arousal, and activation levels, performance, and aggression. It is possible that some drugs would have a detrimental effect on marriage and friendship, or that they would make the person more socially and physically passive, which would have negative effects on her happiness.

The variation in happiness in well-ordered societies is, to a large extent, caused by differences in personality, motivation, and abilities. Factors such as optimism, self-esteem, a problem-focused coping style, an idealistic value-orientation, and mental effectiveness are all conducive to happiness, and it is also beneficial for us to be intrinsically rather than extrinsically motivated, and to be a "satisficer" rather than a perfectionist. Our happiness might of course be increased if any of these psychological determinants could be enhanced, but how would a frequent use of *mood*-enhancing drugs affect these determinants? Again, this depends on the non-hedonic effects; but as far as I can tell, such drugs may well have positive effects on some psychological determinants, for example, it may well give rise to more optimism, self-esteem, and idealism. Some drugs would perhaps have a negative effect on problem-focused coping, however, and a frequent use of mood enhancers may well make us less mentally effective and more emotionally perfectionist, but the net effect on the psychological determinants may well be positive.

These arguments do not show that one should abstain from using mood enhancers altogether, but only that that there are circumstances under which one should (for self-interested reasons) abstain from taking them. We should also note that mood enhancers may have positive effects on the determinants of happiness, and that this would give us a good reason for taking them (in addition to the immediate effect). It has been shown that happiness has positive effects on some of its determinants (which gives rise to virtuous circles), for example, that happy people are healthier, more social, and more active, and that they perform better and tend to be more creative and flexible in their thinking (Argyle 2001; Layard 2005). It is possible that this holds for artificially induced positive mood as well, but the fact that "normally caused" happiness has these effects does not imply that artificially induced mood has similar effects. To what extent this is so needs to be investigated empirically.

ARE MOOD-ENHANCING DRUGS BAD FOR THE IMMEDIATE ENVIRONMENT?

If someone would take mood-enhancing drugs on a frequent basis, how would this affect his partner, children, parents, friends, or colleagues? Again, this depends on what effects the drug has on sociability, arousal and activation levels, and aggression. It is possible that some drugs would have a detrimental effect on intimacy, trust, guilt, or empathy, and that some drugs would also make the person more violent and aggressive, which would have negative effects on the happiness of others. However, we must not forget that happy people are (as a rule) more social, cooperative, agreeable, generous, and helpful, and that they have better relationships (Argyle 2001). This may well hold for artificially induced positive mood as well, but we cannot be certain of this. It can also be argued that, for example, intimate relationships may become less crucial for happiness if there is constant access to good mood enhancers.

NEGATIVE EFFECTS ON THE SOCIAL DETERMINANTS OF HAPPINESS?

There are several large-scale social factors that have good long-term effects on happiness. For example, we are happier if we live in a society that is wealthy (up to a certain point), peaceful, secure, and politically stable. It is good to live in a democracy where people's rights are protected, and in an individualistic culture (where people are free to live their lives in accordance with their own values) characterized by interpersonal trust (Argyle 2001; Layard 2005; Frey 2008). A widespread use of mood-enhancing drugs would probably not have negative effects on these determinants. The only risk I can think of is that political participation might decrease, and possibly also interpersonal trust and solidarity. But if there is constant access to good mood enhancers, these factors may carry less weight.

CONCLUSION

After having offered a brief overview of the ethics of enhancement, I have reflected on whether mood-enhancing drugs might (even if successful) have detrimental effects on our long-term happiness. In my view, this depends on what non-hedonic effects these drugs would have, for example, on sociability, activation, performance, and aggression. But in any case, there is some cause for concern on the following counts: Mood enhancers may have a detrimental effect on marriage and friendship, social and physical activity, the capacity for rational problem-solving and mental effectiveness, and they may lead to an increase in emotional perfectionism. A widespread use of mood-enhancing drugs may also have negative effects on political participation and interpersonal trust, which might affect the average happiness in society as a whole. On the other hand, many determinants of happiness might be positively affected by these drugs. However, it remains to be seen whether the positive effects of "normally caused" happiness occur in the case of artificially induced positive moods as well. We simply need to know more about how mood-enhancing drugs are supposed to work, and about the non-hedonic effects of these drugs. At present, it seems that the best "mood enhancement" is achieved through "natural causes," and that we should help each other to improve our relationships, find more rewarding activities, and learn to think differently.

NOTES

1. The gap between the rich and the poor might also grow for other reasons, for example, if the new technologies are commercially successful, more research will most likely be conducted in this area, rather than in areas that are more important for the poor countries.

2. But note that some forms of neuromodulation (e.g., by means of electrode implants) are quite immediate, precise, and fine-tunable.
3. That such drugs might also increase our happiness by affecting our *life satisfaction* is less relevant in this context, because this variable can most likely not be affected directly, and because I have already decided to restrict my attention to drugs that are designed to increase hedonic level directly.
4. I am aware that this question is rather vague, because I do not specify what I mean by "regular" in this context: a constant, habitual use of enhancement drugs would most likely not affect the user in the same way as an occasional use would. Most of the time, I have a rather frequent use in mind, but I think there is good reason to keep the question somewhat open.

REFERENCES

Argyle, M. (2001). *The Psychology of Happiness* (2nd ed.). London: Routledge.

Brülde, Bengt. (2007a). "Can Successful Mood Enhancement Make Us Less Happy?" *Philosophia*, vol. 79, pp. 39–56.

Brülde, Bengt. (2007b). "Happiness Theories of the Good Life," *Journal of Happiness Studies*, vol. 8, pp. 15–49.

Brülde, Bengt. (2011). "Is Mood Enhancement a Legitimate Goal of Medicine?" In *Enhancing Human Capacities*, edited by Ter Meulen, R., J. Savulescu, and G. Kahane. (Hoboken, NJ: Blackwell-Wiley.) pp. 218–229.

Buchanan, A. (2008a). "Enhancement and the Ethics of Development," *Kennedy Institute of Ethics Journal*, vol. 18, pp. 1–34.

Buchanan, A. (2008b). "Human Nature and Enhancement," *Bioethics*, vol. 23, pp. 141–150.

Caplan, A., and C. Elliott. (2004). "Is It Ethical to Use Enhancement Technologies to Make Us Better than Well?" *PLoS Medicine*, vol. 1, pp. 172–175.

Frederick, S., and G. Loewenstein. (1999). "Hedonic Adaptation." In *Well-Being: The Foundations of Hedonic Psychology*, edited by Kahneman, D., E. Diener, and N. Schwarz (New York: Russell Sage Foundation Press) pp. 302–329.

Frey, B. S. (2008). *Happiness: A Revolution in Economics*. Cambridge, MA: MIT Press.

Layard, R. (2005). *Happiness: Lessons from a New Science*. London: Allen Lane.

Liao, S. M., J. Savulescu, and D. Wasserman. (2008). "The Ethics of Enhancement," *Journal of Applied Philosophy*, vol. 25, pp. 159–161.

Morris, W. N. (1999). "The Mood System." In *Well-Being: The Foundations of Hedonic Psychology*, edited by Kahneman, D., E. Diener, and N. Schwarz (New York: Russell Sage Foundation Press) pp. 169–189.

Rajczi, A. (2008). "One Danger of Biomedical Enhancements," *Bioethics*, vol. 22, pp. 328–336.

Temkin, L. S. (2008). "Is Living Longer Living Better?" *Journal of Applied Philosophy*, vol. 25, pp. 193–210.

18 On Hubris and Hybrids

Ascesis and the Ethics of Technology

Peter-Paul Verbeek

The question of the good life has come to play a substantive role in the ethics of technology. To mention just a few examples, Albert Borgmann convincingly shows how deeply technology has become interwoven with contemporary ways of answering the question of the good life (Borgmann 1984; Higgs, Light, and Strong 2000), Tsjalling Swierstra analyzes how complicated it is to bring questions of the good life into public ethical deliberation (Swierstra 2000), and Philip Brey assesses the impact of new media on the quality of life (Brey 1998).

In this chapter, I will take the current discussion of technology and the good life back to its roots in ancient Greek thinking. By doing this, I will explore how 'good life ethics' can help to further articulate the so-called nonmodernist approaches to technology that have developed over the past decades, especially in actor-network theory (Latour 1994) and postphenomenology (Ihde 1990; Verbeek 2005). The central idea in these approaches is that humanity and technology need to be seen as fundamentally intertwined: they shape each other, and the one cannot be understood without the other. This makes the ethics of technology a complicated affair, because it makes it impossible to 'defend' humanity against technology—even the criteria by which we can assess technology are technologically mediated in this approach. As I will argue, though, good life ethics makes it possible to develop an alternative form of ethical reflection, which aims to *accompany* the development and use of technologies 'from within' rather than *assessing* them 'from outside.'

To elaborate on this, I will first discuss the relation between the classical notion of 'hubris' on the one hand and the contemporary notion of 'hybrids' on the other, which plays a central role in current 'nonmodern' approaches to technology. The central idea in these nonmodern approaches is that human beings and technological artifacts cannot be separated in our understanding of reality; they co-constitute each other, and for this reason, both human beings and technologies need to be understood as hybrid entities. This boundary-blurring notion of hybridity, though, is closely related to the classical Greek notion of *hubris*. Both notions are related to the boundaries of the human being. Whereas hybridity pertains to the *factual* boundaries of the human being, the notion of hubris pertains to its *normative* boundaries, indicating the overconfidence that transgresses the boundary between the human domain and the domain of the gods. As I will show, these two notions together can form the basis of a nonmodern ethics of technology: On the one hand, they make visible how humans are continually shaped by the technologies they develop and use, whereas on the other hand, they show how the hybridization of human beings and technologies continually raises ethical concern.

Having elaborated this, I will address how this ethical concern can take concrete shape. Drawing on Michel Foucault's reading of classical ethical approaches to sexuality, I will develop the outlines of an ethics of technology that focuses on 'governing' the relations between humans and technologies, rather than protecting the boundaries of the human being and its dignity. The notion of *ascesis* will play a central role here—not to indicate how human beings can refrain from using technology, but to show how human beings can actively shape themselves in close interaction with the ways in which technologies mediate their existence.

HYBRIDS AND HUBRIS

Various approaches in the philosophy of technology that have developed over the past decades have shown that an adequate understanding of the social and cultural roles of technology requires that we soften the boundaries between humans and technologies. Don Ihde's post-phenomenological approach to technology, for instance, begins with the idea that philosophy of technology should direct its attention to human-technology *associations*, rather than to technologies themselves. By analyzing the character of the various shapes these human-technology relations can take, Ihde has been able to analyze how technologies help to shape human interpretations and practices (Ihde 1990). Bruno Latour's actor-network theory works along comparable lines. To understand reality, Latour claims we need to give up the a priori separation we make between human and nonhuman beings. Human 'subjects' and nonhuman 'objects' are always intertwined, and the networks of relations between them produce the world in which we live (Latour 1992).

Following Latour's analysis in *We Have Never Been Modern* (Latour 1993), approaches like these can be classified as 'nonmodern.' They do not take the modernist separation of subject and object as their starting point, but think in terms of a mutual constitution of the human and the nonhuman—in the case of technology, of humans and technologies. These approaches actually have a place in a longer tradition in philosophical anthropology, where the human and the technological have always been understood in close relation to each other. Ernst Kapp (Kapp 1877), for instance, approached technologies as projections of the human body—a hammer is a materialization of the fist, a saw of the teeth, and so on. Hermann Schmidt (Schmidt 1954) argued that technologies also embody human *capacities*: while a tool has to be operated both physically and mentally by a human being, a machine takes over the physical part and an automation the mental part. Arnold Gehlen then added to these analyses that these tools, machines, and automata all have specific relations to the human, organic being. Some technologies facilitate our organs, whereas others strengthen them or even replace them (Gehlen 2003).

Hybridization, therefore, has always been a central theme in philosophy of technology. There is no way to understand human beings and technologies separately; we need to understand the human in terms of the technological and vice versa. The French philosopher Bernard Stiegler even speaks about an 'originary technicity' of human existence: Human existence has always taken shape in close interaction with technologies (Stiegler 1998). Not only have we never been modern—we have always been hybrids.

This conclusion has serious ethical implications. First of all, not all human-technology associations can be considered desirable or ethically defendable—don't we need a criterion to distinguish good hybrids from bad ones? Actually, virtually all technological innovations went through a period of societal resistance. To mention an example in the realm of media: while Plato was afraid that the broad introduction of writing would threaten the process of thinking ("Once any account has been written down, you find it all over the place, hobnobbing with completely inappropriate people no less than with those who understand it"; Plato 2002, p. 70), contemporary critics of technology are afraid that social media will have a negative impact on relations of friendship and love and on people's attentiveness and concentration (Carr 2010; Turkle 2011).

Current discussions on human enhancement are an interesting example as well: The very possibility to modify human nature to enhance our performance in specific fields is currently under heavy ethical debate (Bostrom 2005; Fukuyama 2002; Habermas 2003; Kass and President's Council on Bioethics 2003). Technological hybridization can be experienced as a form of hubris—which could actually be a reason to think further about the possible etymological relations between 'hubris'/'hybris' and 'hybrid.' Although technology always puts the *factual* boundaries of the human being at stake by continuously offering a potential to change human existence, at the same time, it brings *ethical* boundaries in view. Hybrids and hubris go hand in hand.

A second ethical implication of the hybrid character of human existence is that ethical frameworks cannot be understood separately from technology. The criteria by which we assess technology are technologically mediated themselves. Our understanding of human privacy, for instance, takes shape in interaction with the specific ways in which technologies put our private sphere into question, just like our understanding of human dignity develops in interaction with the technologies we have to reduce suffering. There is no 'outside' of technology from which we can do ethical reflection; ethics can only take place from within our technologically mediated situation.

The fact that technologies change what it means to be human, including our ethical views, does not necessarily imply, though, that hybridization should be seen as a threat to human beings. To the contrary: history has shown that human beings have continually re-invented themselves in relation to technology, and that technologies that initially appeared as a threat end up being the start of something new. The large-scale introduction of writing, for instance, has not led to the end of history, but is actually often seen as its very *beginning*: everything that happened before is now seen as prehistorical. Human beings are still able to remember things, but they have extended their memory into technologies (Stiegler 1998; Clark 2004). Human beings 'expand' themselves with technologies, or better: They reshape themselves in close interaction with technologies. In this process of hybridization, not all technologies will be accepted, and not all forms of hybridity will be seen as desirable. Although the originary technicity of human beings makes it impossible to define a nontechnological human essence that needs to be protected against technology, human beings continually develop new criteria to determine the quality of their lives-with-technologies. Living with technologies is an active process in which the very character of human existence is continually put at stake.

This close connection between hybridity and hubris, therefore, opens up a very specific ethical space. On the one hand, it makes us aware that there is much at stake in the intertwinement of humans and technology, which gives ethical reflection on technology a crucial importance. On the other hand, the hybrid intertwinement of humans and technologies makes it impossible to give ethics the role of a border guard, protecting the boundaries of the human being. We need to find a nonmodern ethical approach to technology, an approach that moves beyond the modern purification of subjects and objects. Although ethical frameworks to assess technology are never pure but always technologically mediated, the goal of ethical activity cannot be to determine how far technology can be allowed to invade the human sphere. Rather than *assessing* technologies from an external position, the ethics of technology should be directed at *accompanying* the intertwinement of humans and technologies, by focusing on the quality of human existence-with-technology. The central question in the ethics of technology is not so much where we have to draw the line—for humans, or for technologies—but how we are best to give direction to the ongoing process of blurring the line: How to shape the interrelatedness between humans and technology that has always been a hallmark of the human condition? We need an ethics that does not stare blindly at the issue of whether a given technology is morally acceptable or not—as is often the case, for instance, in the current debate on human enhancement technologies—but which looks at the quality of our lives as lived with technology. And this takes us directly to the ethics of the good life.

ASCESIS: THE ETHICS OF SELF-CONSTITUTION

The ethics of the good life, as it took shape in classical antiquity, can offer an interesting alternative to modernist ethical frameworks. Good life ethics does not focus on the modernist question of 'how I should behave,' as a moral subject in a world of objects, but on the nonmodern question of 'how to live.' The good life was directed by aretè—a term frequently translated as 'virtue,' but which is better rendered by the word 'excellence.' Ethics, then, was about mastering the art of living.

Foucault on Ascesis

Michel Foucault's later work offers an inspiring reading of this 'ethics of the good life' (Foucault 1990, 1992). Although his earlier work focuses on power, as the forces and structures in society that produce specific human subjects, his later work investigates how human beings can shape their existence *in interaction with* these power structures. Foucault's ethical work does not approach human beings as the mere 'objects' of power, but as subjects who shape their existence in relation to the powers and structures around them.

To make Foucault's approach fruitful for the ethics of technology, some translation work needs to be done. His ethical work did not focus, obviously, on technology; it primarily addressed the classical ethics of sexuality. Foucault investigated the ways in which sexuality has become an object of ethical activity. He demonstrated that the ethics of sexuality has not always been a matter of commandments and prohibitions. In classical times, the ethics of sexuality rather focused on finding good ways of dealing with lust and passion. Sexuality was not primarily organized via a moral code that determined what was allowed and what was not, but rather in terms of styling one's dealing with pleasure. Passions impose themselves on us, so to speak, and ethics was about how to shape oneself as a sexual subject. Excellence was to be achieved by avoiding the extremes of either denying the passions or becoming their slave. Sexual ethics was about mastering the art of living with sexuality.

This "design" or "styling" of oneself in relation to the passions took place in the form of what Foucault called "self-practices." Such practices were ways to experiment with and give shape to one's sexual existence. Foucault also indicated them as "technologies of the self" (Foucault 1997, pp. 223–252). Rather than simply following one's passions and desires, self-practices are aimed at *governing* them by gaining a productive distance to them. In a variety of ascetic and aesthetic practices, the subject thus took shape in an explicit way, rather than being implicitly given in a moral code that requires people to adopt a specific subjectivity without making this explicit. Ethics was a matter of "care of the self": paying careful attention to one's subjectivity, and shaping one's life in a desirable way.

This ethics of shaping oneself as a sexual subject thus followed a typical classical Greek model for ethical practices: It aimed to find a middle position, a balance between two extremes. It was not focused at the question of whether or not specific forms of sexuality are morally acceptable. Between this 'yes' and 'no' it was looking for a 'how.'

In fact, the Greek myth of Daedalus and Icarus embodies this very ethical approach with respect to technology. Daedalus and his son Icarus were kept caught in Crete. They longed for home, but they could not cross the sea—until they got the idea to escape by flying, just like the birds do. They made themselves wings, out of feathers and wax. Using these wings required a delicate balance, though. Flying too close to the sea would cause the feathers to absorb water, which would make them too heavy to fly. Flying too close to the sun, though, would cause the wax to melt, making flying impossible as well. And while Daedalus managed to keep a good balance, flying neither too high nor too low, Icarus got so much carried away by the fact that he was able to fly and by the beauty of the skies that he wanted to get higher and higher—until his wings fell apart and he crashed into the sea.

The ethics of technology, therefore, can learn from classical ethics of sexuality, and especially from Foucault's reading of it. The work of Steven Dorrestijn (Dorrestijn 2006) illustrates this in a very illuminating way. When technological means force themselves on us incessantly, just like the passions do, then the art of living in a technological culture is the art of shaping our own mediated subjectivity. Rather than assessing which technologies we should let in and which we should keep out, we should ask ourselves *what kind of hybrids we want to be*. Technologies cannot be kept out, because they inevitably change our lives as soon as they come into being—often even without them being used, because *not* using them is normally a decision that has to be made explicitly as well. Foucault's approach can teach us that making visible the fundamental intertwinement of humanity and technology does not imply giving up on ethics. Just like he did not start from an 'external' conceptualization of human beings and sexuality, but took their close intertwinement as a starting point, we can base an ethics of technology on the fundamentally mediated character of human existence. Ethical reflection does not need to be directed at the question to what extent technology should be given access to the realm of the human being. It can also be focused on the question of what are good ways of living with technology.

This approach makes it possible to connect hybridity and hubris. Although acknowledging the radically mediated character of human actions on the one hand, which deprives the subject of its alleged autonomy; on the other hand, it enables the subject to develop a relation to the powers that help to shape it, making it possible to co-shape their impact. In our technological culture, technology plays a role comparable to sexuality in Foucault's approach. Because of their fundamentally mediating role in people's actions and interpretations, technologies help to shape what constitutes a good life. And by developing explicit relations to these technological mediations, human beings can take responsibility for their technologically mediated existence.

Ascesis and Technology

The Foucauldian interpretation developed in the previous results in a very specific reading of good life ethics. It does not reduce the question of 'the good life' to issues concerning the 'quality of life.' Focusing on the 'quality of life,' after all, can easily develop into a modernist framework, using a specific definition of the quality of life as a measure for evaluating technologies. Moreover, a Foucauldian ethical approach to technology is to be seen as a *practice* rather than a framework: It focuses on the practice of shaping one's technologically mediated existence, rather than on determining theoretically if specific technologies are morally acceptable or not.

As previously mentioned, such ethical practices of shaping one's existence can be called 'technologies of the self.' They are ways to get actively involved in the constitution of one's subjectivity. Foucault discusses these practices of the self as forms of *ascesis*. Here, *ascesis* should not be understood as "being moderate," or as "rejecting" the powers that are exerted on somebody. Rather, *ascesis* consists in developing a conscious and explicit relation to the powers that be, not to master or conquer them, but to govern one's existence *with* them.

Foucault elaborated four dimensions of this 'ascetic' self-constitution—and Deleuze (1988) has shown that these dimensions actually follow Aristotle's analysis of causation: the *causa materialis* (the material that causes an entity 'to be'), the *causa formalis* (its form), the *causa finalis* (the goal or *telos* of the entity), and the *causa efficiens* (its 'maker'). For Foucault, the four dimensions of the constitution of one's moral subjectivity are: the moral substance (the 'material' of ethical activity, the part of the self that is taken as a point of application for ethical activity), the mode of subjection (the 'maker' or 'source' of the ethical subject), the practices of the self (the form that shaping one's subjectivity takes, the technologies of the self), and the *telos* (the goal at which self-practices are directed).

In the case of technology, these four dimensions of the moral self take on a specific shape. The moral substance is people's technologically mediated existence: the actions and experiences that now come about in a technologically mediated way. The mode of subjection, the authority that is exerted on someone, is the power of technological mediation. Self-practices then, are the deliberate ways in which people shape their lives in interaction with technologies, whereas the *telos* of these practices is to develop a free relation to technology—a relation in which human beings develop the possibility to shape their lives in good ways in close interaction with the powers of technology.

This technological *ascesis* implies quite a different elaboration of the ethics of technology than many other approaches have. The central question is not if a specific technology is morally justifiable or not, but what a good life in interaction with this technology looks like. The issue is not whether or not the technology itself is morally justifiable (as is the case when focusing on technological risks, for instance), but how we can find good ways of shaping our lives *with* that technology. Current discussions on social media, for instance, tend to focus on the potentially detrimental effects of such technologies on friendship relations, concentrated presence, and attention and engagement with other people and with specific activities (Borgmann 1999; Carr 2010; Turkle 2011).

There is no doubt that such negative effects can indeed occur—thinking in terms of hybrids rather than autonomous human subjects does not imply that every hybrid is a desirable hybrid. But to arrive at more desirable hybrids, the central question in the Foucauldian perspective I propose here is not merely how we can *evaluate* or *assess* these technologies (to determine if they are good or bad), but rather how we can *accompany* their development and social introduction in responsible ways. There is no way to undo the existence of these technologies—we can only make sure that we develop excellent ways of integrating them in our lives and of modifying their design and social embedding. The analyses by Borgmann, Turkle, and Carr can be helpful here—not to reach a verdict about the desirability of social media, but to help people to shape their lives in interactions with the mediating powers of social media technologies.

We should not forget, moreover, that we have gone through such processes of appropriating—in the literal sense of making it one's own—new media already. Plato's resistance against writing, as previously mentioned, bears witness to this. Still, we have developed many ways to make writing 'our own.' It actually became an intrinsic part of what it means to be a human being, even to the point that we now call the period before writing 'pre-history.' Reading and writing have actually become a full-blown part of our existential repertoire, even so deeply that we also learned to maintain relations of friendship or love 'through' them. Writing letters even became a form of art, a genre in itself, which shows that human beings have attempted to achieve excellence in their dealing with writing.

This is exactly the type of ethical relation we could also develop with regard to current media of friendship and love. We will need to engage in 'self-practices' in which we develop new configurations for relations of love and friendship, founded on new ways of dealing with technologies. Moreover, our current awareness of the specificities of *how* technologies have an impact on people's practices and experiences could also inspire practices of technology design. Introducing, removing, or changing specific characteristics of social media can foster specific aspects of love and friendship that we find important.

CONCLUSION: HUBRIS AND THE GOOD LIFE

Good life ethics can play an innovative role in the ethics of technology, connecting contemporary insights in human-technology relations and contemporary issues in the ethics of technology. More specifically, it has the potential to fill the ethical gap that resulted from new conceptualizations of technology, in which humanity and technology are fundamentally intertwined. Although this hybridization of humans and technologies seems to make ethics impossible—if technology always already defines what it means to be human, after all, there seems to be no room for ethical assessment—the ethics of the good life opens up an alternative ethical space. Rather than assessing technologies 'from outside,' it makes it possible to accompany 'from within' how technologies help to shape human practices and experiences. Not the protection of 'humanity' against 'technology' is the core issue on the agenda of this type of ethics, as is often the case in risk-oriented approaches in the ethics of technology, but a careful experimentation with technological mediations, to enable people to shape their lives in conscious and responsible ways, in close interaction with technologies.

But does this then imply that there is no criterion at all? Does the conceptualization of hybridity urge us to forget about the phenomenon of hubris altogether? The answer to these questions is no. For Foucault, after all, the *telos* of subject constitution is freedom. For this reason, the specific Foucauldian interpretation of freedom can provide guidance in governing our relations with technologies. When human beings do not have any room to develop a free relation to technology to shape their existence in interaction with it, we lose a crucial condition for being human: the capacity for self-constitution. This capacity is a first requirement for self-practices to come about, after all. When the power exerted by technological mediation turns into domination, people's subjectivity stops being the outcome of a subtle interplay between technological mediation and human appropriation. Human beings are entirely determined by technology then.

To be sure: Foucault's notion of freedom, as mentioned, should not be confused with the modernist concept of autonomy. Freedom has nothing to do with the *absence* of power; it rather is a specific *relation* to power. As Foucault stated:

> I am sometimes asked: 'But if power is everywhere, there is no freedom.' I answer that if there are relations of power in every social field, this is because there is freedom everywhere. (Foucault 1997, p. 292)

It is not *power* we should reject, therefore, but *domination*, as a "perversion of power" that makes freedom impossible (O'Leary 2002, p. 158). This means that freedom plays a double role in Foucault's work: on the one hand, it functions as the *condition* for ethics; and on the other hand, it is its ultimate *telos* (O'Leary 2002, pp. 154–170). Although ethics is only possible when people are not completely dominated by power relations, ethics itself consists in developing practices of interacting with power, as a basis for constituting one's subjectivity.

The Foucauldian concept of freedom therefore is a rich alternative to the criterion of autonomy that is predominant in ethical theory. Although the concept of autonomy stresses the importance of the absence of 'external influences' to keep the moral subject as pure as possible, the concept of freedom recognizes that the subject gets shape in *interaction* with these influences. The subject is not what remains when all powers and mediations are *stripped* from it; it is what results from an active design and styling of the impact of these powers and mediations. In all this hybridity, one form of hubris remains, and that is giving up the very possibility of developing a free relation to technology.

It must be noted that in practice many technological innovations that might seem oppressive at first glance do indeed offer the possibility to develop a free relation to them. Technological possibilities in anesthesia, for instance, as Gerard de Vries (de Vries 1999) elaborated, were initially seen as immoral. Such technologies intervene in the way God created us: as beings who feel pain when their bodies are opened. By now, we consider it immoral *not* to give people anesthesia when they have to undergo surgery. A technology that at first seemed to threaten human dignity, partly because it 'switches off' human consciousness and therefore seems to annihilate human freedom, has come to be seen as a *contribution* to human dignity and as part of the repertoire human beings have to constitute their subjectivity.

A similar case concerns the brain implants sometimes used for patients with Parkinson's disease. Such devices can have quite serious side effects, as a well-known Dutch case illustrates (Leentjens, Verhey, Visser-Vandewalle, and Temel 2004). A patient in the Leiden Academic Hospital experienced this quite severely: his behavior turned manic, without him being conscious of that. He spent more money than he had, and got involved in extramarital relationships. After a car accident, the implant had to be switched off; suddenly the patient regained his previous self, but at the price of being bedridden and dependent again. Yet, even though this technology seems to dominate human beings so severely that their very freedom seems to get lost, this patient did manage to develop a free relation to it—albeit quite an extreme one. He decided that he neither wanted a bedridden life, nor a life of manic episodes, and therefore he let himself be taken to a closed psychiatric ward where the implant had to be switched on again, under the condition that the hospital would prevent him from getting out when his manic behavior recurred. In freedom, this man decided to use a device that impacts his capacity of being free, and to give up his autonomy by trusting himself to the judgment of the doctors in a psychiatric hospital.

These examples show that we must be very careful when drawing a boundary between power and domination. Every technology puts at stake what it means to be a human being, and only when the relations between human beings and new technologies are starting to crystallize can we develop an explicit relation to their role in our existence. What initially appears as domination can also be the starting point of a new configuration of humans and technologies that is wrongly perceived as dominating because of the fundamental way in which it challenges human nature.

Living a good life in a technological culture, to conclude, is a learning process in which technologies become domesticated and in which human beings actively engage in the technological mediation of their lives. This implies that several conditions have to be met to make an ethics of the good life possible, in a practical way. First, it requires that human beings be equipped with a critical attitude toward technology that allows them to see through their functionality and recognize their mediating role. And second, it requires that the *design* of technologies take place in more reflexive ways. Not only *using* technologies, but also *designing* them can be an important 'self practice' in our technological culture, after all (cf. Verbeek 2011). On a basis like this, the hybrid character of our lives does not appear as a threat to our authenticity, but as its very starting point—with the notions of freedom and hubris as beacons to guide the ways we govern our mediated selves.

REFERENCES

Borgmann, Albert. (1984). *Technology and the Character of Contemporary Life*. Chicago: University of Chicago Press.

Borgmann, Albert. (1999). *Holding On to Reality: The Nature of Information at the Turn of the Millennium*. Chicago: University of Chicago Press.

Bostrom, Nick. (2005). "In Defence of Posthuman Dignity," *Bioethics*, vol. 19, no. 3, pp. 202–214.

Brey, Philip. (1998). "New Media and the Quality of Life," *Techné: Journal of the Society for Philosophy and Technology*, vol. 3, no. 1, pp. 1–23.

Carr, Nicholas. (2010). *The Shallows: What the Internet Is Doing to our Brains*. New York: W. W. Norton.

Clark, Andy. (2004). *Natural-Born Cyborgs*. Oxford: Oxford University Press.

Deleuze, Gilles. (1988). *Foucault*. Minneapolis: University of Minnesota Press.

Dorrestijn, Steven. (2006). *Michel Foucault et l'éthique des techniques: Le cas de la RFID*. Nanterre: Université Paris X (Mémoire).

Foucault, Michel. (1990). *The Care of the Self–The History of Sexuality*, vol. 3. London: Penguin Books. (Originally published 1984)

Foucault, Michel. (1992). *The Use of Pleasure–The History of Sexuality*, vol. 2. London: Penguin Books. (Originally published 1984)

Foucault, Michel. (1997). *Ethics: Subjectivity and Truth*, edited by Paul Rabinow. New York: New Press.

Fukuyama, Francis. (2002). *Our Posthuman Future: Consequences of the Biotechnology Revolution*. New York: Farrar, Straus and Giroux.

Gehlen, Arnold. (2003). "A Philosophical-Anthropological Perspective on Technology." In *Philosophy of Technology: The Technological Condition*, edited by Scharff, R. C. and V. Dusek (Oxford: Blackwell) pp. 213–220.

Habermas, Jürgen. (2003). *The Future of Human Nature*. Cambridge: Polity Press.

Higgs, Eric, Andrew Light, and David Strong. (Eds.). (2000). *Technology and the Good Life?* Chicago: University of Chicago Press.

Ihde, Don. (1990). *Technology and the Lifeworld.* Bloomington: Indiana University Press.

Kapp, Ernst. (1877). *Grundlinien einer Philosophie der Technik. Zur Entstehungsgeschichte der Kultur aus neuen Gesichtspunkten.* Braunschweig, Germany: Verlag George Westermann.

Kass, Leon, and President's Council on Bioethics. (Eds.). (2003). *Beyond Therapy. Biotechnology and the Pursuit f Happiness.* Washington, DC: President's Council on Bioethics.

Latour, Bruno. (1992). "Where Are the Missing Masses? The Sociology of a Few Mundane Artifacts." In *Shaping Technology / Building Society,* edited by Bijker, W. E. and J. Law. Cambridge, MA: MIT Press.

Latour, Bruno. (1993). *We Have Never Been Modern* (C. Porter, trans.). Cambridge, MA: Harvard University Press.

Latour, Bruno. (1994). "On Technical Mediation: Philosophy, Sociology, Genealogy," *Common Knowledge,* vol. 3, pp. 29–64.

Leentjens, A. F. G., F. R. J. Verhey, V. Visser-Vandewalle, and Y. Temel. (2004). "Manipuleerbare wilsbekwaamheid: een ethisch probleem bij elektrostimulatie van de nucleus subthalamicus voor ernstige ziekte van Parkinson," *Ned Tijdschr Geneeskd,* vol. 148, pp. 1394–1398.

O'Leary, Timothy. (2002). *Foucault: The Art of Ethics.* London: Continuum.

Plato. (2002). *Phaedrus* (Robin Waterfield, trans.). Oxford: Oxford University Press

Schmidt, Hermann. (1954). "Die Entwicklung der Technik als Phase der Wandlung des Menschen," *Zeitschrift des VDI,* vol. 96, no. 5, pp. 118–122.

Stiegler, Bernard. (1998) *Technics and Time: The Fault of Epimetheus,* vol. 1. Stanford: Stanford University Press.

Swierstra, Tsjalling. (2000). *Kloneren in de polder: Analyse van het maatschappelijk debat over klonen en kloneren in Nederland.* Den Haag: Rathenau Instituut.

Turkle, Sherry. (2011). *Alone Together: Why We Expect More from Technology and Less from Each Other.* New York: Basic Books.

Verbeek, Peter-Paul. (2005). *What Things Do: Philosophical Reflections on Technology, Agency, and Design.* University Park: Pennsylvania State University Press.

Verbeek, Peter-Paul (2001). *Moralizing Technology: Understanding and Designing the Morality of Things.* Chicago: University of Chicago Press.

Vries, Gerard de. (1999). *Zeppelins–over filosofie, technologie en cultuur.* Amsterdam: Van Gennep.

19 Brave New World

Platonism 2.0

Tsjalling Swierstra

It is not uncommon that new technologies raise new fears. Such fears basically come in two types. The first type of fear concerns the possibility that the technology will not live up to its promise, for example, by breaking down, exploding, or poisoning us and the environment. The other fear concerns the opposite possibility, that is, that the new technology will exactly live up to its promise, that it will indeed fulfill our wishes and grant our desires. But the fear is that in doing so, the technology will keep us from living a good life.

It is this second fear that is particularly manifest in the debate on biotechnology. And nowhere is this fear of detrimental wish-fulfillment expressed more forcefully than in Aldous Huxley's famous 1932 novel *Brave New World*. The novel invokes a future when biotechnology—in combination with other technologies such as mass production and drugs—has helped create a society where every desire is satisfied, where frustration is banned, where everyone is happy, where stability and peace reign supreme—and that appears to our eyes as utterly dehumanized.

Although published more than two decades before the discovery of the double-helix structure of DNA, the book's dark prophecy seems to have lost nothing of its power. Current discussions about the life sciences are interspersed with references to *Brave New World*. Two examples. One of the most influential, as well as thoughtful, voices in the debate is that of Leon Kass, chairman of the President's Council on Bioethics between 2001 and 2005. According to him "the deepest source of public anxiety about biotechnology [is] represented in the concern about 'man playing God,' or about the Brave New World" (2003, p. 10). And in the Preface to his *Life, Liberty, and the Defense of Dignity* he writes: "Aldous Huxley saw it coming two generations ago" (2002, p. 5). Francis Fukuyama simply states at the start of his *Our Posthuman Future: Consequences of the Biotechnology Revolution*, published in the same year as Kass's book: "The aim of this book is to argue that Huxley was right" (2002, p. 7).

The promise of biotechnology is that life itself becomes the object of manipulation, which would include ourselves. The fear is not so much that an evil elite will produce an army of mindless organic robots, but that we will willingly embrace biotechnology because it promises to eliminate common causes of suffering, such as disease and hunger, and even to enhance us humans. A nice prospect? Not according to critics such as Kass and Fukuyama. They warn us that this is a Faustian pact with technology. And in the end we will pay with our soul. If you don't believe them, read *Brave New World*.

But did Huxley indeed "see it coming"? Was he indeed "right" about biotechnology? I think he was not. *Brave New World* does express something essential about the good life, but it neither does justice to technology nor to politics and modern culture.

PLATONISM 2.0

What are Brave New World's essential features? Critics of modern biotechnology usually start by pointing to the famous opening scene of the 'Hatchery,' where new citizens are conceived. Or to put it more evocatively: where they are designed, programmed, and mass-produced as common artifacts. In this way, the dehumanizing consequences of an omnipotent and omnipresent technology are placed in the foreground. But presented like this, we miss the most important thing about Brave New World: its rationale. The problem to which Brave New World provides the solution is the endless cycle of wars that has been wearing humankind down since its infancy.

From the ancient Greeks onward, the dominant strand in political theory has identified frustrated desire as the cause of conflict, war, and chaos. Think of Helen and the Trojan War. In trying to solve this puzzle, philosophers have pointed to frustration's two dimensions: one objective—the availability of goods to quench desire with, the supply side—the other one subjective—the intensity and scope of our desire, the demand side. Frustration is then defined as the result of a mismatch between the objective and the subjective. This definition immediately points to the solution: the problem of frustration—causing unhappiness, instability, conflict, and war—can be solved from two sides: we can *decrease* subjective demand, *increase* objective supply, or do both simultaneously (see for instance McMahon 2005).

Common people, as philosophers have pointed out time and again, typically pick the first option: Their demand is taken for granted, and then they try to increase supply. This strategy is bound to backfire for two reasons. First, ultimately, human beings lack the power to really control supply. Fate strikes down even the rich and powerful—bankruptcy, divorce, accidents, disease, death, a falling out with one's children. Second, as especially the Stoics liked to point out, desire is highly plastic. Right after it has been satisfied, expectations are raised and dissatisfaction takes over again. Desire is an insatiable master, leaving its slaves perennially unhappy.

From Plato onward, wise men have therefore devised a two-layered strategy to become truly happy. First of all: don't attempt to satisfy desire with earthly goods, as these are temporary and not immune to the attacks by Fate. Fortunately, there exist other types of goods that are in principle plenty, attainable, and immune to Fate. Truth, for instance, is a universal, perennial, and attainable good. Similar with Beauty, or, most importantly, Goodness. Or all three in one: God. These can be safely desired. No one desiring God is ever left frustrated. Platonism, says Isaiah Berlin (1999), can be summarized in the form of three theses: (1) all genuine questions know one true answer, (2) these answers are—at least to some—knowable, and (3) they are mutually compatible, forming a harmonious whole. In an ideal Platonist world, eternal and universal truth has replaced error and mere appearance: There is no longer need for conflict or strife or movement, but stability, order, and peace reign supreme. Peace and stability require the firm hand of an enlightened elite in possession of Truth.

This is why a life devoted to *theoria,* contemplation, is much to be preferred above a life devoted to base *praxis.* Unfortunately, because of our bodily nature, is our desire not naturally turned to these safe—because universal and eternal—goods. Therefore, the second part of the strategy entails disciplining and exercising one's own desire to turn it toward these safe (intellectual and spiritual) goods. This is the true meaning of *askesis* (exercise). This lesson, that the wise man should master his desire rather than become desire's slave, is as old as civilization. As a reward, only the wise man is a truly happy man because in him desire and goods, demand and supply, are perfectly in balance.

Sadly, for the common masses, this spiritual and ascetic ideal is not attainable. They are therefore constantly subject to frustration, envy, conflict, and war. The solution to the problem of the common masses can therefore not lie on the level of individual ethics and life style; it can only be political. Lacking self-control, the masses need to be controlled by the elite—through lies, manipulation, religion, force, or whatever means necessary—to safeguard peace and stability. Ironically, it is exactly the fact that this elite of wise men is too elevated to be interested in exerting political power, that makes them perfect, because disinterested and incorruptible, rulers.

This double strategy, disciplining one's appetite for earthly goods and concentrating on universal goods by directing one's mind's eye toward them, constitutes the core of the Platonist conception of the good life—arguably the most influential one in Western history, as it deeply influenced Christian ethics. In its political form, it has provided the kernel of most anti-democratic, elitist, political theories: the masses are not spiritual enough, too much earth-bound, to govern themselves.

But Platonism has one major flaw: it hasn't worked very well, considering our endless histories of instability, conflict, and war. Furthermore, by and large, people don't like to be oppressed, even if it is for their own good. They much prefer to revolt. The Platonist solution to the problem of frustration, conflict, instability, and war, is thus inherently instable.

The genius of *Brave New World* is that in this book, Huxley provides the missing piece to the Platonist puzzle. The piece being technology. Thanks to modern technology, finally the subjective and objective side of wish-fulfillment, supply and demand, can be matched, making oppression of the masses by an elite of philosopher-kings superfluous, because it tackles the problem of lack, frustration, and unhappiness at the root. With regard to the supply side, Truth, Beauty, and Goodness are replaced by mass-produced goods. Fate can only touch particulars, not universals. Mass-produced goods are as universal, and as safe, as Truth, Beauty, Goodness, and God, because they are instantly replaceable with identical exemplars. In *Brave New World,* this is true for artifacts as well as human beings:

> . . . 'after all, what is an individual?' With a sweeping gesture he indicated the rows of microscopes, the test-tubes, the incubators. 'We can make a new one with the greatest ease—as many as we like. . . . (2006, p. 121)

With regard to the demand-side, the role of self-discipline and askesis is taken over by biotechnology and psycho-techniques. As ascetics have always voluntarily tried to stylize their desires, so the inhabitants of Brave New World freely opt for prenatal conditioning in the Hatchery, psycho-conditioning of infants, hypnosis, and drugs to realize the same goal: that their desire is turned toward 'universals' rather than particulars or individuals. Mass production of replaceable identicals + mind control = a perfect match of supply and demand, and thus peace, stability, and happiness:

> The World's stable now. People are happy; they get what they want, and they never want what they can't get. (2006, p. 180)

Brave New World is Platonism democratized, Platonism for the masses, Platonism industrialized, Platonism realized, Platonism 2.0—finally. And all this only thanks to modern technology!

THE ROMANTIC CRITIQUE

When our culture is deeply influenced by Platonism, why then don't we rejoice when reading Huxley's novel? Why will *Brave New World* never be mistaken for a utopia? The answer is: because of our Romantic heritage. We moderns are not simply Platonists, we are also, and by now maybe even more fundamentally, Romantics.

Romanticism is an incredibly rich, heterogeneous, intellectual phenomenon, famous for escaping any coherent definition. I will therefore not attempt one but simply align myself with Isaiah Berlin (1999), who neatly fashions Romanticism as Platonism's opposite. Rather than universal truth(s), romantics stress individual truth, expression, and authenticity. Furthermore, these truths are not readily accessible for others, but remain firmly tied linked to the individual or group. Finally, and most important, these individual truths don't add up to a harmonious, stabile order.

For most Romantics, struggle, strife, change, conflict, competition, victory, and failure are the stuff of life. These are not to be deplored, but to be celebrated. Only under those conditions, humans can acquire depth, individuality, character, a life story, and soul. The human life is a life lived by attaching us, as particulars/individuals, to other particulars/individuals. It is not the life of the anonymous mass consumer of mass-produced replaceables. The Romantic view of the good life is not about being satisfied and calm. There can be no truly human life without frustration, without a gap between supply and demand. The experience of loss, for instance, is intrinsically connected to the experience of love (Nussbaum 1986, 1990). Furthermore, the Romantics opt for an activist stance toward the good life. In the view of the German philosopher Johann Gottlieb Fichte, it is not selfless contemplation of the world by the intellect but the conscious shaping of reality by the will that is paramount. And it is all more about the act, the process, than about the end result. That is why so many Romantic heroes end in tragic defeat—so as to spare them the peace and quiet of old age and retirement.

The role of romantic hero in *Brave New World* is played by 'the Savage.' Having grown up in the unforgiving, masculine world of the Reservation, he gets transported to a world whose morals he abhors and despises as shallow and feminine. The Savage doesn't avoid hardship and suffering, but actively seeks it out, if necessary. If the world denies him conflict, he will find it deep within himself. What appears to the innocent eyes of the inhabitants of Brave New World as masochism in its purest form, is in fact a Romantic living the good, deep, and conflicted life.

The Savage signifies the rejection of the Platonist conception of the good life as one in which objective supply and subjective demand match. And insofar as in Platonism 2.0, this match is due to technological control over the contingencies of life and desire, his life embodies a rejection of the ideal of perfect technological control. In the eyes of the Romantic, if we control reality to such an extent that it ceases to withstand us, to challenge us, to deny us, to resist us—than meaningful life comes to an end.

This seems to me to be the deep ethical truth of the Romantic vision of the good life: Humans are fundamentally ambiguous about controlling reality. We develop technology that is meant to help us to acquire such control, but this endeavor is constantly accompanied by the deep, albeit usually inarticulate, fear that we may eventually in fact succeed—completely. In Fichte's terms: the 'activist' stance where the Ego manifests itself in subjecting reality to its practical will, should ultimately fail, because it is essentially self-negating. If reality no longer resists, we would be condemned to metaphysical solitude because our relationship with it ceases to exist. For example, however much we sometimes wish that our partner fulfilled more of our desires, very few people would choose a perfectly programmable love robot. Recalcitrance is a part of every healthy relationship, because it is a manifestation of alterity, of the fact that the other is in fact other, rather than a manifestation of our own desire. We need opposition just as much as we need to overcome it. A subjugated world may not entail sorrow, loss, humiliation, and defeat, but it no longer has any happiness, gratitude, satisfaction, or surprises in store for us either. If we can make or take everything and do not have to accept or receive anything, life loses its meaning and beauty.

This fear of perfect control of man and world constitutes the heart of *Brave New World*. The novel expresses something that cannot be translated in terms of the accepted moral coinage of not-harming, doing good, autonomy, and justice. It confronts us with the metaphysical loneliness of the human who only meets his mirror image everywhere: a world that obediently yields to all our wishes and desires. That is the underlying fear that overcomes many of us when we contemplate a world without resistance, in which everything, including who we are as humans, has become the object of human intervention. And it is this message that modern commentators of biotechnology such as Kass, Fukuyama, Michael Sandel, and others recognize in *Brave New World*. This is why they think that Huxley's message is timeless, and why they think biotechnology is inherently dangerous—not in the physical, but in the spiritual sense.

ROMANTICISM AND TECHNOLOGY

But I don't think they are right in appealing to Huxley to warn against modern biotechnology. It is not that I do not agree with the Romantic vision of the good life, but I do disagree with Huxley and his present day followers that the world of mass production and consumption, and of modern technology, makes that kind of life hard or impossible. In this aspect, Huxley drew his inspiration from the wrong type of Romanticism.

The Romantic movement started as a reaction in a peripheral part of Europe, 18th-century Germany, against the arrogance and dominance of its modern and successful neighbors: the French, priding themselves for their cold and abstract rationalism, and the British, priding themselves for their empiricism, utilitarianism, liberalism, and commercialism. Because they felt hurt in their pride, the Romantics turned the tables on their neighbors. They felt far, far elevated above the others' base materialism and superficial rationalism, because they themselves were deep, lonely, suffering, torn, and spiritual. Maybe not so happy, but happiness is shallow anyway.

In doing so, they drew on a on a much older repertoire of the good life that opposes the masculine honor of the warrior to the decadent, weak, effeminate character of those who seek luxury and pleasure. This repertoire can be traced from the Greek defiance of the effeminate, luxurious Persians

(Holland 2005) to the present-day Jihadists who are convinced that the rich West will fall because it is corrupt and weak (Buruma and Margalit 2004). It is certainly no coincidence that the characters in *Brave New World* who fullest embody the degeneration caused by a regime of constant and immediate bodily satisfaction are women: Lenina and the Savage's mother.

There is something deeply elitist about Romanticism, in that it thrives on an opposition of spiritual and materialistic—whether we are talking about the exceptional bohemian versus the mass of happy slaves, or of a deep German spirituality opposed to crass English materialism, French rationalism, and Jewish cosmopolitanism. In fact, there exists a hidden complicity with Platonism in this respect: in both approaches, the sheeplike, lazy, shallow, hedonistic, and materialistic masses are contrasted with the 'deep' elite—even thought the Romantic and Platonist conceptions of 'deepness' differ considerably. This same elitism is evident in *Brave New World*. Take for instance the character of Mustapha Mond—one of Brave New World's philosopher-kings. He rules the witless, happiness-driven masses on the basis of his superior knowledge of the Truth. He is not power-hungry; he is rather sacrificing himself for the common good of the herd. Huxley makes it clear that Mond only unwillingly turned away from a scientific career to take upon himself the burden of political power. But although this is all still in accordance with Platonism, Mond is also a deeply Romantic hero. His life is marked by loneliness, loss, and tragedy. He is mild and wise, but his is the wisdom acquired by self-denial. As opposed to the masses, Mond knows suffering and depth. He is the Platonic philosopher-king and the tragic Romantic hero all in one.

This elitism informs Huxley's vision of Brave New World. Never is there any hint that ordinary people could still lead meaningful lives. Given the choice, they would always choose happiness—which in this context equals shallowness, materialism, hedonism, and the lazy recourse to technology rather than to discipline and askesis: "whenever the masses seized political power, then it was happiness rather than truth and beauty that mattered" (2006, 187). And it is this same distrust of the 'lower selves' that informs the pessimistic assessments of his present-day followers.

But I think these Romantics were then, and are now, wrong in their disgust for the world of mass production, consumption, and technology, and in their distrust of the masses. For a start, feeding and clothing the population is a good thing, even if it doesn't rank high on the scale of Romantic ideals. And we have mass production, and thus technology, to thank for that progress. But even more interesting: rather than constituting its antithesis, basic security has provided the conditions for the Romantic ideals of the good life to spread. In terms of Abraham Maslow's pyramid: only when people are warm, fed, and safe, the Romantic urge to self-actualize can take flight. An urge, one may add, that is protected and fed by the liberalism many Romantics despised because it was not heroic enough—trying to avoid violent conflict rather than celebrating it. And mass production also creates, besides a lot of conformism and uniformity, an enormous opportunity for self-expression and individuality through the sheer amount of products and consumer choices. So, Huxley was wrong to simply denounce the world of mass production.

I think the same is true for the Romantics' perception of the world of commerce and trade. Many Romantics simply follow their Platonist counterparts in denouncing commerce as base, apt for people who care for nothing but material gain. A domain where honor counts for nothing, where everything is up for sale, where people don't fight but bargain, and where the masses determine the value of things. Of course, there certainly is truth in this picture, but it is only half the truth. Think of entrepreneurship, or of the thrill of competition and conflict. Capitalism has simply proven to be too dynamic, exciting, dangerous, and destructive to create a mass of happy slaves. Nor has the world evolved into anything as ordered and controlled as Brave New World. Look at the architectural dreams of someone like le Corbusier and you have to admit: these dreams, utopian for some and dystopian for others, have remained just what they were: dreams, not reality. The docile, satisfied world that was so much feared during the first half of the twentieth century has simply not materialized. In sum: Huxley was mostly wrong when he prophesized a world where mass consumption would produce uniform and shallow mass consumers. The modern world allows for a large degree of individualism and diversity. Or to put it bluntly: there is far too much cause for frustration, fear, and unhappiness to start worrying about what would happen in their absence.

But this all concerns the supply side. How about the demand side? This is what the modern critics of biotechnology are most worried about. They fear that biotechnology will bring about the ultimate objectification and manipulation of the human subject. This is what they see foretold in *Brave New World*. But should we really take Huxley's warnings seriously? I doubt it. Indeed, in *Brave New World,* biotechnology plays a pivotal role in adapting people to what life has to offer them. But it is not itself the cause of this manipulation: It is an instrument wielded by a tiny elite of benevolent Controllers who are human subjectivity in a highly concentrated form. Biotechnology enthusiasts never fail to point this out: under democratic conditions, such as ours, there is nothing to fear from biotechnology, because in that situation, people can freely choose to use biotechnology or not. Critics then object that 'freely' is a relative concept when one is, for example, forced by social competition to genetically enhance one's offspring or by social pressure to avoid children with genetic defects. This is an interesting debate, but my point is different. My point is that in *Brave New World,* biotechnology is still an *instrument*, a *tool.* And that is not realistic.

If we have learned one thing in the previous century, it is that technology is not a simple tool at all. Neither is it an impersonal force, towering over us, and dictating to us who to be and how to live. First, after technology leaves the lab for the real world, it never functions quite as flawlessly as the blueprints and prototypes promised. Second, it is a creative or destructive force that rarely obeys our wishes and desires. It may solve existing problems, but while doing so, it usually creates some new ones. Moreover, technology is a highly dynamic force, constantly destabilizing the ways we perceive the world and act in it, closing some doors while simultaneously opening others. These general observations also apply to biotechnology. Now the hype has faded somewhat, we see that many of the radical hopes and fears have been inflated. It is extremely unlikely that in the foreseeable future, disease and hunger will be eradicated thanks to biotechnology, nor that a eugenic logic has already set in. There may be such tendencies, but these invoke immediate reactions that highlight the unique value of any human life. Yes, progress in the domain of biotechnology will affect our lives, as computers have done. But there is no reason to fear that biotechnology will take the drama out of our lives.

Stated in more general terms: Technology will not take the challenges, conflicts, dynamism, unpredictability, and resistance out of our lives, because technology itself is the source of new challenges, conflicts, dynamism, unpredictability, and resistance. The Romantics were right: too much control over the world is not good for humans. The struggle to gain control may give meaning to our lives, but this struggle should never finish. *Brave New World* evokes a world where this struggle has been finished, and human life as we know it is over. When critics of biotechnology refer to *Brave New World,* they refer to this Romantic wisdom. And in this, they are right. What they miss, however, is that a capitalist, liberal, technological world like ours, provides much more room for Romantics than Huxley believed.

REFERENCES

Berlin, Ian. (1999). *The Roots of Romanticism.* Princeton, NJ: Princeton University Press.

Buruma, Ian, and Avishai Margalit. (2004). *Occidentalism: The West in the Eyes of Its Enemies.* New York: Penguin Press.

Fukuyama, Francis. (2002). *Our Posthuman Future: Consequences of Biotechnology Revolution.* New York: Profile Books.

Holland, Tom (2005). *Persian Fire: The First World Empire and the Battle for the West.* New York: Little, Brown.

Huxley, Aldous. (2006). *Brave New World.* New York: HarperPerennial. (Originally published 1932)

Kass, Leon. (2002). *Life, Liberty and the Defense of Dignity.* San Francisco: Encounter Books.

Kass, Leon. (2003, Spring). "Ageless Bodies, Happy Souls: Biotechnology and the Pursuit of Perfection," *The New Atlantis*, pp. 9–28.

McMahon, Darrin (2005). *Happiness: A History.* New York: Atlantic Monthly Press.

Nussbaum, Martha C. (1986). *The Fragility of Goodness.* Cambridge: Cambridge University Press.

Nussbaum, Martha C. (1990). *Love's Knowledge: Essays on Philosophy and Literature.* New York: Oxford University Press.

Sandel, Michael. (2007). *The Case Against Perfection: Ethics in the Age of Genetic Engineering.* Cambridge, MA: Harvard University Press.

20 Care Robots, Virtual Virtue, and the Best Possible Life

Mark Coeckelbergh

Growth of the elderly population and shortage of caregivers place increased pressure on health care systems in North America, Europe, and Japan. Aging rapidly, these societies are on their way to what the U.S. President's Council on Bioethics called 'mass geriatric' societies with 'more long-lived individuals than ever before in human history' (President's Council on Bioethics 2005, p. xvii). The problem is not so much the absolute numbers of elderly people but the ratio between older and younger people. This poses a problem for elderly care. Of course many of us welcome a longer life span. But although we age, we do not always age healthy: the 'oldest' old (80+, 85+) often suffer from chronic diseases, such as dementia, before they die. Hence, there is a need for more caregivers for long-term care. But it is unlikely that children take up these tasks:

> precisely because many individuals have taken advantage of modern freedom's opportunities for education, careers, and geographic mobility, many elderly persons will live in greater isolation from loved ones, separated from children and grandchildren who have settled elsewhere or whose lives are defined primarily by work and school. (President's Council on Bioethics, p. 2)

At the same time, state funding—which pays for professional caregivers—is hardly rising. Thus, we need more caregivers, while their availability decreases for various reasons (see also Lynn 2004). Hence, we find ourselves 'on the threshold of a crisis in long-term care' (President's Council on Bioethics 2005, p. ix–x).

If this analysis is right, then one possible response is the following future scenario: let *robots* take over care tasks from humans. First, we delegated care from family, neighbors, and friends to professionals and institutions. This professionalization of care created all kinds of problems (Illich 1976) as well as benefits. But if this solution is inadequate in the light of the demographical and political situation, then why not delegate care to robots?

This scenario is not as exotic as it may appear at first sight. It is seriously considered or even promoted by some governments (e.g., Japan) and seems to be in line with developments in other fields. We increasingly delegate tasks to automated systems in all kinds of domains of life, from coffee machines to autopilot systems. Medicine and health care is no exception; although machines usually assist rather than *replace* human care. Human lives depend on heart-lung machines, electronic monitoring devices, pace makers, robotic surgery, and other systems. Moreover, in response to social-economic problems with the current health care systems, some suggest that we should use electronic monitoring and care equipment to enable people to live longer in their own homes and to save hospital costs.

Is this care robot scenario desirable? First, we need to specify what kind of robots and care tasks we are talking about. Care robots could appear human in some way (humanoid robots) or look and act like a pet (pet robots). As *personal* robots they would "share physical and emotional spaces with the user" (Arras and Cerqui 2005, p. 13). Care tasks would range from therapy to entertainment and companionship: robots could assist ill and elderly people by monitoring them, delivering drugs, moving them around, helping them with domestic tasks, entertaining them, talking to them, or keeping them company.

In robotics, concrete steps are being taken toward the development and use of such personal or 'social' robots, as well as less interactive assistive robots in health care. Consider Paro (a baby seal therapeutic robot), My Spoon (a feeding robot), or HAL (assistive limb suit), and ongoing research projects such as the Companionable project,[1] which aims to enhance the independence and quality of life of elderly and disabled people. Therapeutic and companion robots could be regarded as more social and personal than an assistive limb suit or feeding robot, which, in turn, are more social and personal than, for example, a cleaning robot or a car manufacturing robot.

This chapter will focus on intelligent, autonomous, and interactive robots that take on 'personal' or 'social' dimension and perform the previously mentioned therapy, monitoring, entertainment, and companionship functions. Let me also assume for the sake of argument that in these scenarios, care robots would *replace* real humans or pets.

The replacement assumption invites one of the primary ethical worries: is it right that robots would replace humans in care? In the robot ethics literature, this is called the *replaceability* question (Decker 2006, 2008). There are various approaches to the question. For instance, Decker has analyzed the question from a Kantian perspective; and Robert Sparrow and Linda Sparrow have understood the problem as being about "genuine social interaction" and respecting people as "ends in themselves" (Sparrow and Sparrow 2006, p. 149).

In this chapter, I reformulate the research question concerning evaluation of the 'care robot' in terms of the good life: Can interaction with care robots contribute to the good life, and what is the difference with human care?

THE GOOD LIFE AND THE BEST POSSIBLE LIFE

What is 'the good life'? There are many theories of well-being (Crisp 2008). Usually, the term 'good life' or the 'flourishing life' refers to Aristotle's view that we are to realize our potential and our aim (*telos*) that lies in our nature as humans. His theory, which gave rise to the virtue ethics tradition in moral philosophy, can be understood as implying that one of the things we need for a good life is friendship. I will explore the applicability of the Aristotelian notion of complete friendship, which he thought was constitutive of the good life, and consider some alternative models of interaction in which human good can appear.

But first we should ask what the good life means given the specific situation of ill and elderly people. Human flourishing, *the* good life, seems easier when one is young, healthy, and in touch with one's nearest and dearest. But these conditions are not always fulfilled. Ill and elderly people cannot aspire to the same level of human flourishing as others and have specific needs and vulnerabilities. It is more appropriate, then, to speak about *the best possible life* (BPL) in their situation.

On reflection, however, this term seems appropriate with regard to all of us. How absolute can the criteria for human flourishing be? Are they relative to one's personal situation and one's unique individual character, or are there needs and aspirations shared by all humans? Let me respond to this familiar problem in ethical theory by distinguishing between three equally important 'good life' questions that must be raised to evaluate care robots and other technology, related to three kinds of norms: one concerns general criteria that apply to all humans (criteria for *the* good life, that is, the best possible life conceivable without taking into account contextual or other limitations: the good life for a healthy human being able to use and develop all her abilities), one concerns universal criteria that apply to the particular situation, context, and technology (criteria for the *BPL* given that situation), and one concerns criteria that are unique to the person and that person's interaction with the technology (criteria for the life of person X). This approach amounts to what I coin a *vertical moral pluralism* of norms: there are different kinds of moral norms that range from general to personal levels of application, but none of them takes priority by definition.

Of course, such an approach raises the question how to balance the different criteria in case of tension between them. I believe this is a question that is difficult (if not impossible) to solve *a priori*, disregarding the particular situation. Moreover, this scheme is not by itself able to guide an ethics of (elderly) care; we need to specify the criteria and what they mean for elderly people in specific situations. In the remainder of the chapter, I shall start with Aristotle's account of friendship (formulated at the most general level) as a way to flesh out the BPL criterion at a less general (but not personal) level: the BPL for elderly people that find themselves in a situation of needing care (that is, more and perhaps to some extent different care than a young and healthy person needs) and of receiving professional care (in an institution or in their own home).

FRIENDSHIP

What kind of interaction between humans is constitutive of the good life, that is, makes us flourish as humans? In contrast to human rights theory or the capabilities approach, the good life approach does not ask about *minimum* criteria for human life, but asks about how to *maximize* the quality of that life. Part of Aristotle's answer to that question is his account of friendship. In the *Nicomachean Ethics*, he argues that the most worthy relationship between people is what he calls 'perfect friendship,' which requires that one desires the other for the sake of the other and one desires the good for the other (book 8, p. 1156b and p. 1157b25; see also Badhwar 1987). It is a relationship between moral equals (p. 1156b); we acknowledge and enjoy virtue in the other. In imperfect friendship, by contrast, the other is a source of advantage or pleasure (p. 1156a15–20). It is an *instrumental* relationship.

Perhaps this notion of friendship could guide nonprofessional, noninstitutional care by family and friends, provided that there is a relation between moral equals (this is far from obvious: our contemporary understanding of friendship seems closer to pleasure-based friendship than to moral equality). However, this Aristotelian criterion is certainly far too demanding to guide relations between care professionals and care receivers. Professional caregivers and care receivers (and maybe nonprofessional caregivers also) need not be friends in this Aristotelian sense to achieve a good care relation which contributes to the BPL. Yet we do not want to resort to 'minimum' morality embodied in ethical codes, contracts, and care protocols; we need a concept that refers to 'warmer' and less instrumental relationships that concerns the good rather than (only) the (legal or deontological) right.

A further problem is that Aristotle's theory of friendship has little to say on human-technology relationship. Using the example of a bottle of wine, he argues that the notion of friendship is not applicable to lifeless objects because we do not wish good to them for their own sake but only for our own use (we want the bottle of wine to remain good for us to drink it), whereas to friends we wish good because of themselves (book 8, p. 1155b30). Therefore, *insofar as* robots count as 'lifeless objects,' Aristotelian theory is not very helpful. Within its framework it is inappropriate to talk of moral symmetry between humans and robots, or to discuss a robot's contribution to the good life. (Unless, of course, robots are *not* viewed as lifeless objects. This would trigger further discussion of 'friendship' with robots. But even if we create this opening, it seems unlikely that the relation could be defined in terms of moral equality as opposed to use or pleasure. Generally, we tend to regard robots as a kind of slaves. This excludes the possibility of perfect Aristotelian friendship. Or is there another possibility? I will discuss this further in my section of 'virtual virtue'.) Moreover, given its historical roots in the culture of the ancient Greek city-state, Aristotle's theory appears to be addressed to active, rational, independent, and healthy male citizens of high social rank, whereas care receivers embody dependency, vulnerability, inactivity, and nonparticipation in the public life. Perhaps they even lost their rational capacity, for example, when they suffer from dementia.

COMPANIONSHIP

Some of these problems could be avoided by appealing to *companionship* rather than Aristotelian friendship. This notion has something of the 'warm' connotation of friendship, but is less demanding. It does not require moral symmetry, without supposing a totally instrumental relationship. The aim is social: being together, doing things together. The (virtue of the) other is not the end, although a somewhat Aristotelian argument can be made that there is a noninstrumental dimension to the extent that we are social beings and that therefore being together realizes that *social* dimension of what we already are, the social *telos* of humans. We are social by nature. Furthermore, the term lacks the erotic element: desire for the other is not needed; desire for company suffices. Nevertheless, the end of companionship cannot be reduced to pleasure or utility: companionship is 'more' to the extent that it realizes the social end of humans.

Given these features, we can apply the term to relations between humans and nonhumans, such as 'social' animals (pets) and 'social' robots. People sharing their physical and emotional space with pets or personal robots do not typically experience their relation to the pet robot or care robot in instrumental terms. Rather, empirical research (e.g., Turkle 2005, see following) suggests that these nonhumans *appear* as their companions and are treated and interacted with accordingly. At the same time, humans—if mentally healthy—are aware of the 'real' asymmetry. They usually know that the robot is not really human or not a biological pet. Therefore, the contribution of such interactions to human flourishing as the BPL must be judged on the basis of what people experience. Appearance turns out to be ethically relevant. If people *experience* companionship, then this contributes to the BPL in the given situation and personal circumstances. I will expand this argument in the next sections.

VIRTUAL VIRTUE

However, if appearance matters so much to the good life, can we push this argument further and consider the possibility of moral symmetry in human-robot relations? Are robots necessarily like Aristotle's bottle of wine? They sometimes appear to us as being 'more' than lifeless objects. Perhaps an advanced humanoid robot could appear to us as a morally *equal* companion, despite us knowing that it is a robot. In that case, we would ascribe 'virtue' to the robot on the basis of its appearance. A robot with *virtual* virtue—whatever its designed or intended functions may be—we would call a 'good' robot; and human good can emerge in relations with such a robot. A robot with virtual virtue, therefore, could contribute to the BPL in a way that differs qualitatively from 'mere' companionship. It would provide virtual friendship and thereby potentially contribute more to the BPL of people.

The ascription of virtual virtue is less odd than it may seem. After all, we are used to virtual characters in novels, films, and computer games. And very little is necessary for us to ascribe a personality to an object such as a computer (Reeves and Nass 1996). Moreover, we judge human virtue on the basis of habits, that is, on the basis of consistently good behavior. If robots acquired 'good habits,' then by the same moral-social 'mechanism,' we would probably ascribe virtue to them.

Of course, Aristotle was not a behaviorist and he would require that virtuous actions are done for the right reasons. This means that in addition to observation of virtuous behavior of the robot, the ascription of virtual virtue to the robot would require that there is also the appearance of reasoning and intention. It requires that we 'forget' that the robot is following an algorithm, that we 'forget' that it is a machine. On the part of the designer, it demands the creation of the *illusion* of virtue and its requirements in terms of behavior and virtual motivation. Is this deception? And if so, is it morally wrong?

Using the term virtual virtue enables us to redeem the Aristotelian notion of friendship in the context of human-care robot relations, at least if we are content with *virtual* friendship with care robots and disregard potential moral concerns this concept might raise. (Next I will discuss the deception objection.)

CARESHIP

But why take great pains to apply notions such as friendship and companionship to health care when we already have the concept of *care* itself? Using that concept, we could try to develop criteria specific to the kind of life ill or elderly people lead: criteria specific to those who do not meet the Aristotelian assumptions regarding autonomy and independence and can only attain the *best possible* life in their vulnerable situation. Such a BPL ethics would require that ill and elderly people enjoy interaction with those who care for them or who appeal to their capacity to care. They participate in, and contribute to, a care relation. Here 'care' does not have the narrow, more instrumental meaning of 'medical care' or 'health care,' and it is not required that one desire the other (as in Aristotle's perfect friendship); rather, one desires *the best* for the other in the given situation of illness or old age and given that person's unique character and identity. I believe this criterion of 'careship' can be met by others than friends and relatives only, but is more demanding than what protocols and codes require from health professionals. (Hence, the criterion is in tune with the intuition that professional care is only a 'second best' option—the best option being that relatives, friends, and others who care *about* the person are the caregivers.)

However, while this criterion for achieving the BPL or model of care could in principle be met by health professionals if certain contextual conditions are fulfilled (e.g., sufficient staff and time available, good intentions, the possibility to develop a care relationship, etc.), it cannot be met by robots. If robots cannot be conscious or cannot have (real) emotions, they can 'deliver' care but not 'really' *care* about their patients. Only if we remove the reality demand, can we create an opening: care robots could *appear* to care. But this invites the deception objection: It would amount to making the person believe that the robot really cares, while in reality it does not.

This objection relies on making an ontological distinction between appearance and reality, an assumption that may be contested (for example—but not exclusively—by behaviorists). But even if one accepts the distinction, one might argue in response to the objection that apparent care does not count *as deception* if care receivers are aware that the robot is a robot. They would *know* this but relate to the robot on the basis of its *apparent* care and this could contribute to their BPL. At most, there is temporary *self-deception*, but one that enhances the BPL. Personal robots are never 'mere' robots in this experiential, perceptual sense. They are what Sherry Turkle, Will Taggart, and others call 'relational artefacts' (Taggart, Turkle, and Kidd 2005). In their study of human-robot interaction in nursing homes, a pet robot looking like a baby seal was treated by elderly people in a nursing home *as if* it needed nurturing. This suggests that if people *feel* that the robot needs their care or that the robot cares about them, this helps them achieve their BPL, depending on their situation and character. The emphasis on the moral significance of feeling and appearance lends support to the claim that both human care and robot care should be adapted to the specific situation ('case') of the person and to the human-robot relation that emerges between the unique, individual person and the robot. For example, Turkle's research shows that people respond very differently to the same robot according to their character. The stress on appearance renders this care ethics of BPL sensitive to individual and contextual differences.

THE DECEPTION OBJECTION

Nevertheless, it seems that there is something morally problematic here. Does virtual friendship and virtual care respect people? Sparrow and

Sparrow admit that people can feel happy when interacting and forming 'relationships' with robots, the authors take issue with the cause, which they call *delusion*:

> In most cases, when people feel happy, it will be because they (mistakenly) believe that the robot has properties which it does not (. . .) It is these delusions that cause people to feel loved or cared for by robots and thus to experience the benefits of being cared for. (Sparrow and Sparrow 2006, p. 155)

Why is this problematic? According to the authors, failure to apprehend the world accurately is a moral failure. Moreover, they claim that our well-being is not served by 'illusions:'

> What most of us want out of life is to be loved and cared for, and to have friends and companions, not merely to *believe* that we are loved and cared for, and to *believe* that we have friends and companions, when in fact these beliefs are false. (Sparrow and Sparrow 2006, p. 155)

The authors conclude that using robots in this way would express a gross lack of respect for older people.

I agree that *if* virtual friendship and virtual care were to constitute a delusion, the use of robots for this purpose would be morally wrong. However, provided that the elderly people or patients in question are mentally healthy, the authors mis-describe what happens in virtual care. Just as in interactions with virtual agents in a computer-generated virtual world, most mentally healthy people have a twofold awareness: They are aware of the virtual other, but they are also aware that this other is not real. If we really wish to take seriously ill and elderly people as autonomous beings, which the authors want, then we should not assume that they live in delusion. Rather, we should assume that they can enjoy interaction with their robot nurse and robot companion without believing that they are interacting with a human being.

Of course, some people can no longer distinguish between the real and the virtual. But this implies they *already* live in 'delusion' or illusion (for example, they may mistake the *human* nurse for someone else or mistake their spouse for someone else), and it is unclear why care robots would worsen their condition or not respect them.

Moreover, in *human* health care such people are *already* treated as non-autonomous persons and for a good reason: they *lack* the capacity for autonomy. But that does not mean that we necessarily disrespect them. Paternalism can be justified under certain conditions. And respect can also be granted on the basis of recognizing a person as human or on the basis of the particular relation we have to that person. The questions whether or not care robots can help mentally confused elderly people and how we can best respect such people and enhance their well-being are different questions that must be separated from the deception issue.

Sparrow and Sparrow are right to point to subconscious processes that go on when humans interact with robots. But surely these processes do not stop when we interact with humans. In human-human interaction too, we can be deceived about the other—and worse: *by* the other. 'Worse' because I assume that intended deception is morally worse than nonintended deception, which may not always be the case.[2]

Finally, human-human interaction often involves nonintended deception and partly *depends* on appearance in the sense that as social beings we always play roles to *some* extent (roles of a nurse, a doctor, a partner, a friend, etc.) and it is difficult to make a strict distinction between appearance and reality in human social life. At least it is not *obvious* what reality is as opposed to appearance (e.g., what is the real self or real person?). Hence, given the role of appearance in human-human relations, including care relations, putting all emphasis on the deception objection when evaluating care robots is misplaced. As it happens, thinking about human-robot relations and human-robot care is a lens through which we can look afresh at what we mean by good human care and relations. Perhaps there is something dubious about the role appearance plays in human-robot care relations, but it is not obvious why it is morally wrong *per se*.

However, there may be other, better objections against delegating care to robots not discussed in this chapter.[3] Moreover, is the problem well-defined? Are robots the problem?

NON-TECHNOLOGICAL SOLUTIONS? REFORMULATING THE PROBLEM

Why consider robots at all in response to problems in health care and elderly care? Discussing a technological solution may bypass reflection on the causes of the problem as outlined in the beginning of this chapter. Must we not ask why there is a shortage of care workers? Must we not ask why we should have professional care at all instead of care by relatives and friends? How could our society end up with this problem in the first place?

Once we pursue this line of inquiry, it is tempting to move immediately to a normative view. For instance, with Ivan Illich, one may argue against the professionalization and institutionalization of modern health care by referring to the ethic of the 'Good Samaritan' (responding to people, feeling dis-ease at the sight of suffering, caring reciprocity as opposed to following rules or conventions) and to more communal forms of living together (see, for example, Illich and Cayley 2005; Kahn 2009).

I hesitate to embrace this normative view without further qualification. A weak reply to the 'Samaritan' objection is that we cannot turn back time—that we have to start from (late) modern society as it is. This is a weak reply because however wide the gap between norm and practice is, philosophers have a duty to imagine and discuss normative ideals. A stronger reply is that both the objection and the weak reply assume (1) that we already have, and agree about, a conception of good care and good community, and (2) that there was a time when that conception of good care and community was realized. But we must doubt if the past was really better, be critical of the current health care system, and scrutinize conceptions of care that allow us to evaluate that system.

The 'Good Samaritan' model of care is yet another model that, like the Aristotelian, is not simply applicable to robot care or to current professionalized care. It is based on the model of reciprocal love in *human* relations, an ideal that is close to the Aristotelian model of perfect friendship. Robots can be programmed to help any human person in need, regardless of how they otherwise appear, and thus act 'impartially' as good rule followers,[4] but they cannot love. They cannot really feel the suffering of the care receiver. However, this, by itself, is not an argument against using robots in health care. Moreover, professionalization and the preservation of systematic knowledge do not only have disadvantages. The intuitive 'Samaritan' response to the care problem invites us to further reflect on what good care is, what a good community is, what love is, and what the differences are between human-robot relations and human-human relations. Therefore, normative evaluation of robots in health care and elderly care should involve more conceptual and empirical analysis of the care problem and of the moral ideals that might help us to create better care futures. This includes being more critical of our current norms and practices that do not involve robots. Philosophy of technology is not (only) about technology, but about us.

CONCLUSION

I started this chapter with a methodological proposal: evaluate care robots in terms of the good life. In particular, we should shift our attention to criteria for BPL for unique and vulnerable people in particular situations. I have discussed friendship, companionship, and careship as models for evaluating BPL in relation to human-robot interaction. More empirical research is needed to evaluate how people in health care contexts could attain the BPL. This notion of BPL could also be used to evaluate other human practices.

But what is my normative conclusion? There is little doubt that robots could contribute to better lives for ill and elderly people. Therefore, there is nothing intrinsically wrong with giving them *a* place within health care. However, there is disagreement about what place that should be if our aim is the BPL of ill and elderly people—that is, the BPL of all elderly people, not just those who happen to like robots. At first sight, it seems unlikely and undesirable that robots would replace human caregivers. They cannot be our 'Aristotelian' friends and they cannot *really* care about us. A way to get around these problems is to lower the Aristotelian standard and take seriously what I have called 'virtual virtue' and 'virtual care.' In virtual friendship and virtual care, human good could emerge that is experientially similar in form to human friendship and human care. If this happens, it contributes to people's BPL, does not necessarily violate the moral requirement that people are not deceived, and if it did, this would be less problematic than appeared at first sight given the mental condition of the people in question and given the importance of appearance in human social life.

In the current technological-social situation, the conditions for this kind of good to arise are seldom fulfilled. Robots cannot produce much 'virtual virtue' yet, and our current health care systems do not stimulate the development of virtual virtue—or real virtue for that matter. To the extent that these conditions are not fulfilled, therefore, we should make sure that the use of care robots does not lead to a reduction of human contact, as Sparrow and Sparrow rightly fear (Sparrow and Sparrow 2006, p. 152), and use robots only as a supplement to human caregiving, not as its replacement.

However, if robots cannot always provide what ill and elderly people need to reach BPL, it must be noted that professional caregivers cannot necessarily do this either. This is not just a matter of individual ethics; we should assume that health care professionals usually have good intentions. But good intention is not enough. As previously suggested, current health care systems may not stimulate the kind of relations we hold most worthy or most conducive to BPL. Therefore, a comprehensive discussion about using robots in health care should involve a reassessment of the way we deal with the ill and elderly in our society. For instance, in contrast to most robots, humans *can* intentionally deceive. If they are human, we want caregivers to *really* care. If our current social institutions do not support that kind of care, we should not blame the robots but imagine better, *best possible* ways of organizing care and society.

NOTES

1. This is an acronym for 'Integrated Cognitive Assistive and Domotic Companion Robotic Systems for Ability and Security.' The project is funded by the EU 7th Framework Programme. The robotic system would be able to recognize emotional states of people, distinguish between normal and exceptional behavior, remind people to take their medication, and so on.
2. Sometimes deception by robotic and other caregivers may be required (Wagner and Arkin 2010).
3. For example, Adam Briggle thinks delegating care to robots is like the ostrich sticking its head in the sand, because it gives us the illusion that we live in a world without ageing. In other words, we deceive ourselves if we let technology take away the 'burden' of care. (Compare this situation also to the initial life of the Buddha before his departure: He lived in the illusion of a world without suffering.)

4. Note that even robotic rule-following is not straightforward: as Asimov's stories informed by the 'Laws of Robotics' show, rules may conflict.

REFERENCES

Aristotle. (1908). *Nicomachean Ethics* (William David Ross, trans.). Oxford: Clarendon Press.

Arras Kai O. and Daniela Cerqui. (2005). "Do we want to share our lives and bodies with robots? A 2000 people survey," Technical Report Nr. 0605–001, Autonomous Systems Lab, Swiss Federal Institute of Technology Lausanne (EPFL).

Badhwar, Neera Kapur. (1987). "Friends as Ends in Themselves," *Philosophy and Phenomenological Research*, vol. 48, pp. 1–23.

Crisp, Roger. (2008, December). "Well-Being." In *Stanford Encyclopedia of Philosophy*. Stanford: Stanford University Press. Retrieved September 21, 2010, from http://plato.stanford.edu/entries/well-being/Decker, Michael. (2006). "Replaceability of Humans by Autonomous Robots? Ethical Reflection within the Framework of an Interdisciplinary Technology Assessment." Paper presented at the ETHICBOTS workshop, Naples, Italy, October 18, 2006.

Decker, Michael. (2008). "Caregiving Robots and Ethical Reflection: The Perspective of Interdisciplinary Technology Assessment," *AI & Society*, vol. 22, pp. 315–330.

Illich, Ivan. (1976). *Medical Nemesis: The Expropriation of Health*. New York: Random House.

Illich, Ivan, and David Cayley. (2005). *The Rivers North of the Future: The Testament of Ivan Ilich*. Toronto: Anansi.

Kahn, Richard. (2009). "Critical Pedagogy: Taking the Illich Turn," *The International Journal of Illich Studies*, vol. 1, pp. 37–49.

Lynn, Joanne. (2004). *Sick to Death and Not Going to Take it Anymore!* Berkeley: University of California Press.

President's Council on Bioethics. (2005). *Taking Care: Ethical Caregiving in our Aging Society*. Washington, DC: U.S. Printing Press.

Reeves, Byron, and Clifford Nass. (1996). *The Media Equation: How People Treat Computers, Television, and New Media Like Real People and Places*. Cambridge: Cambridge University Press.

Sparrow, Robert, and Linda Sparrow. (2006). "In the Hands of Machines? The Future of Aged Care," *Minds and Machines*, vol. 16, pp. 141–161.Taggart, Will, Sherry, Turkle, and Cory D. Kidd. (2005). "An Interactive Robot in a Nursing Home: Preliminary Remarks." Paper presented at the 27th Annual Conference of the Cognitive Science Society Android Science workshop, Stresa, Italy, July 2005.

Turkle, Sherry. (2005). "Relational Artefacts/Children/Elders: The Complexities of CyberCompanions." Paper presented at the 27th Annual Conference of the Cognitive Science Society Android Science workshop, Stresa, Italy, July 2005.

United Nations Programme on Ageing. (2010). "The Ageing of the World's Population," Retrieved November 25, 2010, from http://www.un.org/esa/population/publications/worldageing19502050/

Wagner, Alan R., and Ronald C. Arkin. (2010). "Acting Deceptively: Providing Robots with the Capacity for Deception," *International Journal of Social Robotics*, vol. 3, pp. 5–26.

Part VI

Technology Design and Policy

21 Can We Design for Well-Being?

Ibo van de Poel

Because engineering design is aimed at the creation of useful products, it is always in some sense aimed at creating value for users. In the last two decades, designers have started to pay more systematic and theoretical attention to customer satisfaction or user value (e.g., Akao 1990; Woodruff and Gardial 1999; Holbrook 1999). Still, what exactly constitutes user value, and how design can contribute to this is not yet well-understood (Boztepe 2007). Moreover, the notion of well-being, on which I will be focusing, is usually not discussed in this literature.

In this chapter, I will propose an objective list account of well-being and explore how, on such an account, considerations of well-being can be integrated into the design of new technologies. Design may be described as the process in which certain desiderata for a technological product, including relevant values, are translated into a blueprint for that product that can then be produced and subsequently be used. The reason to focus on design is that it is a central activity in the shaping of new technologies and an activity in which value considerations already explicitly play a role. In further exploring how considerations of well-being can be integrated into design, I will make use of the basic framework of value sensitive design (VSD) (Friedman and Kahn 2003; Friedman, Kahn, and Borning 2006; Flanagan, Howe, and Nissenbaum 2008).

If one wants to integrate considerations of well-being into design, one needs to specify considerations of whose well-being. Following the opinion in most of the current design literature, I will assume this to be primarily the well-being of the expected users of the technology. I admit that this is a rather big assumption. After all, it can and has been argued that design should also take into account the well-being of other stakeholders; this is, indeed, one of the main aims of the VSD approach. For the purposes of this chapter, I have two reasons to restrict my discussion to user well-being. One is that I think that the *philosophical* problems for integrating stakeholder well-being are rather similar to those of those of integrating user well-being.[1] So once we have solved the latter, we at least have a good idea how to deal with the former. The second reason is that I doubt whether there is a moral imperative for designers to *increase* the well-being of other stakeholders besides users. Surely, designers should take into account the well-being of stakeholders, but quite often this will be restricted to making sure that the well-being of such stakeholders is not diminished by a design, and this imperative would, I think, translate into paying attention to such values as safety, sustainability, privacy, and justice but not necessarily to considerations of overall well-being of stakeholders. In contrast, increasing or maximizing user well-being is often mentioned or assumed as goal in design (e.g., Hauser and Clausing 1988; Bakker II and Loui 1997).

Although I focus on user well-being rather than stakeholder well-being, this means that designers have to design for the well-being of somebody else. This raises at least two philosophical problems for integrating considerations of well-being into engineering design: an epistemological problem and an aggregation problem.

The *epistemological* problem is that, because the designers are usually not the users of the product they design, the designers need to know what constitutes well-being for the (expected) users, and how their product may affect that well-being. This knowledge may be difficult to gather because design typically concerns products that do not yet exist; in fact, design is largely an open-ended process that relates to creating a product.

The *aggregation* problem arises due to the fact that a design does not affect the well-being of just one person, but rather that of a range of people. This raises the question of how the well-being of these people should be aggregated so that it can be taken into account in the design process. If one believes that well-being constitutes plural and incommensurable prudential values, as some philosophers like Joseph Raz (1986) have suggested, then an aggregation problem arises with respect to how these values can, or cannot, be aggregated into an overall measure of well-being.

In this chapter I will discuss both the epistemological and aggregation problem if one adopts an objective list account of well-being. I will suggest ways of dealing with both problems, so that neither problem is a reason to doubt whether we can actually integrate considerations of user well-being into engineering design. I start with a brief discussion of well-being and the VSD approach.

WHAT IS WELL-BEING?

Most studies and approaches in the design literature assume a subjective notion of human well-being. They presuppose what philosophers would term a desire-satisfaction account of well-being.[2] Such accounts conceive of well-being as the satisfaction of human desires. Desire-satisfaction accounts have a number of problems, as is well-documented in the philosophical literature (e.g., Crisp 2008; Griffin 1986). One problem is that people might desire things that do not contribute to their well-being. I might have a great longing to eat an entire pie every day, but on closer reflection, it is likely that I will come to the conclusion that this will not contribute to my well-being. Well-being then is not so much about satisfying the desires I have here and now but rather about how my life is going overall over a longer period of time. Another reason why desires may not be able to appropriately track the components of human well-being is the phenomenon that when people are deprived of basic rights or needs for a long time, they may lose their desire for such rights or for fulfilling those needs (Nussbaum 2000, p. 136–142). It would, however, be wrong to conclude that fulfilling those rights and needs no longer contributes to their well-being. In fact, most of these people would start to appreciate those rights and the fulfillment of those needs again once they were no longer deprived of them.

One philosophical alternative to desire-satisfaction accounts are so-called objective list accounts of well-being. Such accounts assume that it is possible to list a number of values that together constitute well-being. We might call the values on the list prudential values. Prudential values may thus be understood as those values that contribute to someone's well-being, without necessarily contributing to or taking into account the well-being of others, as moral values typically do. Objective list accounts also have their problems. First, the very process of arriving at a list of objective prudential values and how we know when it is complete seems rather obscure. Second, such accounts seem to ignore reasonable differences between what constitutes well-being for different people. After all, my well-being is highly dependent on my ability to set my own goals in life and to accomplish these (Raz 1986). One possible way to try to avoid these problems is to base the list not only on certain features of human nature but also in part on so-called informed desires (Griffin 1986, p. 70). These are basically the desires that people would have if they were fully (or at least sufficiently) informed and took a reflective attitude toward their own life. In addition, it could be argued that even if the resulting list outlined the basic components of human well-being, the specific content that these abstract values represent in the life of individual people and their relative importance may differ from person to person (and maybe also between cultures).

Several authors have proposed a list of general prudential values. James Griffin (1986, p. 67), for example, provides the following list:

1. Accomplishment
2. The components of human existence. This includes values such as autonomy, liberty, and the basic capabilities to act ("limbs and senses that work, the minimum material goods to keep body and soul together, freedom from great pain and anxiety" (p. 67).
3. Understanding
4. Enjoyment (including perception of beauty)
5. Deep personal relationships

The list from Griffin provides us with a basic notion of central prudential values that is good enough to be a starting point for designing for human well-being. Let me now turn to a possible approach to integrating prudential values in design.

VALUE SENSITIVE DESIGN

VSD is an approach that aims to integrate values of ethical importance systematically into the design of information systems and software (Friedman and Kahn 2003; Friedman, Kahn, and Borning 2006). The central ideas of the approach are, however, more broadly applicable. That is, the basic tenets of the approach apply to engineering systems, and they apply not only to moral values but to prudential values as well. In fact, well-being is a value of ethical importance.

The original VSD approach integrates three kinds of investigation: conceptual, empirical, and technical. Conceptual investigations aim to clarify the values at stake. Empirical investigations "involve social scientific research on the understanding, contexts, and experiences of the people affected by technological designs" (Friedman and Kahn 2003, p. 1187). Technical investigations "involve analyzing current technical mechanisms and designs to assess how well they support particular values, and, conversely, identifying values, and then identifying and/or developing technical mechanisms and designs that can support those values" (Friedman and Kahn 2003, p. 1187). Recently, Mary Flanagan, Daniel C. Howe, and Helen Nissenbaum (2008) proposed a more detailed methodology for 'embodying' values in technology. This methodology consists of three steps:

1. *Discovery.* This activity will result in a list of values that are relevant for the design project. If we are interested in prudential values and adapt an objective list account of such values, the list by Griffin may provide a good starting point for identifying relevant values and applying these to the design project under consideration. In addition, empirical investigations may be useful to discover what users value in a certain product and to decide which of the values on the general list are relevant and which are not for the product being designed.
2. *Translation.* Translation is "the activity of embodying or expressing [. . .] values in system design" (Flanagan, Howe, and Nissenbaum 2008, p. 338). It consists of two subactivities: specification of values and the implementation of values into technical features. In this phase, tradeoffs also have to be made between the relevant prudential values.
3. *Verification.* This is the stage of assessing (e.g., through simulation, tests, or user questionnaires) whether the design indeed has implemented the values that it aimed to implement.

I will not outline in detail these different phases if we want to design for well-being. Rather, I will focus on the two problems identified in the beginning of this chapter: the epistemological problem and the aggregation problem. Both problems mainly relate to the translation phase. The epistemological problem especially surfaces in the specification of the list of general prudential values for specific users in specific situations. The aggregation problem is relevant for how acceptable tradeoffs can, or cannot, be made.

THE EPISTEMOLOGICAL PROBLEM

Specifying prudential values in design raises epistemological problems for two distinct reasons. One is that the specification will depend on the goals of the users whose well-being is concerned. The other is that the specification will depend on the technology being designed and the context in which this technology is used, including, for example, existing physical circumstances, user skills and knowledge, cultural habits, existing social institutions, and so on.

Raz has argued that our well-being is to a large extent determined by our ability to set our own goals and to achieve these (Raz 1986, chapter 12; Scanlon 1998, pp. 118–123). According to him, we adopt goals because we have good reasons to consider them valuable. Might we nevertheless adopt goals for which such reasons are lacking, they will not contribute to our well-being (Raz 1986, pp. 301–304; Scanlon 1998, p. 121). So, attaining reasonable goals can contribute to our well-being, but only if it is we who attain those goals rather than others or, one can also add, rather than technology.[3] If a specific technology attains our goals without any activity or effort on our side, it might be argued that it does not really contribute to our well-being. The relation between technology and the value of accomplishment thus seems indirect: technology best increases our capability to attain our goals rather than attaining those goals directly. The same seems to apply to prudential values such as deep personal relations, understanding, and enjoyment. For example, technologies such as e-mail, Web chats, and social networks enable us to engage in new types of social and personal relations. In doing so, they may contribute to our well-being, but they do so by enabling us to engage in new relations rather than just providing us with new relations. An imaginary social network site that would automatically provide each participant with 200 new friends, without any choice or effort on the side of that participant, would arguably not really contribute to that person's well-being.

The specification of prudential values also depends on the context of use. This is because the question of whether or not a technology promotes or constrains certain prudential values or capabilities is not solely determined by that technology but also on how it is used. An example may be helpful here to illustrate this point. Selinger (2008) discusses the Village

Phone program in Bangladesh, which offers women the opportunity to become entrepreneurs by renting call time on mobile telephones to illiterate villagers who cannot afford a phone. The program has been so successful that mobile phones have been characterized as "weapons against poverty." The program has been praised not least because it offers women employment opportunities, which are denied by their traditional culture, as they are often marginal in Bangladesh and have no access to credit. The program also creates an opportunity for women to speak with males outside the family. Moreover, it is argued, the women gain authority and respect from their spouses and the community, and—through increased income—they can take a more active role in their children's futures. The phones then seem to contribute to what Griffin calls the "components of human existence."

However, some critics contend that the actual effects of the Village Phone program are far less positive. It has been suggested, for example, that some men force their wives to take loans and then appropriate the revenues. More generally, it seems naive to assume that patriarchal social structures and norms can be challenged without considering the forces that sustain them. Selinger himself draws attention to what he calls the techno-economic script of the Village Phone program. Women are provided "with the opportunity to offer only one type of service, and that service invariably requires a default protocol to be followed—a simple, yet strict script whereby a phone is traded for a fee" (Selinger 2008, p. 35). As a consequence, '[w]hen phone ladies do their job, they are passive, invisible, deferential, and unremarkable," they "are viewed more as a service than as a human being" (Selinger 2008, p. 37).

This example contains important lessons with respect to two aspects of the VSD approach: (1) verification of the design, and (2) the specification of values. First, regarding the verification process, the example shows that verifying whether a design indeed embodies certain values cannot be done in the abstract, but requires detailed knowledge and study of how the technology functions in the life of actual people and how it realizes or inhibits certain values. Second, with respect to the identification and specification of values, the example shows that current practices are an important point of departure. In the example, to design and implement the Village Phone program, it is crucial to understand the role patriarchal social structures and norms play in Bangladesh society. Only in this way does it become possible to specify the relevant values or capabilities for the design process in such a way that they have a fair chance of being realized in actual user contexts. I do not wish to suggest that this process is infallible. On the contrary, design is usually an iterative process. Failure to realize certain values may be one reason for revising the previous specification of the values at play and to redesign the product. This iterative character of design therefore underscores the reliance of the specification of values on concrete user contexts and user experiences.

The epistemological challenge of gaining in-depth knowledge of user practices and user contexts for design is of course not new. In fact, in the design literature, a large number of methods and approaches have been developed to this end (e.g., Woodruff and Gardial 1999). However, it is worth bearing in mind that, according to the situation I have previously outlined, VSD requires more than just an identification of user demands by means of surveys or marketing research. Rather, it requires insight into how technologies and values function in the life of actual users. This fits with design approaches such as 'applied anthropology' and 'empathic design' that try to understand how people use products.[4] Another appropriate approach may be participatory design, whereby users (and other stakeholders) are involved in the design process (e.g., Schuler and Namioka 1993).

THE AGGREGATION PROBLEM

Objective list accounts assume that well-being can be assessed (or even measured) in terms of a common set of values, thereby easing interpersonal comparisons (Griffin 1986, pp. 106–124). Nevertheless, they face an important challenge with respect to aggregation: namely, the (alleged) incommensurability of prudential values. Two or more values are incommensurable if they cannot be expressed or measured on a common scale or in terms of a common value measure. Incommensurable values cannot be traded off directly. Raz (1986) has suggested that incommensurability and a resistance to trade-offs is constitutive of certain values. Consider, for example, the following trade-off: For how much money are you willing to betray your friend? It may well be argued that accepting a trade-off between friendship and financial gain undermines the value of friendship. On this basis, it is constitutive of the value of friendship to reject the trade-off between friendship and financial gain. Values may resist trade-offs because they are 'protected' or 'sacred' (Baron and Spranca 1997; Tetlock 2003). This seems especially true of moral values and values that regulate the relations between, and the identities of, people. Trade-offs between protected values create an irreducible loss because a gain in one value may not always compensate for, or cancel out a loss in the other. The loss of a good friend cannot be compensated for by having a better career or more money.

Not all philosophers, however, agree that incommensurable prudential values can exist. Griffin (1986, pp. 89–92), for example, denies the incommensurability of prudential values. Rather, Griffin sets out a plurality of prudential values, as his list testifies, while also conceding that there is no substantive super value that encompasses all those prudential values. Nevertheless, he also believes that all prudential values can be expressed on a common scale, which is somewhat equivalent to the idea of giving 'worth to one's life' (Griffin 1986, p. 90). Although Griffin denies incommensurability, he seems to recognize that different people may reasonably disagree about the specification, prioritization, and (acceptable) trade-offs between the various prudential values. It seems, then, that the aggregation problem returns through the back door: How are we to aggregate the different tradeoffs that different people are willing to make between different prudential values?

If we design for well-being, we need to account somehow for the fact that people have different goals. The point is not just that people have different goals at different moments in time or at different stages of their life, but also that people have different comprehensive goals. Examples of such comprehensive goals are 'being an artist' or 'being a mother.' Generally speaking, comprehensive goals contribute more to our well-being than short-term goals. Moreover, many, although not all, of our short-term goals derive their meaning and importance from comprehensive goals. Our goals in this sense have a hierarchical nature.

Raz has rightly argued that comprehensive goals are often incomparable, that is, that we often cannot reasonably judge one comprehensive goal to be better overall than another (Raz 1986, pp. 340–345). I nevertheless think that there are minimal conditions that a comprehensive goal should meet to be reasonable and acceptable. A comprehensive goal is reasonable if, and only if, the person who adopts it has good reasons to adopt it and if these reasons are not overridden by other reasons.[5] In addition, we might want to require that comprehensive goals should be morally acceptable.[6]

Comprehensive goals are likely to be shared by certain groups or subcultures in a society. One reason for this is that comprehensive goals often depend on what Raz calls 'social forms.' An example of such a social form is marriage; marriage requires certain social institutions and shared norms and values. This is not to say that each marriage exactly reflects what might be called the 'conventional marriage,' yet variations such as an open marriage or a gay marriage also depend at least in part on the social form of marriage.

People can have several comprehensive goals related to, for example, marriage, friendship, career, and political and religious commitments and ideals. In addition, people may have a single comprehensive goal that unites all those goals but that by no means is necessary and it is not obvious that having such a uniting goal would add to one's well-being. People may also have what I would like to call a 'vision of the good life.' Such visions of the good life may be religiously motivated but they are not necessarily so. One might, for example, think of the ideal of living an adventurous life, in which one is always looking for new, exciting experiences. A 'vision of the good life' entails an evaluative framework according to which certain comprehensive goals are more worthwhile than others (although it does not necessarily order all possible comprehensive goals) and which articulates certain, more specific, values and virtues. Those who want to live an adventurous life would probably prefer a career as entrepreneur over that as civil servant. They may adopt mountaineering or paragliding as hobby rather than collecting stamps. They may consider taking risks a virtue rather than a vice. As with comprehensive goals, visions of the good life will often be shared. Our modern society is characterized by a diversity of competing visions of the good life that are largely incomparable beyond minimal conditions of reasonableness and moral acceptability.

Like comprehensive goals, visions of the good life give meaning and priority to many of our more specific goals and values. They will often articulate not just single values and goals but also specific combinations of these. In such cases, well-being is not simply the 'sum' of the attainment of specific goals and prudential values, but is to be found at least partly in the specific combination of a range of nested goals and values.[7]

The above description suggests the following basic approach if one wants to design for well-being:

1. Select a user group that shares a comprehensive goal or a vision of the good life, or both;
2. Analyze this comprehensive goal (vision of the good life) in terms of how it articulates a specification and mix of the general prudential values in terms of more specific goals, values, and virtues;
3. Identify how the technology that is being designed may contribute to this mix in the user context for which is designed;
4. Design technology that enables this mix as much as possible.

The first step avoids the need to aggregate the well-being of people who have different, incompatible comprehensive goals or visions of the good life. This step does not necessarily mean that people with other comprehensive goals or visions of the good life are neglected; however, they may be better served by another design that takes their comprehensive goal as a point of departure rather than a messy compromise between two or more incompatible comprehensive goals. For people whose vision of the good life includes the idea that it is essential to always and continuously keep up with what is going on in the world and with their friends, it may be absolutely essential to have a mobile phone with Internet access, twitter, social networks, and the ability to make and send around photos at any time. For other users, such multifunctional mobile phones may not just be unnecessary, they might refuse to use them because they represent a vision of the good life they reject. They may be better served by a much simpler type of mobile phone.

It should further be noted that step four does not imply a maximizing approach to well-being. The focus is on the mix of values rather than on maximizing an overall measure of well-being. The focus is also not on maximizing each of the prudential values in isolation, because it is usually the mix of values that contributes to the overall goal rather than the values considered in isolation.

CONCLUSION

I have considered the epistemological and aggregation problems related to integrating considerations of well-being into design on an objective list account and suggested ways for dealing with them. In particular, I have underlined the importance of in-depth studies of how technologies function in the life of users (and other relevant stakeholders) and how these may help realize certain values (or not). I have also suggested an approach to deal with the aggregation problem that focuses on distinguishing different user groups with different visions of the good life or different comprehensive goals. Neither the epistemic nor the aggregation problem is thus a bar to designing for well-being, and the title question of this chapter needs to be answered in the affirmative.

ACKNOWLEDGMENTS

I am grateful to NIAS, the Netherlands Institute for Advanced Study, for providing me with the opportunity, as a Fellow-in-Residence, to write this chapter.

NOTES

1. However, it must be admitted that the practical problems may be much bigger for stakeholder well-being.
2. It might be argued that some accounts, especially those that emphasize the importance of experience for user value, can be better seen as mental state accounts. Here I will assume that mental states accounts are close to or can be integrated into desire-satisfaction accounts.
3. In Raz's terminology, goals that are important for our well-being provide us with action reasons rather than with outcome reasons (Raz 1986, p. 305–307).
4. Applied anthropology approaches were pioneered by Lucy Suchman; for examples of this approach, see Blomberg (1987) and Lie and Sorensen (1996); for emphatic design, see, for example, Ilpo Koskinen, Katja Battarbee, and Tuuli Mattelmäki (2003). For an overview of these and other relevant approaches to human-centered design, see Marc Steen (2008).
5. This does not mean that the reasons he or she has for adopting the goal should override all possible reasons against it. This may simply be impossible to decide. This is precisely the reason why comprehensive goals are incomparable. The situation is as follows: for each of the comprehensive goals, reasons can be given that are not overridden, but these positive reasons do not usually override the reasons that can be given for other comprehensive goals.
6. Care should be taken in further specifying what is meant by 'morally acceptable' here. As Raz (1986, p. 406) notes, from the perspective of those who have certain comprehensive goals (or what others would call a vision of the good life), other comprehensive goals may seem morally unacceptable. The reason for this is that comprehensive goals are often incompatible in the sense

that they articulate different conflicting virtues. This, however, seems insufficient reason to reject those other goals as morally unacceptable. A better criterion might be whether the adoption of a certain comprehensive goal is likely to harm others (with other comprehensive goals), although it may still be difficult to agree on what constitutes 'harm' and what not.

7. Such specifications and mixes give raise to what Raz (2003) has called genre-specific evaluations.

REFERENCES

Akao, Y. (Ed.). (1990). *Quality Function Deployment: Integrating Customer Requirements into Product Design*. Cambridge, MA: Productivity Press.

Bakker II, W., and M. C. Loui. (1997). "Can Designing and Selling Low-Quality Products Be Ethical?" *Science and Engineering Ethics*, vol. 3, no. 2, pp. 153–170.

Baron, Jonathan, and Mark Spranca. (1997). "Protected Values," *Organizational Behavior and Human Decision Processes*, vol. 70, no. 1, pp. 1–16.

Blomberg, J. L. (1987). "Social Interaction and Office Communication: Effects on User Evaluation of New Technologies." In *Technology and the Transformation of White-Collar Work*, edited by R. E. Kraut. Hillsdale, NJ: Lawrence Erlbaum.

Boztepe, Suzan. (2007). "User Value: Competing Theories and Models," *International Journal of Design*, vol. 1, no. 2, pp. 55–63.

Crisp, Roger. (2008, Winter). "Well-Being." In *The Stanford Encyclopedia of Philosophy*, edited by E. N. Zalta. Stanford: Stanford University Press.

Flanagan, Mary, Daniel C. Howe, and Helen Nissenbaum. (2008). "Embodying Values in Technology: Theory and Practise." In *Information Technology and Moral Philosophy*, edited by Van den Hoven, J. and J. Weckert. Cambridge: Cambridge University Press.

Friedman, Batya, and Peter H. Kahn, Jr. (2003). "Human Values, Ethics and Design." In *Handbook of Human-Computer Interaction*, edited by Jacko, J. and A. Sears. Mahwah, NJ: Lawrence Erlbaum.

Friedman, Batya, Peter H. Kahn, Jr., and Alan Borning. (2006). "Value Sensitive Design and Information Systems." In *Human-Computer Interaction in Management Information Systems: Foundations*, edited by Zhang, P. and D. Galletta. Armonk, NY: M. E. Sharpe.

Griffin, James. (1986). *Well-Being: Its Meaning, Measurement, and Moral Importance*. Oxford: Clarendon Press.

Hauser, John R., and Don Clausing. (1988). "The House of Quality," *Harvard Business Review*, vol. 66, no. 3, pp. 63–73.

Holbrook, M. B. (Ed.). (1999). *Consumer Value: A Framework for Analysis and Research*. New York: Routledge.

Koskinen, Ilpo, Katja Battarbee, and Tuuli Mattelmäki. (Eds.). (2003). *Emphatic Design: User Experience in Product Design*. Helsinki: IT Press.

Lie, Merete, and Knut H. Sorensen. (Eds.). (1996). *Making Technology Our Own? Domesticating Technology into Everyday Life*. Oslo: Scandinavian University Press.

Nussbaum, Martha C. (2000). *Women and Human Development: The Capabilities Approach*. Cambridge: Cambridge University Press.

Raz, Joseph. (1986). *The Morality of Freedom*. Oxford: Oxford University Press.

Raz, Joseph. (2003). *The Practice of Value*. Oxford: Oxford University Press.

Scanlon, T. M. (1998). *What We Owe to Each Other*. Cambridge, MA: Harvard University Press.

Schuler, D., and A. Namioka. (Eds.). (1993). *Participatory Design: Principles and Practices.* Hillsdale, NJ: Lawrence Erlbaum.

Selinger, Evan. (2008). "Does Microcredit 'Empower?' Reflections on the Grameen Bank Debate," *Human Studies*, vol. 31, pp. 27–41.

Steen, Marc. (2008). *The Fragility of Human-Centered Design.* Amsterdam: IOS Press.

Tetlock, Philip E. (2003). "Thinking the Unthinkable: Sacred Values and Taboo Cognitions," *Trends in Cognitive Sciences*, vol. 7, no. 7, pp. 320–324.

Woodruff, Robert B., and Sarah Fisher Gardial. (1999). *Know Your Customer: New Approaches to Understanding Customer Value and Satisfaction.* Oxford: Blackwell.

22 The University, Metrics, and the Good Life

Robert Frodeman, J. Britt Holbrook, and Kelli Barr

Knowledge and money have no common measure.—Aristotle

In the *Nicomachean Ethics,* Aristotle argues that the best of all lives is the life of contemplation. Outside of the cash nexus, indeed separate from any societal function, contemplation is the godlike activity of thought thinking itself. With the decline of monastic life, the academy might be thought of as the last bastion of reflection. But whether Aristotle would find much contemplation there today is doubtful. Driven by economic crisis and the need for greater and greater output, academic life is increasingly characterized by impatience and anxiety. In his or her daily activities, the contemporary academic is more likely to be found with shoulders bent over a keyboard, answering a daily avalanche of email, and hurrying from teaching to personnel meetings to the pursuit of grants.

But what is contemplation? Is it a distinctive form of thinking? Isn't thinking just a matter of getting mentally to work? If we seek help from etymology, the English 'contemplation' comes from the Latin *contemplatio*, the silent gazing on the divine. Of long use within Christian theology, the term itself contains *templum*, which identifies the marking out of a sacred district or sanctuary. *Contemplatio* was itself a translation of the Greek *theoria*, which in its nontechnical sense simply meant 'seeing.' Overall, 'contemplation' points to a slow, recursive type of thinking—not the rock dropped in the pond, but rather the waves that move out across the water.

It is a type of thought that is now decidedly out of favor—commonly viewed as unproductive and indulgent. David Fincher's film *The Social Network* (2010) highlights our alternative reality. Facebook inventor Mark Zuckerberg exemplifies our understanding of intelligence today: swift, technological, and leaning toward the autistic. The opening scene, a conversation between Zuckerberg and a soon-to-be former girlfriend, shows what has been lost. Where the girlfriend displays an ambient intelligence, sensitive on a variety of registers, Zuckerberg's intelligence is narrowband. Processing information at the speed of light, he lacks the capacity for empathy or a feeling for differences of tone and nuance. Irony escapes him: as the girlfriend increasingly voices her irritation he asks several times "is this real?" For Zuckerberg, her slower way of thinking is the sign of inferior intelligence—a fact codified by her attending Boston University rather than Harvard.

If contemplation is in retreat today, the drivers of its abandonment are both technological and philosophical in nature. New technologies such as Facebook seduce us by their ease, speed, and brilliance. They also lessen the friction for both sending and receiving information. The cell phone, now a misnamed device, combines text, music, video, scheduler, alarm clock, and more. And in the era of cloud computing, all knowledge is instantly available—giving us everything, except perhaps the time for patient reflection. One result is what Alvin Toffler (1970) has called overchoice. Like the consumer market with too many brands of toothpaste or too many TV channels to choose from, academics face a crisis of overproduction. The volume of material ingested every day through emails, websites, journals, conferences, and publishing venues is impossible to digest.

The traditional strategy for managing this overflow of information has been specialization—the creation of areas of expertise within disciplines and subdisciplines. But this approach falters today. The endless proliferation of specialties has become absurd and unaffordable, a modern Tower of Babel. Moreover, it is far from clear that new strategies, such as the appeal to interdisciplinary modes of knowledge, do much to address the problem. Interdisciplinarity itself faces the unpleasant choice of either becoming its own area of specialization, thereby contradicting itself, or staying at a general level and being labeled dilettantism (Frodeman 2011).

Moreover, like all overproduction, the result has been devaluation: Knowledge and its cognates (information, data) today count for less and less. We see these effects in the way we 'surf' information—skimming books, articles, and emails to extract the meat (or the skin) of the matter, and demanding that complicated issues be reduced to the soundbites of executive summaries. The same is true for teaching, which reacts to today's student by providing tomorrow's soon-to-be replaced technological fix—for instance, clickers in the classrooms that required faculty to adjust their pedagogical style away from lecture or discussion and toward multiple-choice questions that could be answered without excessive effort (Bugeja 2008). As Gleick already noted in 2000, ours is an age characterized by 'faster.'

But beyond the new information technologies and the generally harried nature of contemporary life lurks a philosophical commitment that steadily extends its sway. Call it the technologizing of existence: Everything is subject to a pre-established standard, arrived at via a method, with results that are expressed in terms of quantities. This has been characteristic of the larger culture for quite some time (McDonald's, and more recently Starbucks; Club Med vacations, and 'preworn' jeans). The academy once offered a critique and a bulwark against commodification. No more. What matters today across the academy are numbers—of articles or books, dollars brought in, student ratings of professors, citation indices, and patents.

Somehow, academic numbers have come to count as something 'tangible'—even if the value of such numbers remains a mystery. An article has been cited a given number of times. Has it been understood? Is its argument profound or even correct? Sponsored research increases. But is the work it funds valuable, or does it merely contribute to the mountain of information under which we are buried, the creation of trivial consumer items, or to the growing inequality of American society? The fact that the U.S. spends more than any other nation on health care research, while faltering in so many health indices (life expectancy, infant mortality, percentage of those who are obese), ought to raise questions (Anderson and Chalkidou 2008). But Einstein's comment that not all that can be counted counts, and that not all that counts can be counted, remains little attended to.

Quantity counts as our definition of the real. Driven by the economic crises of the 21st century, quantitative approaches to thinking are spurred on by new technologies. It adds up to a technologized version of 19th century positivism. Positivism was a philosophy that limited the real to the data of experience, excluding what it labeled 'metaphysical' speculations, while emphasizing the achievements of science. Today this attitude more commonly trades under the label of 'neoliberalism,' where the language of markets, efficiency, consumer choice, and personal autonomy is applied to all human institutions and relationships. But it comes to the same: the abandonment of any serious defense of significant social values that lie outside the quantitative (scientific and monetary) nexus. Dickens's Mr. Gradgrind—"a man of realities. A man of facts and calculations"—has found new life, but this time armed with a more rigorous method (Dickens 1854).

This chapter explores the assumptions underlying the increasing focus on developing metrics for research and scholarship, as well as the concomitant marginalization of contemplation, or indeed of anything that falls outside the grasp of the methodological and quantitative. In contrast, we offer a refurbished notion of contemplation—revising *theoria* so that it can have a crucial place within both the academy and society at large. What is at stake here, along with our idea of the university, is the question of what is the best sort of life for human beings to live.

ACADEMIC ACCOUNTABILITY

The question of what role reflection plays within thinking, and whether academic production can be rendered quantitative, calculable, and methodical, is relevant to all the disciplines.[1] But the question presses most deeply on the humanities. Can the values that the humanities provide to society be reduced to quantifiable measures?

Stanley Fish rejects the premise. In his column of January 6, 2008, in the *New York Times*—a public, rather than a scholarly venue—Fish argues that the humanities must be divorced from all that is useful:

> To the question "of what use are the humanities?" the only honest answer is none whatsoever. And it is an answer that brings honor to its subject.

> Justification, after all, confers value on an activity from a perspective outside its performance. An activity that cannot be justified is an activity that refuses to regard itself as instrumental to some larger good. The humanities are their own good.[2]

Now, such claims are not limited to the humanities. Across the sciences, a similar point is often made in terms of 'curiosity.' But Fish's concern is to buttress the purity of humanistic inquiry.

Fish echoes Aristotle here, but it is unclear what his claim actually comes to. That the humanities are their own good in the sense of private amusement, personal religious ecstasy, or contemplation? And with no larger effect on the character of the individuals who study them, or on their students or associates? To paraphrase Fish, an activity that cannot be justified is an activity that had better have a rich patron.

The idea that any type of knowledge is its own justification—Stanley Fish notwithstanding—is not likely to gain much purchase in today's environment of accountability, in which 'intrinsic value' is a contradiction in terms. Academics have been aware of the need to justify their existence for some time. In fact, the term 'interdisciplinarity' was brought to prominence in 1972 by the Organization for Economic Co-operation and Development (OECD), as part of making the case that the pursuit of academic knowledge could be relevant to solving problems in the real world (Apostel, Berger, Briggs, and Michaud 1972). Despite its weaknesses, 'interdisciplinarity' remains the university's best attempt so far to come up with a language of its own to prove its worth: 'interdisciplinarity' is university-speak for what decision makers call 'accountability.'

But interdisciplinarity is not the real issue. In Europe especially, the term has largely been replaced by calls for 'transdisciplinarity'—connecting academic disciplines to those outside the university, rather than merely connecting those inside the university with each other.[3] Yet we may wonder whether talk of any sort of '-disciplinarity' makes sense outside of the academy. Today, university presidencies are increasingly staffed by corporateers, veterans of the business world who are bringing their own notions of accountability to bear on the academy. It is not by chance that they bring with them an affinity for—and a fluency in—the language of numbers.

METRICS—THE TECHNOLOGIZING OF THEORETICAL VALUE

The quest for objective standards, when wedded to a love of numbers, leads inexorably down the path toward metrics. Metrics allow us to compare different things by abstracting out a quality that both share. Different citrus fruits, for example, can be compared in terms of their acidity on the pH scale. Although something is always lost in translation, 'taste' in this instance, the mark of a good metric is that it allows for comparison while retaining some of the important characteristics of the thing in question.

This is the point of 'commensuration': transforming dissimilar qualities to a common quantitative value.

Despite the fact that it relies on methods and techniques, commensuration is more than a simple technical process. However objective these measures may seem by virtue of their general scientific quality or the particular psychological appeal of numerical devices, one key point remains true of all acts of commensuration: Metrics inevitably embody particular values. Often when metrics are applied to measure what has previously been deemed incommensurable, such as the quality of a particular journal article compared to another, the intended object of measure (i.e., scientific quality) is precisely what gets lost in translation. These acts of exclusion, and their underlying normative assumptions, call for a degree of attentiveness largely lacking in today's trends toward mechanization (Espeland and Stevens 1998).

Quantitative metrics of academic performance are now being used for evaluative purposes, often in a haphazard manner (van Raan 2005). Metrics of all kinds drive decision making in higher education with respect to a number of important collegiate and university functions—assessing academic quality, hiring, and the granting of tenure and promotions, to name a few. This trend is especially troubling in the context of research evaluation for the purpose of research funding. While questions concerning their use—what exactly do metrics measure, and how fair an account do they give?—are becoming pressing matters within the academy, concerns of their accuracy and meaningfulness have done little to slow their rise in popularity.

The rise in popularity of quantitative metrics challenges academic tradition. Academics themselves have long relied on peer review, rather than quantitative metrics, to assess academic performance. The ubiquity of peer review within the academy has also forced decision makers outside the academy to rely on academic judgment to inform their own decision making. In the case of science policy, for instance, peer review remains the gold standard for evaluation used to guide decision making (Valdez and Lane 2008). In our time of economic and budgetary crisis, however, peer review is being called into question (Chubin and Hackett 1990). But it is not only the objectivity of peer judgment that is called into question; it is also its relevance. And as disciplinary standards of expertise lose purchase, the question of who counts as a peer is reopened (Holbrook 2010).

There are those within the academy who claim that the superior objectivity of quantitative metrics makes them a more fitting and fair measure for performing evaluations (van Noorden 2010). This forgets, however, that the most popular measures of individual and group contribution to academic literature, bibliometrics, themselves depend on the peer review process. The construction and implementation of any metric relies on expert judgment concerning its merit. The inputs for many metrics, such as journal quality or grant money awarded, additionally rely on peer evaluation (Donovan 2009). No metric can be more objective than the peer review judgments on which it relies—which is to say, metrics are no more 'objective' than peer review (Weingart 2005, Donovan 2009).

The policy implications arising from the use of quantitative metrics, like the assumptions and values on which those metrics are based, are not always clear. A number of studies have noted unintended steering effects that influence the character of whole academic disciplines. Several document behavioral changes induced in researchers attempting to 'game' the numbers (David 2008; Erne 2007; Hicks 2006; Hicks and Potter 1991). There are also consequences for the academy as a whole. In what amounts to an epistemological Gradgrinding, the more interpretive and humanistic knowledge contributions of nonscientific disciplines are devalued (and sometimes ignored) in favor of 'positivistic' scientific knowledge (Donovan 2009).[4]

It is evident today that knowledge production has begun to undermine itself, especially insofar as rigor has become the ultimate academic value. In a knowledge ecosystem that rewards scientifically rigorous methodologies, "expert judgment" is a loser. Peer review, which used to guarantee the autonomy of the academy, now appears as biased self-policing (Holbrook 2010). Policymakers increasingly search instead for a metric that does not rely on the input of experts in order to render a judgment—metrics such as publications authored, patents awarded, or grant money secured. They seek, in other words, a number, and this number needs to be arrived at in some fairly simple way—a black box, a piece of technology for rendering judgment.

THE USES OF *THEORIA*

In *Frames of Mind* (1983) Howard Gardiner identified seven forms of intelligence: logical, linguistic, spatial, musical, bodily kinesthetic, interpersonal, and intrapersonal. It wasn't an entirely new thought: In the *Ethics*, Aristotle described *episteme*, *techne*, *phronesis*, and *nous* as the four intellectual virtues (the fifth, *sophia*, was formed from the combination of *episteme* and *nous*.) Setting aside the details, both of these accounts claim that human beings are capable of knowing the world in a variety of ways. Indeed, for Aristotle it was the mark of an educated person to understand that different domains of reality require different epistemological approaches. If *episteme* referred to types of reasoning akin to what we would today call calculation, *phronesis* and *nous* respectively referred to practical reasoning and the flash of intellectual insight.

This chapter has sought to simultaneously move forward on two fronts. On the one side, we have pushed back against the claim that only what can be counted counts. Of course, criticisms of positivism, scientism, calculative reasoning, and one-dimensional thinking are legion. But our point has been less about deconstructing metrics—they *are* useful, after all, and in any case they are here to stay—and more about appreciating that the creation of quantitative measures is always underpinned by acts of judgment and value.

On the other side, we have also challenged the notion that contemplation is of no practical use. Arguments such as Fish's manage to be both epistemologically incoherent and politically obtuse. But if metrics thinking has suffered from one dimensionality, contemporary humanistic thinking has been two dimensional—seeing the humanities as either proudly useless—*l'art pour l'art*—or as captured by mercenary values.

This is a childish way of thinking. Any real world ethics has to struggle with the better and worse rather than with a simple right or wrong. And any realistic expression of the humanities has to risk some degree of compromise with what former senator Tim Worth once called "the dark art of politics." The problem of 'dirty hands' is a regular refrain of 20th century cinema—think Brian De Palma's *Untouchables* (1987). But while humanists occasionally explore the theoretical dimensions of 'dirty hands' (e.g., Walzer 1973), they express disdain at the idea of actually working on the grubby problems of everyday life.

Part of this has to do with the tarnished status of rhetoric in the 20th century. Shaping an argument for the needs of an audience was of the utmost importance for Aristotle, and as important as the process of reasoning itself. But in the 20th century, claims of objective truth left little room for appreciating the art of contextualizing truths. Rhetoric became a term of dismissal. Moreover, in the 20th century, the humanities lacked the confidence to enter into the public realm, make claims, and take chances, assured in their ability to play a valuable public role.

In *Spectacles of Truth in Classical Greek Philosophy: Theoria in its Cultural Context*, Andrea Wilson Nightingale finds a political motivation behind Aristotle's notion of *theoria*: setting out a specialized field for philosophers, and motivating the polis to produce the conditions necessary for leisure. In contrast, Nightingale suggests, Plato's *theoria* involves a "journey" to another realm, a witnessing of "foreign" spectacles, followed by a return to one's original destination, where one is called to give an account of what one has seen. *Theoria* is then an inherently social and socially useful activity, a means for enlightening one's fellow citizens. Insofar as it involves giving an account of what one has seen, this Platonic sort of *theoria* is also necessarily interpretive, discursive, and dialectical.

If, following Plato, we take *theoria* to be a process, and an interpretive, dialectical, and useful process at that, then it is a mistake to understand thinking in terms of methodologies and mechanisms rather than as a fundamentally political process. Rather than existing in splendid isolation, calculative reasoning instead requires contemplation to provide orientation and governance to its measure.

Practically, this means matching the use of numbers with dialogue about the means of their construction. Now, this suggestion faces its own set of problems. Such dialogue takes time, an unlikely concession in an age of media soundbites and perpetual motion. Moreover, there are any number of political and economic interests that will not find it advantageous to make it clear that acts of judgment and value underlie appeals to the 'objectivity' of numbers. And finally, such conversations are undercut by the continuing subjectivism of our times, where terms such as 'politics' and 'values' are rendered insubstantial.

In the face of these obstacles, we can do nothing other than to look for every opportunity to make our arguments, again and again, at every occasion that arises. And seek out real world examples where humanists can demonstrate on the level of case studies that interpretation, reflection, and discussion add to rather than undermine the pursuit of the common weal.

NOTES

1. For an account of the interpretive and reflective nature of the hard sciences, see Frodeman, R. (2003). *Geo-Logic: Breaking Ground Between Philosophy and the Earth Sciences*. Albany, NY: State University of New York Press.
2. Fish, S. (2008). "Will the Humanities Save Us?" *New York Times*, Opinionator blog. Retrieved January 7, 2011, from http://opinionator.blogs.nytimes.com/2008/01/06/will-the-humanities-save-us/. See also "The crisis of the humanities officially arrives", Opinionator blog. Retrieved October 12, 2010, from http://opinionator.blogs.nytimes.com/2010/10/11/the-crisis-of-the-humanities-officially-arrives/?hp.
3. See Klein (2010) for a taxonomy of interdisciplinarity.
4. Recent mass academic downsizing in a number of universities has been justified on the basis of predominantly economic metrics. Examples include Middlesex University (England) cutting its philosophy department (a decision which was later reversed) (June, 2010), SUNY Albany eliminating five degree programs across the humanities (CBS, 2010), and Texas A&M's model of evaluating departmental contribution according to bottom-line dollar value (Doward, 2010).

REFERENCES

Anderson, Gerald F., and Kalipso Chalkidou. (2008). "Spending on Medical Care: More Is Better?" *Journal of the American Medical Association*, vol. 299, no. 20, pp. 2444–2445.

Apostel, Leo, Guy Berger, Aja Briggs, and Guy Michaud. (1972). *Interdisciplinarity: Problems of Teaching and Research in Universities*. Organization for Economic Cooperation and Development: Paris, France.

Barnes, Jonathan (Ed.). (1984). Nicomachean Ethics. In *The Complete Works of Aristotle*, vol. 2 (Princeton, NJ: Princeton University Press) pp. 1729–1867.

Bugeja, Michael. (2008, December 5). "Classroom Clickers and the Cost of Technology." Chronicle of Higher Education. Retrieved from http://chronicle.com/article/classroom-clickersthe/6009/

CBS. (2010, October 1). "Budget Cuts at SUNY Albany to Hit Academic Programs," CBS Broadcasting. Retrieved from http://www.cbs6albany.com/articles/university-1278894-programs-suny.html

Chubin, Daryl E., and Ed J. Hackett. (1990). *Peerless Science: Peer Review and US Science Policy*. Albany, NY: State University of New York Press.

David, Miriam E. (2008). "Research Quality Assessment and the Metrication of the Social Sciences," *European Political Science*, vol. 7, pp. 52–63.

Dickens, Charles. (2006). *Hard Times*. New York: Oxford University Press. (Originally published 1854).

Donovan, Claire. (2005). "The Governance of Social Science and Everyday Epistemology," *Public Administration*, vol. 83, no. 3, pp. 597–615.

Donovan, Claire. (2007). "The Qualitative Future of Research Evaluation," *Science and Public Policy*, vol. 34, no. 8, pp. 585–597.

Donovan, Claire. (2009). "Gradgrinding the Social Sciences: The Politics of Metrics of Political Science," *Political Studies Review*, vol. 7, pp. 73–83.

Doward, Jannie. (2010, May 8). "Middlesex University Cuts Spark International Protest from Philosophers," *The Guardian*. Retrieved from http://www.guardian.co.uk/world/2010/may/09/middlesex-university-cuts-protest-philosophers

Erne, Roland. (2007). "On the Use and Abuse of Bibliometric Performance Indicators: A Critique of Hix's 'Global Ranking of Political Science Departments,'" *European Political Science*, vol. 6, pp. 306–314.

Espeland, Wendy N., and Mitchell L. Stevens. (1998). "Commensuration as a Social Process." *Annual Review of Sociology*, vol. 24, pp. 313–343.

Frodeman, Robert. (2011). "Interdisciplinary Thinking and Academic Sustainability: Managing Knowledge in an Age of Accountability," *Environmental Conservation*, vol. 38, no. 2, pp. 105–112.

Gardner, Howard. (1983). *Frames of Mind: The Theory of Multiple Intelligences*. New York: Basic Books.

Gleick, James. (2000). *Faster: The Acceleration of Just About Everything*. New York: Hachette Book Group.

Habermas, Jurgen. (1968). *Knowledge and Human Interests* (J. J. Shapiro, trans.). Boston: Beacon Press.

Hicks, Diana. (2006). "The Dangers of Partial Bibliometric Evaluation in the Social Sciences," *Economia Politica*, vol. 23, pp. 145–162.

Hicks, Diana., and J. Potter. (1991). "Sociology of Scientific Knowledge: A Reflexive Citation Analysis or Science Disciplines and Disciplining Science," *Social Studies of Science*, vol. 21, pp. 459–501.

Holbrook, J. Britt. (2010). "Peer Review." In Robert Frodeman (ed.). *The Oxford Handbook of Interdiscilinarity* (Oxford, UK: Oxford University Press) pp. 321–332.

June, Audrey W. (2010, August 31). "U. of Mississippi Plans to Cut Programs and 29 Faculty Jobs," *Chronicle of Higher Education*. Retrieved from http://chronicle.com/article/U-of-Southern-Mississippi/124217/

Nightingale, Andrea W. (2009). *Spectacles of Truth in Classical Greek Philosophy: Theoria in its Cultural Context*. Cambridge, UK: Cambridge University Press.

Toffler, Alvin. (1970). *Future Shock*. New York: Bantam Books.

Valdez, Bill, and Julia Lane. (2008). "Prospective Evaluation: Lessons Learned from the OSTP Science of Science Policy Interagency Working Group." PowerPoint lecture delivered to the Washington Research Evaluation Network, George Washington University Marvin Center, Washington, DC, June 6, 2008.

van Noorden, Richard. (2010). "Metrics: A Profusion of Measures," *Nature*, vol. 465, pp. 864–866.

van Raan, Anthony F. (2005). "Fatal Attraction: Conceptual and Methodological Problems in the Ranking of Universities by Bibliometric Methods," *Scientometrics*, vol. 62, pp. 133–143.

Walzer, Michael (1973). "Political Action: The Problem of Dirty Hands," *Philosophy and Public Affairs*, vol. 2, no. 2, pp. 160–180.

Weingart, Peter. (2005). "Impact of Bibliometrics upon the Science System: Inadvertent Consequences?" *Scientometrics*, vol. 62, no. 1, pp. 117–131.

23 Science Policy and the Expectation of Health

The Case for Reforming Peer Review at the National Institutes of Health

Adam Briggle

As tho' to breathe were life!—Alfred Lord Tennyson, "Ulysses"

Ivan Illich (1972) defines "hope" as trusting faith in personal goodwill or the goodness of nature and "expectation" as reliance on results that are planned and controlled by humans. Health used to be something humans hoped for; now it is something we expect.

In his 1626 utopia, *New Atlantis*, Francis Bacon rendered the first depiction of a society where health is a rational expectation of an institution.[1] A ship carrying several sick passengers arrives by chance at the shores of the island of Bensalem. The citizens of Bensalem take in the strangers and give to their sick a "box of small gray or whitish pills" to hasten their recovery. Within a few days, the narrator tells us that the sick sailors "thought themselves cast into some divine pool of healing, they mended so kindly and so fast."

Salomon's House, the "noblest foundation" and lantern of the kingdom, makes Bensalem a place where health is an expectation. It has as its purpose "the knowledge of causes, and secret motions of things; and the enlarging of the bounds of human empire, to the effecting of all things possible." Salomon's House is a massive network of scientific laboratories that make novel metals, fertilizers, infusions, foods, drinks, engines, and artificial animals, plants, and trees. Its scientists perform experiments on animals to "take light what may be wrought upon the body of man." They have developed foods and medicines that help them "live very long" and make "their strength far greater than otherwise it would be."

In Bensalem, scientific research is the engine of social progress. Its citizens can expect health from a reliable, though mysterious and inexplicable, process of discovery and development designed to understand and control the states of the human body and mind. Salomon's House has a complex division of labor that includes "pioneers or miners" who "try new experiments, such as themselves think good" and "benefactors" who look "into the experiments of their fellows, and cast about how to draw out of them things of use and practice for man's life and knowledge." It is a primitive form of our current spectrum of basic, translational, and clinical or applied research.

The idea that progress in human affairs results from progress in science and technology was first given explicit policy formulation by Vannevar Bush in his 1945 report to U.S. President Harry Truman titled *Science—The Endless Frontier.*[2] Bush opened his report with the assertion that "Progress in the war against disease depends upon a flow of new scientific knowledge." The federal government, he argued, "should extend financial support to basic medical research in the medical schools and in universities." Scientific research is essential to health and for this reason "the flow of new scientific knowledge must be both continuous and substantial." Bush's policy has since been termed the "linear model" of science-society relations, because it presumes a linear function—more funding for science equals more science equals more social good (Pielke and Byerly 1998).

The present-day incarnation of Salomon's House in the U.S. is the National Institutes of Health (NIH), with its strikingly similar mission to "seek fundamental knowledge about the nature and behavior of living systems and the application of that knowledge to enhance health, lengthen life, and reduce the burdens of illness and disability." The optimistic spirit of Bush's vision fueled the post-World War II growth of NIH as its budget skyrocketed from $4 million in 1947 to $100 million in 1957 to $1 billion in 1974 to $32 billion in 2010.[3]

After the war, NIH adopted a method of awarding grants and fellowships to nonfederal scientists based on a peer review system that had been pioneered by the National Cancer Institute. Peer review is essential to the linear model, because it assumes that knowledge that has been certified by peers will automatically advance social goals. Good science, as judged by scientific peers, is good for society. All that needs to be done, according to NIH policy, is "to fund the best science by the best scientists." But the "best science" can be defined in two ways. First, it can be defined as that which a scientific peer group finds most intellectually promising for advancing a given field of research. Second, it can be defined as that which is most likely to contribute to the goal of improved health. The linear model would have us believe that the two definitions are equivalent. But of course they are not, because rather than furthering social goals, scientific research can, at times, be irrelevant to those goals or even contradict them.

At first it would appear that the two-stage NIH peer review process assesses both definitions of "best science." In the first stage, scientific peers judge proposals in terms of five criteria: Significance, Investigators, Innovation, Approach, and Environment. This is designed to assess the first definition of best science as "the likelihood for the project to exert a sustained, powerful influence on the research field(s) involved" (National Institute of Health Peer Review Process n.d.). NIH then conducts a second stage of review done by an advisory council composed of scientists and public representatives that is designed to determine "best science" in the second sense of the term, that is, as most likely to contribute to improved health. Yet, stage one is far more developed and rigorous than stage two. In contrast to the elaborate discussion of stage one, the NIH Website only states that advisory council members will examine proposals and "consider these against the IC's [Institute's or Center's] needs."

The NIH mission hinges on the connection between scientific research and health, and thus one would expect that stage two review would be the fulcrum of the entire agency. The second stage review is the place where progress in science (as judged by stage one review) is explicitly assessed in terms of the goal of improved health. Yet, judging from its publically accessible documents, NIH does not put near the emphasis on second-level review as on the first. The first stage has criteria, metrics, and procedures outlined in detail, but this is not true of the second stage.[4] Rather than a well-structured process designed to assess proposals in light of their ability to advance the goal of health, stage two seems to be an underdeveloped afterthought.[5]

This would not be a problem if the assumption of the linear model was adequate. In that case, we truly would inhabit a utopia like the citizens of Bensalem who reap all of the benefits, *and none of the costs or harms*, from the autonomous workings of peer-controlled scientific research. All of our health problems would be best addressed by the advancement of new knowledge. We could go about our way, confident that we can expect nothing but the best health outcomes from the opaque yet beneficent workings of an institution and its experts.

In what follows, I paint a picture of reality that suggests we are far from this utopia. Scientific research is costly, and its rewards are dubious. This being the case, NIH needs to concentrate its energies on the second stage of peer review. Such an effort will better serve society by developing a mechanism whereby public dollars are more explicitly guided toward public health outcomes. It will also serve the interests of NIH, because in an age of accountability and budget woes, an agency armed with an evidence-based explanation of how it serves the public good will fare better than one that assumes more science equals a healthier society.

EXPECTING HEALTH: EXPENSIVE AND INEFFECTUAL

Health-as-expectation finds its strongest voice with those who, like U.S. Senator Ted Kennedy (2009), believe that everyone has a "fundamental right" to "the best medical care money (and a good insurance policy) can buy." The problem is that the "best medical care" is a moving target that requires more and more money to buy. In the U.S., spending on health care tripled between 1965 and 1985, from $187 billion to $666 billion. Then it tripled again to $1.9 trillion in 2005. In 1965, health care spending accounted for 5% of the U.S. GDP. In 2005, it accounted for 15%. In 1965, annual per capita spending on health care was $1,000. In 2005, it was nearly $7,000.[6] The U.S. does spend more on health care than other industrialized nations. But "most industrialized countries—even those with a financing system quite different from that in the United States—have experienced a substantial long-term rise in real spending on health care" (Congressional Budget Office 2008, pp. 3–4).

As I write this, we expect a 2012 level of health care, not a 1965 level. The relevant difference between the levels is the increased capabilities of medicine, that is, the health care system's use of new medical services made possible by scientific discovery and technological advance. In the 1960s, caring for patients with coronary artery disease was inexpensive, because there was little that physicians could do for them. Now we have coronary angiography, heart-lung machines that make bypass surgery possible, angioplasty, and thrombolytic therapy. Neonatal intensive care, joint replacement, and diagnostic imaging are other examples. Victor Fuchs (1999) notes, "Advances in medical technology have made it feasible and desirable to do more for each patient and to intervene with more patients" (p. 13). Yet unlike in Bensalem, where medical marvels seem to come at no cost, in reality, new capabilities come with a steep price. The U.S. Congressional Budget Office (CBO) estimates that "about half of all growth in health care spending in the past several decades was associated with changes in medical care made possible by advances in technology" (CBO 2008, p. 1).[7]

Expectation is infinite, as the wording of the NIH mission statement makes clear: Can we ever enhance health too much, extend life too much, or reduce the burdens of illness and disability too much? Health-as-expectation operates by a logic of beneficence first voiced by René Descartes. As soon as we make scientific discoveries, Descartes argued that we cannot "keep them concealed without greatly sinning against the law which obliges us to procure, as much as in us lies, the general good of all mankind." New knowledge will "render ourselves the masters and possessors of nature," which is to be desired above all "because it brings about the preservation of health, which is without doubt the chief blessing and foundation of all other blessings in this life" (Descartes 1637, book 6). Once health is conceived both as something in our control and as the highest good in life, we are obliged to pursue it ad infinitum.

This puts us into a policy bind: we cannot continue the trend and we cannot discontinue it. We cannot continue it, because 100% of all funds will eventually be spent on health.[8] Technological advances will continue to yield services that we will come to expect, fueling further spending and "imposing difficult choices between spending on health care and spending on other priorities" (CBO 2008, p. 1). Fuchs (1999) argues that "Over the long haul, there is only one reliable way to slow spending growth: slow the growth of services to patients" (p. 15). Yet we cannot do this, because it would mean relinquishing the humanitarian logic of infinitely improving health. It would mean saying "enough," even though people continue to suffer and die.

Elizabeth G. Nabel (2009), director of the National Heart, Lung, and Blood Institute at NIH, articulated the linear model assumption: "Biomedical research provides the basis for progress in health and health care" (p. 2858). When challenged, the assumption is defended by a reaction of the "look out the window" variety: Just look, now we can treat coronary artery disease, replace joints, save premature babies, transplant organs, even synthesize replacement organs from a patient's own cells, and so much more that we could not do in the past, and all of this can be traced back to investments in scientific research.

The "look out the window" hypothesis implicitly defines improved health as increased consumption of the institutional products of modern medicine. It is analogous to arguing that we are better educated now because we consume more of the institutions of modern schooling, or that we are happier now because we consume more of the products of the modern economy. Institutions have a way of defining the goals (health, education, happiness) in such a way that their achievement is synonymous with increased dependence on said institution. This is, however, to confuse process with substance. It is as if progressing through the school system was the same as learning or as if having a cabinet full of the latest pharmaceuticals was the same as being healthy. Are we really healthier as a result of our dependence on medical institutions?

Answering this requires some metric for health. The problem with metrics is that they can at best point in the general direction of what really matters. GDP growth is our favorite metric for "progress," but it is a notoriously poor proxy for goods such as time with family and friends, meaningful opportunities, and natural beauty. The converse problem is that these goods are harder to measure. When it comes to health, we also have a couple of favorite metrics. In the U.S., infant mortality rates dropped between 1950 and 2004 from nearly 30 to under 7 deaths per 1,000 live births. From 1900 through 2004, life expectancy at birth increased from 46 to 75 years for men and from 48 to 80 years for women (Centers for Disease Control 2007).

There are two possible reactions to these figures. First, we can question the adequacy of such indicators. As a U.S. Center for Disease Control (CDC) report notes, "even as progress is made in improving life expectancy, increased longevity is accompanied by increased prevalence of chronic conditions and their associated pain and disability" (CDC 2007, p. 3). People may now enjoy added years of healthier life, but at the price of protracted debility, dementia, dependence, and loneliness "stacked up at the end before they eventually die" (President's Council on Bioethics 2005, p. xviii). Further, a society that expects health becomes health obsessed, and even those not ill become the "worried well," forever insecure about their ability to attain the products of the medical establishment. Expecting ever-more from the technological capacity to fend off mortality means that we are, in Epictetus' words, "thwarted, miserable, and upset" whenever our expectations are not met. Living longer does not necessarily mean living well.

Second, we can take the indicators as genuine signs of improved health. But then the next question is one of attribution: what causes the upward trends? We cannot assume a causal relation between biomedical research and improvement in these indicators; because several countries that spend far less on health care outperform the U.S. in terms of infant mortality and life expectancy (see WHO 2009).[9] As Daniel Sarewitz (1996) notes, the steep rise in expenditures on biomedical research and health care in the U.S. did not occur until the mid-1950s, which is about the same time that improvements in mortality and life expectancy began to taper off. Indeed, education, economic opportunity, nutrition, and sanitation are better predictors of good health than health care expenditures. It is far more effective to prevent the release of toxic chemicals in the first place than to liberally permit them and then fund a $700 million Institute of Environmental Health Sciences to research their health effects.

Fuchs (1979) argues that health improvements are "much more dependent on non-medical factors than on the quantity of medical care" (p. 155). He goes on to argue that the marginal contribution of medical care to health is "very small," and that those who argue for eve- more research, hospitals, and physicians "are either mistaken or have in mind objectives other than the improvement of the health of the population" (p. 155). Sarewitz concludes that the direction of causation is actually opposite of the linear model: only once high levels of affluence, public health, and education are achieved did the U.S. begin to invest in high-tech institutionalized medicine. Low-cost, low-tech alternatives, such as education and prenatal care, are often more effective in terms of health outcomes than high-tech interventions. Furthermore, greater capacities often bring greater risks, a fact seen with the prevalence of iatrogenic diseases (those caused by medical interventions) and antibiotic-resistant bacteria.

In a more concrete example about the questionable value of medical research, Gayle Sulik (2010) points to disturbing evidence about breast cancer. A woman now has a 1-in-8 chance of getting breast cancer in her lifetime; in 1975, that figure was 1-in-11. Despite billions of dollars invested in research, the risk of dying from the disease, upon diagnosis, decreased just 0.05% from 1990 to 2005. Indeed, a woman with breast cancer today will receive many more treatments and spend a lot more than her grandmother would have on care, but she will have about the same chance of dying from the illness as women did 50 years ago. Meanwhile, cancer drugs are the fastest growing class of pharmaceuticals, which poses the question one skeptic raises: "Who needs a cure if you can make so much money without one?" (Rich 2010).

Daniel Callahan and Sherwin Nuland (2011) draw a similar conclusion from a more global survey of biomedical research: "We have arrived at a moment where we are making little headway in defeating various kinds of diseases. Instead, our main achievements today consist of devising ways to marginally extend the lives of the very sick." Putting this in the context of a major debt crisis and soaring health care costs, David Brooks (2011) writes that America is a nation willing to spend itself "into bankruptcy to extend life for a few more sickly months."

In summary, biomedical research is a mixed blessing. Along with, or sometimes instead of, the intended good outcomes, we get a lot of bad unintended ones. Progress in health outcomes results from a mixture of behavioral, cultural, and economic factors—scientific research needs to be seen as just one component of this ecosystem of health determinants, rather than as a straight golden road to a healthier society (see Crow 2011). The problem is that the linear model is so politically entrenched: "more research" is a much safer political tactic and far more profitable than social policies that target habits, cultures, and values. Americans eat poorly, are overweight, inactive, and pollute their environment. But that is their choice, whereas medical care is their right.

REFORMING PEER REVIEW AT NIH

NIH has just completed a major overhaul of its peer review process with the primary objectives of reducing the administrative burden of reviewing grants and improving review quality. The goal is to streamline the process whereby the "best science" is identified and funded. In general, scientists have reacted favorably to the changes, but some argue that they do not do enough. For example, the editors of the journal *Infection and Immunity* lamented the state of science where investigators spend half their time writing grants with only a 10% chance of being funded. The real problem with NIH, they claim, is the lack of resources: "There is no escape from the relentless math of an unprecedented number of applications . . . chasing a shrinking number of grant dollars" (Fang and Casadevall 2009, p. 930). This "funding imbalance," they argue, is harming science, innovation, economic growth, morale, and collegiality.

Tellingly, the editors do not list "health" among the casualties of the funding imbalance. They beat the linear model drum demanding more money for more science, not for public health, but to save scientists' jobs and to grow the economy. But NIH is tasked with improving health, and the previous section cast doubts on the connection between biomedical research and health. It highlights a problem that cannot be solved by increasing the budget or streamlining the peer review process. NIH peer review reform has missed the boat in supposing that its new goal of "fund the best science, by the best scientists, *with the least administrative burden*" is an adequate fix. The real goal is determining the "best science" according to the standard of improved health. This problem requires an overhaul to the meager second-stage process of peer review.

A remark from the father of Salomon's House can be the inspiration for reforming the second stage. In describing their peer review process, he notes "this we do also: we have consultations, which of the inventions and experiences which we have discovered shall be published, and which not; and take all an oath of secrecy for the concealing of those which we think fit to keep secret." Their peer review process is self-censoring: With a beneficent eye to the common good, even high-quality research may be squelched if the members of Salomon's House judge that it is not in the interest of society.

Fuchs gave contemporary voice to the idea of aborting scientifically promising research. We will, he argues, continue to see increasing marginal costs per life saved the more we come to rely on high-tech medical services. "To the extent that we deem this an undesirable outcome, the way to guard against it is to rule out the *possibility* of relatively high-cost interventions. If the intervention is unknown, society may, in some sense, be better off" (Fuchs 1979, p. 165). Society would be better off, because it would opt for lower cost preventative measures or other more efficacious means toward improved health.[10] The decisive strike must be abortive. Once a high-tech intervention is brought to fruition, comparative effectiveness analyses would have to convince doctors and patients to opt for fewer and less intensive services than are actually available. This seems unlikely given the logic of expectation: we opt for more and more of what the medical institutions have to offer, because we have come to define health as increased consumption of their wares.

One way to make this beneficent censorship both practical and democratic is to transform the second stage of peer review into a science court. This court would judge the societal worth of all proposals that pass the first stage with a fundable score. The principal actors would be the prosecutor, the defendants, and the jury. Its essence would be an open challenge, wherein the prosecutor throws down the gauntlet by arguing that the proposed research will not improve public health. The prosecutor would represent the *interdisciplinary* moment in the peer review process, because his or her expertise would not be that of the research proposal under consideration. It would, rather, be a mixture of economics, ethics, sociology, and whatever else is needed to make a reasonable, well-founded, pessimistic case regarding the interface between the proposed research and society.

The defendants would be the principal investigators (PIs) listed on the research proposal in question who will be tasked with constructing a reasonable, well-founded optimistic case about the prospects of their research contributing to the social goal of improved health. The PIs represent the *disciplinary* moment of peer review. They are, thus, a continuation of stage one peer review, but now they are compelled to make the case for their project in terms accessible to those who are not experts in their field. In this task, they may enlist the advisory support of any other field of knowledge, but the PIs must themselves do all the speaking in the courtroom. This is designed to foster within scientists a breadth of thought, an interdisciplinary skill set, and an awareness of social contexts.

The jury would be composed of "lay citizens" and would be selected via much the same process as is used in civil litigation. They represent the *transdisciplinary* moment of the peer review process, that is, the perspective that begins from the interests and values of society rather than from the interests and values of knowledge production. After listening to testimony from both sides, the jury would issue its ruling and direct the appropriate course of action.

There would be three possible rulings entailing different actions: guilty, innocent, or provisions. If the prosecutor is successful in obtaining a guilty verdict, the money that would have gone to the grant in question becomes property of the prosecutorial staff. They may use 15% of such monies to fund their work and are required to devote the remaining 85% to low-tech public health projects, such as educational and preventative health programs, environmental remediation for low-income neighborhoods, and vaccinations for poor children. If the defense is successful, then their grant is funded. If the jury rules a finding of provisions, then they specify changes that must be made to the grant for it to receive funding.

This science court would likely increase the administrative burdens of peer review, but so be it: NIH is not funding science for the sake of science, but to "enhance health, lengthen life, and reduce the burdens of illness and disability." Assessing the relationship between such goals and scientific research is a task that demands expanding the "peer" category in a way that mirrors the complex realities at the interfaces between science and society. If we are going to "expect" science to be in service of the good life, we can no longer just "hope" that what's good for science will be good for society.

NOTES

1. Full text can be found here: http://oregonstate.edu/instruct/phl302/texts/bacon/atlantis.html.
2. Full text can be found here: http://www.nsf.gov/od/lpa/nsf50/vbush1945.htm.
3. NIH is comprised of 27 centers and institutes that research cancer, heart disease, neurological disorders, aging, and much more. Private industry nearly doubles NIH, expending roughly $57 billion on biomedical research in 2008.
4. The NIH "peer review process" Webpage includes five pages of material on the first stage of peer review, including a detailed scoring system and criteria, whereas the second stage is described in one short paragraph: http://grants.nih.gov/grants/peer_review_process.htm. In a Power Point presentation by an NIH scientific review officer, 13 slides were devoted to an account of stage one peer review; one slide to stage two review (Allen, 2010).
5. For more evidence that NIH takes scientific and technical merit to be more important than societal considerations, see the National Institute of Allergy and Infectious Diseases (NIAID) document about funding decisions: http://

www.med.illinois.edu/administration/research/resources/NIH/writing/2a.pdf. It states that "Scientific merit is the most important factor for NIAID's funding decisions." And that the "biggest hurdle is initial peer review. Once you've cleared that, the second-level review is a small step, which ensures there are no administrative problems."

6. That is 2005 dollars.
7. CBO notes that very few technologies actually reduce costs. Most increase health care spending because they make treatment possible for conditions that were previously untreatable. Improvements that decrease mortality increase costs, because surviving patients live longer and therefore use health care services for more years. W. B. Schwarz (1994) argues that many technologies raise costs, because they create new expensive interventions that extend life but are not curative.
8. CBO predicts that the U.S. will spend 49% of its GDP on health care in 2082.
9. For example, Japan exceeds the U.S. on these indicators, yet spends roughly one-third per capita on health care.
10. There is an analogy here with American football. Players may in fact be safer without the very helmets that were supposed to make them safe, because those helmets have given them a sense of invincibility and changed the way the game is played (e.g., tackling by leading with the head).

REFERENCES

Allen, Janice. (2010). "The NIH Scientific Review Process," CAPR Digital Repository. Retrieved from http://csid-capr.unt.edu/fedora/repository/capr:1063

Brooks, David. (2011). "Death and Budgets." New York Times, July 14,. Retrieved from http://www.nytimes.com/2011/07/15/opinion/15brooks.html.

Bush, Vannevar. (1945). *Science—The Endless Frontier*. Washington, DC: United States Government Printing Office. Retrieved from http://www.nsf.gov/od/lpa/nsf50/vbush1945.htm

Callahan, Daniel, and Shwerin Nuland. (2011, May 19). "The Quagmire," *The New Republic*. Retrieved from http://www.tnr.com/article/economy/magazine/88631/american-medicine-health-care-costs?passthru=ZDY1YmFhOTg2YzhmMWE2Njg2MTMzNmM1OWUyYmUyMDQ

Centers for Disease Control. (2007). *Health, United States, 2007: With Chartbook on Trends in the Health of Americans*. Hyattsville, MD: National Center for Health Statistics. Retrieved from http://www.cdc.gov/nchs/data/hus/hus07.pdf

Congressional Budget Office. (2008). *Technological Change and the Growth of Health Care Spending*. Washington, DC: Author. Retrieved from http://www.cbo.gov/ftpdocs/89xx/doc8947/01–31-TechHealth.pdf

Crow, Michael. (2011). "Time to Rethink the NIH," *Nature*, vol. 471, pp. 569–571.

Descartes, René. (1998). *Discourse on Method* (Donald A. Cress, trans.). Indianapolis: Hackett. (Originally published 1637)

Fang, Ferric, and Arturo Casadevall. (2009). "NIH Peer Review Reform—Change we Need, or Lipstick on a Pig?" *Infection and Immunity*, vol. 77, no. 3, pp. 929–932.

Fuchs, Victor. (1979). "Economics, Health, and Post-Industrial Society," *Milbank Memorial Fund Quarterly*, vol. 57, no. 2, pp. 153–182.

Fuchs, Victor. (1999). "Health Care for the Elderly: How Much? Who Will Pay for It?" *Health Affairs*, vol. 18, pp. 11–21.

Illich, Ivan. (1972). *Deschooling Society*. New York: Harper and Row.
Kennedy, Ted. (2009, July 18). "The Cause of My Life," *Newsweek*.
Nabel, Elizabeth G. (2009). "Linking Biomedical Research to Health Care," *Journal of Clinical Investigation*, vol. 119, no. 10, p. 2858.
National Institute of Health Peer Review Process. (n.d.). http://grants.nih.gov/grants/peer_review_process.htm
Pielke, Jr., Roger, and Radford Byerly. (1998). "Beyond Basic and Applied," *Physics Today*, vol. 51, no. 2, pp. 42–46.
President's Council on Bioethics. (2005). *Taking Care: Ethical Caregiving in Our Aging Society*. Washington, DC: Government Printing Office.
Rich, Katherine Russell. (2010, October 29). "Sink Pink," *MSN Slate Magazine*. Retrieved from http://www.slate.com/id/2272767/
Sarewitz, Daniel. (1996). *Frontiers of Illusion: Science, Technology, and the Politics of Progress*. Philadelphia: Temple University Press.
Schwarz, W. B. (1994). "In the Pipeline: A Wave of Valuable Medical Technology," *Health Affairs*, vol. 13, no. 3, pp. 70–79.
Sulik, Gayle. (2010). *Pink Ribbon Blues: How Breast Cancer Culture Undermines Women's Health*. Oxford: Oxford University Press.
World Health Organization. (2009). *World Health Statistics, 2009*. Geneva: Author. Retrieved from http://www.who.int/whosis/whostat/EN_WHS09_Full.pdf

24 Neutrality and Technology

Ortega Y Gasset on the Good Life

Jeroen van den Hoven

New technologies such as human enhancement, robotics, GM food, and geo-engineering give rise to a good deal of controversy and disagreement. It is difficult to reach closure or consensus on these matters. Scientists and policymakers tend to cast the issues in a vocabulary of risk, probabilities, uncertainties, costs, and benefits. The general public, on the other hand, tends to react emotionally to new technologies. Both approaches, separately or combined, usually fail to deliver unanimous decisions and clear direction to public policymaking. The first approach has the appearance of an objective and scientific way of dealing with the problems, but a closer inspection may reveal that the assignment of probabilities, distribution of risks, the choice of discounting rates, the way of dealing with uncertainties, and the economic valuing of costs and benefits and their allocation are based on choices and moral assumptions that can be reasonably disputed by reasonable and informed people. Disputes about the moral parameters of risk and cost-benefit assessments are added to extant disputes about what the relevant facts of the matter are. When it comes to the second approach, emotional reactions, and the public scares and panic cascades to which they often give rise, enhanced by the new media, and prompt people to accept beliefs that are often unsupported by the best available scientific evidence, logic and mathematics.

It seems fair to say that we can neither simply calculate nor simply feel our way out of the problems concerning cloud seeding, enhanced soldiers, lettuce with lady bug genes, and companion robots for the elderly. There is reason to think that even a fairly nuanced and regimented applied ethics of technology that produces utilitarian or Kantian analyses of technologies and policy options will not prevent us from getting bogged down in endless and fruitless debates. The reason is not difficult to understand. Disagreements concerning high technology are often occasioned by what John Rawls has called "the burdens of judgement" (Rawls 1993, p. 56), such as different weights assigned to value considerations, different conceptual frameworks, vague terms, and uncertainty about the scientific facts. Furthermore, these disagreements are beset by another formidable catalyst of controversy about science and technology: a plurality of views about the good life and the recurrent, unavoidable, and fundamental choices about what (kind of) things to bring into existence and what (kind of) life we ought to aspire to.

It is this set of fundamental questions about ultimate goals and ideals that are often only marginally discussed in the applied ethics of technology. Given the prominence of debates about science and technology, we need to learn how to deal with them in terms of the basic question about the good life that they raise, because no amount of instrumental reasoning and deliberation about means can replace reflection on our ultimate ends and purposes. The rapid development of technology will eventually make it necessary to pose and answer questions about final destinations. This has become more pressing now that almost everything has become the object of design and engineering, including the very basic things, such as the living cell and its components, the human genome, and patterns of atoms and molecules and combinations thereof. As a result, the consequences of our engineering interventions in the world are potentially greater than they have ever been. As Edward O. Wilson has put it, Homo sapiens is about to decommission natural selection and is about to enter the stage of "volitional evolution" (Wilson 1998). Furthermore, the pace, scale, impact, irreversibility, and complexity of our technological interventions and engineering efforts require that we address the issues at an early stage and on a fundamental level, lest our ruminations about them will be superfluous and obsolete before we can hope to act adequately in accordance with them.

At least two features of our thinking make it difficult to engage in political discussions about what we *ought* to make and what we ought to make of our selves. First, there is the aforementioned technical discourse of risk assessment, which often obfuscates reflection on fundamental moral choices about what ought to be our substantive goals and replaces them by discussions about probabilities, utility, costs, and benefits. The other impediment to fundamental debates about technology and the good life is the culture of neutrality of the dominant liberal tradition in Western political thinking. According to the doctrine of liberal neutrality, government and government policies should be neutral with respect to controversial conceptions of the good life. In Ronald Dworkin's classical formulation: "the government must be neutral on what might be called the question of the good life (. . .) political decisions must be, so far as is possible, independent of any particular conception of the good life, or of what gives value to life" (Dworkin 1985, p. 191).

Neutrality as defended by Rawls, Dworkin, Bruce Ackerman, and others has a respectable pedigree in Western political philosophy and has served as a means of protection against oppressive and perfectionist governments that meddled with religion and religious practice. The ideal of neutrality, however, is now becoming an obstacle to dealing with the defining issues of our age in science and technology. I will present an argument against a construal of neutrality that would make it impossible to have systematic and careful political discussion and decision making about fundamental new technologies, such as nanotechnology, robotics, human enhancement, and geo-engineering. The key element for this argument I take from the Spanish philosopher Jose Ortega Y Gasset, who equates technology and conceptions of the good life: To think about technology is to engage in thinking about the good life and designing or using technology amounts to implementing a conception of the good life.

His essay "Man the Technician" (Ortega 1961) provides a philosophical account of the human condition and argues that there is an internal relation between technology and the good life. I will first discuss Ortega's views on technology and expound how he construes their interrelation. Second, I argue—drawing on Ortega's account of technology and the good life—that public decision making concerning technology and applied science in the 21st century cannot be neutral, in any interesting sense, regarding conceptions of the good life. Finally, I argue that if public decision making about technology cannot possibly proceed in a neutral way, then we are in need of institutions that allow us to conduct political discussions concerning technology that can accommodate disagreement about conceptions of the good life. This requires as—Dennis Thompson and Amy Gutmann have argued—a shift from preclusion to accommodation of disagreements.

ORTEGA Y GASSET ON TECHNOLOGY

According to Ortega Y Gasset, man is perhaps the sole being that is capable of annihilating himself. If he lives, he lives because he wants to. He lives in a world that is not completely hostile, nor completely accommodating. The world provides a mix of opportunities and constraints in which he can decide to continue to live or to discontinue his life. Only if he chooses to live, are shelter, warmth, clothing, and spears seen as necessities (1961, p. 90). A peculiar feature of our lives is that pretty much everything we want, pursue, and tend to think of as necessities are, strictly speaking, not necessities. We could survive without shelter, agriculture, fire, cooking, and utensils. Our lives would be nasty, cold, uncomfortable, or even miserable, but that's about it. We would probably be confined to happier and warmer latitudes.

When man* is unable to satisfy these perceived necessities immediately, when he finds no cave, no water, or hot springs, he engages in a very peculiar activity found nowhere else in nature, Ortega observes. Man mobilizes a second line of defense (1961, p. 91). If he cannot find warmth and protection against the elements, he builds a shelter that provides him with warmth and protection, now and in the future. He shifts gear and moves from a level of activities that seek to satisfy his immediate needs—as perceived—directly to a level of activities that seek to satisfy them indirectly. In doing so, he creates an environment—a second or artificial environment—that allows him to satisfy his needs.

*Hereinafter, 'man' stands for humans, yet I tend to follow Jose Ortega Y Gasset's somewhat archaic terminology.

Ortega emphatically draws attention to the fact that "keeping warm" and "making a fire" are very different acts. In the second-order type of activity, humans engage in planning and design (1961, p. 94). They invent and make things (tools, implements, procedures) that are not available in their environment, and they thereby transform nature to make it suitable to their purposes. These acts are technical and exclusively and specifically human. Technology is man's answer to nature and circumstance. In technology, human beings adapt nature, instead of adapting themselves to nature: "Man without technology—that is, without reaction upon this medium—is not man" (1961, p. 96).

Under the label of human necessities there is an often disregarded conflation of what is objectively necessary and what is superfluous (1961, p. 98). Human beings are "incredibly elastic towards these necessities which (. . .) would seem (. . .) indispensable" (1961, p. 98). Man has a tendency to consider as necessary only the objectively superfluous. (. . .) And to know this is essential for our comprehension of technology." This leads Ortega to his famous characterization of technology as "the production of the superfluities, today as in the Paleolithic age." Ortega points out that "(. . .) man's desire to live (. . .) is inseparable from his desire to live well" (1961, p. 98). Man "(. . .) conceives of life, not as simply being, but as well-being; he regards the objective condition of being as necessary only because it is the necessary condition of well-being" (1961, pp. 98–99). According to Ortega "Not being, but well-being, is the fundamental necessity of man, the necessity of necessities." "Man," "technology" and "well-being" are in the last instance synonymous" (1961, p. 100).

Life in the human sense, that is, as 'good life,' 'human flourishing'—is variable. Human necessities are a function of this and therefore technology—as a set of actions—accommodates the variety and is therefore 'protean in nature' (1961, p. 101); "(. . .) life in the human sense of good life is always mobile and infinitely variable." If we supposed that "the kind of well-being we seek today changed its character perceptibly (. . .) would not our present technology be thrown out of gear (. . .)?" (1961, p. 104). Technology and engineering aim at the good life, the good life is the *terminus ad quem* (goal) of technology, and therefore they presuppose acquaintance with a more or less articulate conception of the good life. A corollary of this claim is that 'to be an engineer it is not enough to be an engineer" (1961, p. 103).

Now humans have been incredibly successful in innovation, invention, and have managed to save themselves considerable time and energy by engaging in technology and engineering. This is therefore essential to technology; that it frees up time and energy to do something else. What is this something else? In the freed up time—unexposed to the exigencies of nature—man starts to invent himself, he shapes his life as he aims at "well-being" and the good life. Man has to fight for his life, not only economically and physically, but also metaphysically. He is an "ontological centaur," half nature, half product of his own imagination and fantasy. He has to decide who and what he wants to be. That part of him that is given and part of nature is not a problem and it is taken for granted and perhaps even thought to be of lesser value. That part of himself that he has to work hard for to bring about matters most to him; it is his self, his life of his own making. Man is a program of what he is not, a program of what he wants and attempts to be. Man is the engineer of his own self and life. Life is about finding the means to be able to be what one wants to be.

Technology allows us to do so in two ways: first, by freeing up time and energy, which can be devoted to crafting one's biographical life; and, second, by providing us with the means to express and realize ourselves. "The particular activities which deserve and have received the name of technology are only concrete specializations of the general character of the autofabrication inherent in human life" (1961, p. 118). Without that extra-natural aspiration, no technology would exist.

Technology is thus preceded by a desire and is just the instrumentation of the preceding desire that forms the blueprint for a human life. Although this sounds easy, Ortega points out that this desire and program for life is not simple to come by. Some people do not know what to want and just want what others want. Therefore, it often seems that the technology and technical innovations that stem from such a common place desire seem ordinary and trite. Life requires conceptions of well-being, human flourishing, or the good life to give direction. If nature is transformed through technology, one needs to know into what it has to be transformed in which direction and why.

There are many conceptions of the good life that have provided guidance and direction to technologies in the past, or of which technologies have been the expression. The Buddhist ideal of the bodhisattva, the individual who wants to be like Buddha, the ideal of the athlete and well-rounded aristocrat of Ancient Greece, the Republican Public Man of Rome, the medieval knight or medieval ascetic, the Dichter und Denker of the early 19th century, the ideal of the gentleman, the Schoene Seele in the end of the 18th century, and the hidalgo. These are some of the examples Ortega provides of life programs (and one could add many more: the samurai, the urban professional of the late 20th century, uomo universale in the Renaissance, the Prince, the scholar of the 21st century, the modern CEO, and the bourgeois bohemian).

These life programs and conceptions of the good life come with technological assumptions, requirements, and implications. The bodhisattva, the Buddhist ideal of man, he who aspires to become like Buddha, aims at contemplation, reflection; on economy of movement, he focuses on techniques of breathing, concentration, and meditation. He protects all living creatures, and it is very unlikely that the invention of the combustion engine, the car, intensive farming, and weapons of mass destruction could have arisen in a community where the bodhisattva is the established ideal mode of being (see Teschner 2009). The gentleman needs leisure, sports, lots of running water, a water closet, clean shaves, and so on.

Ortega draws our attention here to something important, which has been insufficiently discussed in the philosophy of technology and is hardly acknowledged in political philosophy and moral philosophy, namely that there is a special relation between technology and the good life. In the contemporary literature on the philosophy of technology, there is little discussion of Ortega's view. There is discussion of Ortega's work by Carl Mitcham (1994), and occasional others, but it does not go into the relation of technology and the good life. A similar view of the internal relation between conceptions of the good and technology may be found in Friedrich Dessaur, an engineer and philosopher of technology, who was writing in 1950s Germany. Just like law is oriented toward justice and medicine aims at improving and restoring health and well-being, so is technology, according to Dessauer, oriented toward human "freedom from subordination to nature and freedom to design one's own life" (see Van der Wal 1994, p. 221).

There is one notable exception in the recent literature. Basalla's evolutionary historical account of innovation and technological development was heavily influenced by Ortega's main idea that our ideas about the good life and well-being are at the origin of the incredible variety and abundance of artifactual reality. The idea that we desperately *need* the things for which six million patents have been filed since 1790 in the U.S. alone and that we cannot do without them does not make sense. They are all strictly speaking unnecessary as Ortega observed. Basalla concludes that "artifacts are material manifestations of the various ways men and women throughout time have chosen to define and pursue existence," they "(. . .) began to cultivate technology and in the process created what has come to be known as human life, the good life, or well-being" (1988, p. 14). According to Basalla

> Artifacts are uniquely identified with humanity—indeed they are a distinguishing characteristic of life; nevertheless, we can survive without them. That is why Jose Ortega Y Gasset declared in 1933 that technology was the production of the superfluous. Fire, the axe, or the wheel are no more items of absolute necessity than are the trivial gadgets that gain popularity for a season and quickly disappear. Biological necessity is not the reason that so much thought and energy are expended on the making of novel artifacts. People make new kinds of things because they choose to define and pursue human life in this particular manner. The history of technology is not a record of the artifacts fashioned in order to ensure our survival. Instead, it is a testimony to the fertility of the contriving mind and to the multitudinous ways the peoples of the earth have chosen to live. Seen in this light, artifactual diversity is one of the highest expressions of human existence. (Basalla 1988, p. 208)

The diversity of technologies and artifacts, the fact that there is no clear progress in technological development, the fact that recent studies in science and technology, value sensitive design have shown that technology is not neutral but comes with inscribed and embedded values are all accounted for by Ortega's account of the internal connection between technology and the good life.

All technologies are imposed on users with the choices made by designers and developers with the implicit or explicit suggestion of their inventors and designers to the effect that life would be better, more interesting, if we had them and used them. Sometimes we are misguided, mistaken, or deceived about technology's purported contribution in this respect. This may apply to thumbscrews, atom bombs, ferry boats, RFID tags, or nanoparticle sunscreens. Still, the people who invent them and introduce them suggest that it would be good to have them.

If this account of technology is correct, and conceptions of the good life and human flourishing are invariably—although sometimes marginally—implicated in our technical inventions and innovations, then the first question to be posed regarding technology, artifacts, devices, instruments, designs, and infrastructures is to which conception of the good life are these technologies supposed to contribute, and do they really accomplish what they implicitly or explicitly promise to deliver? Every introduction needs to be accompanied by the question "why are we doing this, what good does it serve?" Does it really make the contribution in the suggested ways to the realization of a particular view of the good life and human flourishing, or are we mistaken, misguided, or deceived? And are we satisfied with the conception of the good to which they lend support?

NEUTRALITY AND TECHNOLOGY

Much of contemporary liberal political theory is preoccupied with neutrality regarding substantive and controversial conceptions of the good life. A more or less articulated conception of the good life gives a person a sense of what it means for her life to go well and provides her with the sentiment of her existence. According to some of the main theorists of the liberal state of the past decades—Rawls, Dworkin, Robert Nozick, and Ackerman and more recently Charles Larmore and Will Kymlicka—conceptions of the good life and human flourishing should be kept off the political agenda. Liberal neutrality guarantees that every citizen can pursue his or her own conception of the good life and is respected as an autonomous chooser in this respect. According to Rawls, a state that is neutral in this sense is also more stable and sustainable. The substantive views of the good life or "comprehensive doctrines," as Rawls termed them, are thought to be too divisive to allow us to reach a consensus, or even sustain a *modus vivendi*. We need, therefore, to start from the acceptance of the fact of radical value plurality and the existence of incommensurable views of life and value and work with a minimal (or "thin") theory of the good, one that is acceptable to all, so as to establish an "overlapping consensus" (Rawls 1993, p. 150)

The ideal of state neutrality has been very useful in times when religious persecution, oppression, and dictatorship were still fresh in our memories. It certainly has provided us with an antidote to the imposition of substantial views of the good life by those in power at the expense of freedom and choice of those over whom the power was exercised. What is exactly meant by neutrality by political liberalists? A useful and widely accepted distinction is made between neutrality of *consequences or effects*, neutrality of *justification*, neutrality of *aim*, and neutrality of *fundamental justification*. Fundamental justification refers to the neutral justification of neutrality itself. Even staunch defenders of neutrality agree that, strictly speaking, liberal neutrality is not neutral itself and cannot be neutrally justified. It refers to values such as liberty, autonomy, respect, and equality, and not every reasonable person would give them the same central and foundational place in the moral governance structure of the state as the political liberalist would.

There is also agreement that the requirement of neutrality of effect or consequences with respect to comprehensive doctrines is impossible. Neutrality of effect would imply that a particular intervention or policy would, in effect, treat all conceptions of the good life the same. A law that does not restrict embryo selection *in vitro* techniques would favor, for example, non-Christians. If, on the other hand, we would have restrictions on the use of such techniques imposed by law, this would, in effect, favor the Christian part of the population, who strongly oppose the technology. With or without the law, we are failing to be neutral in effect between rivaling proponents of these conceptions of the good. George Sher thus observes that "if someone were to insist on neutrality of effects, his proposal would be doomed from the start" (Sher 1997, p. 22).

Second, there is *justificatory* neutrality. Justificatory neutrality requires that no political action may be undertaken or justified on the grounds that it promotes an ideal of the good (as superior to other conceptions) nor on the grounds that it enables individuals to pursue an ideal of the good (at the expense of others). According to Sher for instance, a law, institutional or other political arrangement is "neutrally justifiable if and only if at least one possible argument for it has only neutral normative premises, is reasonably strong and contains no obvious mistakes" (Sher 1997, p. 26). Such a principle of justificatory neutrality can form a basis for the scrutinizing and constraining of the practical thought and intentions motivating public policy.

This could work just fine in the case of excluding interventions and restriction in religious debates. For example, one could introduce Sunday closing laws (in effect, favoring the Christian position on Sunday rest) on the neutral ground that it gives a uniform day of leisure to the population.

This could, in principle, be acceptable to non-religious groups of the population. Government could also decide not to restrict Sunday trading (and in effect advantage the non-Christians) but do so again on the relatively neutral ground that it is economically beneficial to the country. This may be acceptable to Christian groups, especially in times of economic crisis. Whichever way one goes on Sunday trading laws, it will result in a non-neutral intervention as far as *effects* on conceptions of the good life are concerned, but the intervention could at least be justified on neutral grounds and could therefore, in principle, be acceptable to both parties.

This solution satisfies Sher's definition of a neutral justification, because "there is" as the previous example has shown "one possible argument which has only neutral normative premises." But this argument need not be the one that was operative in the minds of those who implemented the policy. It would therefore be open to an objection of the following sort: "the government says it allows shops to be open on Sundays because it is good for the economy, but they are just a bunch of atheists," or alternatively; "they say they restrict Sunday shopping because we need a uniform day of leisure, but they are just a bunch of religious fanatics."

It is not so clear that neutrality of justification could work in matters that are nonreligious but technological. There are some relevant differences between religion and technology to note here. First, it seems much easier to relegate religion to a private or semi-private sphere, whereas for technology that is much harder. All significant technology is public in a relevant sense. One cannot keep any significant technology all to oneself. Technology's typical service value can only be enjoyed in networks of users (telephone, Internet), producers (electricity), and those who maintain (cars and computers), govern (transport), and disseminate it.

Second, because technology is designed with a specific purpose, functionality, and service value in mind, this teleological aspect of technology must therefore be referenced in any adequate justification of its introduction in society. A criterion of adequacy of justification here would be that the reason why a technology works and is valued needs to be the same as the reason for which it's adoption is publicly justified. Another aspect of technology's service value is that it helps to shape our daily lives, experiences, expectations, and intentions of people who use it. An adequate public justification of the introduction of the technology must therefore also refer to the way it is expected to affect our lives, ways of thinking, and experiencing. If Ortega's account is adequate, then the public justification of technology must refer to its underlying conception of the good life, which is its *terminus ad quem*.

A case where neutrality of justification cannot work with respect to technology policy is human enhancement. The application of relevant techniques and technology comes with the inscribed value that improvement beyond what is naturally given to man is good. Some people are of the opinion that we should not move beyond natural selection, curative and preventive medicine, hygiene, and healthy food. They think that human enhancement is a bad idea and that we are playing God, or tinkering with nature, if we start to design babies, chimeras, and engage in transgene engineering. Both enhancement and non-enhancement or anti-enhancement policies and their associated medical programs are undeniably and inexorably vehicles of a distinct view of the good life. Even enhancements that would lead to greater autonomy or freedom could be opposed by people who argue that these should be the result of our own natural endeavors.

A similar problem runs across sustainability debates. Some think of the nonhuman world in terms of resources that can be utilized for human purposes and that need to be managed prudentially. Others have a worldview that opposes this instrumental view of nature. They construe nature not as a resource that can be depleted and used, but as something intrinsically valuable, God given, something that not just happens to be out there, but something to which we are in important and meaningful, albeit poorly understood, ways related. Fundamental environmental policies are bound to upset either of these parties.

Sher (1997), Joseph Raz (1986), William A. Galston (1991), and Amy Gutmann and Dennis Thompson (1996) have explored the viability of strands of (perfectionist) liberalism that are more generous with respect to the good and would allow us to have fruitful public and political debates about them. Even a defender of justificatory neutrality Charles Larmore (1996) argues that although rivaling conceptions of the good are in some sense incommensurable, they can be sensibly and potentially fruitfully compared by parties in public discourse. Equal respect and preparedness to justify one's views to each other seem required and possible. Gutmann and Thompson argued in the mid-1990s for the relaxation of principles of preclusion, which regulate the admission of issues on the political agenda and stressed the importance of the design of principles of accommodation that would allow us to conduct debates in the public sphere about conceptions of the good. Like Larmore, they advocate that parties should construe positions as moral positions, show mutual respect, magnanimity, open mindedness, and attempt to find common ground in moral argument and in science. These are important conditions for accommodation.

A different route out of the problem of deep disagreement over perfectionist views is suggested by Richard Thaler and Cass Sunstein (2008). Their approach of soft paternalism or libertarian paternalism (a form of paternalism that salvages freedom of choice) by design implies that controversial conceptions of the good could be presented and incorporated in institutional frameworks, incentive structures, and technology in such a way that preferred—but controversial—options are suggested; whereas other options and choices would not be precluded and freedom is, strictly speaking, not limited and individuals' own responsibility for their own lives and achievement of the good are not interfered with. This position also tries to identify a "third way" between perfectionism and paternalism and liberalism.

It is interesting to note that Robert Nozick revisited his initial strong beliefs in liberalist neutrality of government in *The Examined Life* (1989). He states that "The libertarian position I once propounded now seems to me seriously inadequate, in part because it did not fully knit the humane considerations and joint cooperative activities it left room for more closely into its fabric" (Nozick 1989, p. 286). He observes that liberalism has a tendency to "neglect the symbolic importance of an official political concern with issues or problems, as a way of marking their importance or urgency, and hence of expressing, intensifying, channeling and encouraging, and validating our private actions and concerns towards them" (Nozick 1989, p. 287). Nozick can be read to suggest that there is value in discussing issues in public and politics that would be filtered out by stringent construals of justificatory neutrality. By putting them on the political agenda, governments signal that (1) these issues must be seen by us as issues that occasion debates about the good life, (2) that they can no longer be avoided, and that (3) it is important that we collectively learn how to deal with them given (1) and (2).

Whichever way we go on particular techniques in genomics, nanotechnology, the Internet, and human enhancement, they are bound to have effects on the way people may be able to realize their life programs and their very conceptions of the good life. The justification of their introduction in society will refer to the way they will be put at the service of the realization of a particular way of life that almost certainly will be controversial. Because neutrality concerning technology is not a feasible ideal, we are well-advised to learn and prepare to deal with non-neutrality and discussions about perfectionist goals and comprehensive conceptions of the good expressed in and supported by technical artifacts. Instead of sweeping these issues under the carpet of neutrality, we need to start to think about the design of the institutional and conceptual frameworks that can help us to make sense of them. We therefore need to relax our principles of admission for controversial technology issues on the political agenda and provide ourselves with the opportunities to become good at something that the future will need more of: public discussions about technology and the good life.

REFERENCES

Arneson, Richard J. (2003). "Liberal Neutrality on the Good: An Autopsy." In *Perfectionism and Neutrality: Essays in Liberal Theory*, edited by Klosko, George and Steven Wall (Lanham, MD: Rowman and Littlefield) pp. 191–208.

Basalla, George. (1988). *The Evolution of Technology*. Cambridge: Cambridge University Press.

Dworkin, Ronald. (1985). *Liberalism*. Cambridge, MA: Harvard University Press.

Galston, William A. (1991). *Liberal Purposes*. Cambridge: Cambridge University Press.

Gutmann, Amy, and Dennis Thompson. (1996). *Democracy and Disagreement*. Cambridge, MA: Harvard University Press.

Larmore, Charles. (1996). *The Morals of Modernity*. Cambridge: Cambridge University Press.
Mitcham, Carl. (1994). *Thinking Through Technology*. Chicago: University of Chicago Press.
Nozick, Robert. (1989). *The Examined Life*. New York: Simon and Schuster.
Ortega Y Gasset, J. (1961). "Man the Technician." In *History as a System*. New York: Norton Library.
Rawls, John. (1993). *Political Liberalism*. New York: Columbia University Press.
Raz, Joseph. (1986). *The Morality of Freedom*. Oxford: Oxford University Press.
Sher, George. (1997). *Beyond Neutrality*. Cambridge: Cambridge University Press.
Thaler, Richard, and Cass Sunstein. (2008). *Nudge: Improving Decisions about Health, Wealth and Happiness*. New Haven, CT: Yale University Press.
Teschner, George. (2009). "Technological Paradigm in Ancient Taoism," *Techne*, vol. 13, no 3.
Van der Wal, Koo. (1994). "Technology and the Ecological Crisis." In *Ecology, Technology and Culture*, edited by Zweers, Wim and Jan J. Boersema (Cambridge: White Horse Press) pp. 215–246.
Wilson, Edward O. (1998). *Consilience: The Unity of Knowledge*. New York: Knopf.

25 Technological Change and the Destabilization of Liberal Politics

Govert Valkenburg

Most of the time, our ideas of the *good life* are private business, not a matter of politics. Most of the time, perhaps, but not during an episode that captivated Dutch politics beginning May 26, 2008. That day, Deputy Minister of Health, Jet Bussemaker, announced that genetic (PGD), or embryo selection, would be permitted in prevention of familial breast cancer (Bussemaker 2008a). PGD is a technique to identify genetic predispositions for congenital diseases in embryos, which enables the selection of embryos without (known) diseases for implantation in the womb, as part of an *in vitro fertilization* (IVF) cycle. Mrs. Bussemaker was convinced that the announcement was a mere formalization of what was already common practice: Maastricht University Medical Center, the only hospital with a PGD permit, already applied PGD in cases of familial breast cancer. Moreover, she considered breast cancer a condition just as burdensome as Huntington's and Steinert's diseases, so extension would only be consistent. In fact, PGD had hitherto only been permitted formally in prevention of genetic or heritable diseases, such as Huntington's and Steinert's, but the practice is a typical example of Dutch 'toleration policy.' In the Netherlands, legal changes are often preceded by a rearrangement of the actual practice. In this period prior to legal change, conduct in line with the envisioned policy is not sanctioned, even though it is formally illegal. (Dutch marijuana regulation is a case in point, showing that the period can be several decades, and moreover not always produces a final, formal solution.)

However, rather than being a minor formality, the announcement generated great dispute. The antagonist was vice-prime minister and Christian's Union leader André Rouvoet. He declared that the matter was so ethically delicate, that it would require extensive discussion in the Cabinet prior to legislation. In his eyes, Bussemaker had made a grave mistake bypassing the Cabinet.

The issue is in the *incomplete penetrance* of familial breast cancer. Unlike breast cancer, the diseases for which PGD was already permitted show *full penetrance*: a person carrying the mutation is (almost) certain that the disease will develop. Yet with familial breast cancer, this chance is between 60% and 90% (Niermeijer, De Die-Smulders, Page-Christiaens, and De Wert 2008).

Crossing the line between complete and incomplete penetrance was one bridge too far for Rouvoet: as the disease might well *not* develop, PGD risks the *de facto* discarding of embryos that would otherwise have developed into healthy individuals. According to Rouvoet, this makes a morally significant difference. He and his party argue that a society that discards even potentially healthy embryos is a cruel society. Instead of showing solidarity with the less fortunate, such a society simply accords them no place (ChristenUnie 2010, sections 3.2, 7.1, and 7.5). Rouvoet's party promotes approaches aimed at care rather than prevention, especially if this means elimination while the disease has—as does breast cancer—good treatment prospects. It promotes a society in which people receive due care, rather than being denied the right to existence (Herderschee 2008).

Of course, the Christian's Union disapproves of discarding embryos in general. Nevertheless, it had hitherto decided not to take up this issue politically. This abidance with the majority opinion must not be mistaken for endorsement.

To the accusation of immorality, Wouter Bos, vice prime minister on behalf of Labor, replied that the Christian's Union does not hold a 'prerogative of moral affairs.' In reproductive freedom for prospective parents, including the freedom to use PGD, he recognizes a different moral value that also deserves defense (Bos 2008). Although this freedom is commonly defended in Dutch politics, the explicit appeal to the underlying moral ideals (reproductive freedom vs. the unconditional right to existence) was exceptional. These usually remain implicit for the very reason that they are obvious for most people in Dutch culture—most indeed, but not all.

The vast majority, both in parliament and in public debate, sided with Labor. For example, Mark Rutte, leader of the Liberal opposition, argued that the Christian's Union would not have the right to impose its morality on others and restrict their freedom (Lower House 2008). On the op-ed pages of national newspapers, it was argued that "a minority imposes a backward, Christian morality, by appealing to a God that a majority does not believe wants us to suffer from breast cancer" (Etty 2008). Moreover, some argued that the morality itself was inconsistent: the Christian's Union approves of other treatments, such as IVF, that entail the disposal of embryos. In addition, it would be a caricature to state that the life of an unaffected carrier is unproblematic: having to live with a disease that may or may not set on remains a Sword of Damocles for the carrier of the disposition (Evers 2008).

The aftermath is only briefly relevant here. The announcement was recalled, and some months later, a new announcement was issued (Bussemaker May 30, 2008, June 27, 2008). Remarkably, in the second announcement, PGD was admitted for a less specific range of diseases than in the first announcement. An important difference between the announcements was in the emphasis: It was not shaped as a list of diseases that qualify for PGD, but as a policy that warrants utmost prudence. Consultation between the parents-to-be and their doctor is required, and each individual case must receive due ethical consideration, in which a considerable number of professionals are involved, while the choice is ultimately left with the parents. Peace and quiet had returned, and politics could return to relying on its routines of privatization of ideas of the good. The ideas of the good that had risen in the case of PGD now silently returned to the private sphere.

POLITICS IN THE NETHERLANDS: LIBERAL PRIVATIZATION OF THE GOOD

A few details of Dutch liberal-democratic politics are indispensible to grasp the intricacies of this controversy. Dutch politics has, in modern history, always been characterized by multiple parties being represented in Parliament, none of which ever acquires absolute majority. Thus, parties must form coalitions. Usually, a majority coalition is created, and the Cabinet is appointed from within the ranks of the coalition partners.

The elections of November 22, 2006, had resulted in a coalition of the Christian Democratic Appeal, the Christian's Union, and the Labor Party. Bussemaker belonged to the Labor Party. Rouvoet, on the other hand, belonged to the Christian's Union. If not together in the coalition, their disagreement might have been just another disagreement between political opponents. However, given their joint responsibility, the challenge was daunting: a solution was needed that both could accept and defend in face of their respective constituencies. (The third coalition partner, the Christian Democratic Appeal, played only a marginal role in this controversy, even though it was the largest party in the coalition.)

Essential to any liberal-democratic system is the recognition that there are things on which it is acceptable and even desirable to remain divided rather than seeking agreement at any cost. Religion, taste, and worldview: these are things that people hold a host of different opinions about. Moreover, for the order and stability of society, agreement on them is simply not necessary. John Rawls argues that therefore, public reason, being the style in which we conduct politics and public debate, should only concern the public good and fundamentals of justice in such a way that they are arguably both acceptable and necessary for all citizens (Rawls 1993). In particular, claims should be able to survive without reference to religion, worldviews, taste, ideas of the good, or in short: *comprehensive doctrines* (Rawls 1993). For a society to be good and stable, no agreement on religion and the like is needed.

Clearly, Rawls's account is a normative one and not an empirical description of Dutch political culture. Moreover, the Dutch case we are discussing here is not a matter of the constitutional essentials that Rawls elaborates on, but of common legislation. Nevertheless, the resemblance is striking. Post-WWII Netherlands has been characterized by a 'peaceful segregation': Socialists, Protestants, Roman Catholics, and Liberals each had their own parties, institutions, time slots on public broadcasting services, schools, and so on. Proposed policy only has a chance to convince if it grants that all different walks of life can still be pursued, and if the arguments supporting the policy do not need specific comprehensive doctrines to be convincing. The lines between traditional group identities have become less articulated and have made place for new lines of segregation from the 1960s onward. Against this background of a dynamic pluralism, it is still the case that political claims mostly are made such that their justifications are acceptable to people from *any* comprehensive background, even if their consequences may be bemoaned by some.

POLITICS AS PURIFICATION

The PGD controversy was not just another political disagreement, but escalated to the level of a collision between comprehensive doctrines. One doctrine holds that the human embryo, being a potential human individual, deserves full protection. No condition, however severe, can ever justify terminating its life. Abandoning this principle would, in the end, lead to a society in which imperfection has no place. Another doctrine holds that, although we must pay a decent minimum of respect to the human embryo, there is no obligation to let it evolve into a human individual, especially if its life will be heavily burdened. Informed choice of the prospective parents should be the ruling principle.

These doctrines are not normally on the agenda of political discourse, no matter how important the discussants involved believe they are. My central questions are why and how these doctrines made it to the agenda in the case of PGD. In the remainder of this section, I will explain this case as an exception to how political routines normally deal with ideas of the good. I will then offer a conceptualization of technologies, including their role in these routines. Finally, I will suggest how liberal democracy can improve its ability to cope with the consequences of technological change.

First, remaining divided was not an option, because a coalition bears the responsibility of making decisions. 'Agreeing to disagree' would maintain the existing practice to which part of the coalition objected. Second, either decision—allowing or disallowing PGD in prevention of familial breast cancer—would be *immoral* according to one of the specific comprehensive doctrines the parties held dear. Therefore, the constituencies of the 'losing' party would never accept the terms of their loss. Both the burden of decision and the alleged immorality compelled the parties to voice arguments from their comprehensive doctrines.

The controversy was a contestation of *what is political*, with two levels of discussion made visible. First, argument took place at the level of whether a certain issue should or should not be regulated by politics. I call this the level of *scope*. Second, there was a debate about whether certain arguments were appropriate in the context of public policymaking. This I call the level of *justification*.

The level of justification is easiest to recognize. Appealing to God is a straightforwardly unacceptable justification for most discussants. However, two remarkable issues surface here. First, Arie Slob, a leader of the Christian's Union in Parliament, argued in a national newspaper that *every* politician is necessarily guided by his or her personal beliefs. It would be unfair to only discredit the Christian's Union for adducing Christian beliefs, while, say, Liberal convictions should apparently be admitted into the debate unconditionally. For the same reason, the accusation that the Christian's Union was abusing its position in the coalition to enforce its morality would be hypocritical: *any* party will use its coalition membership to stage and promote its ideology and moral beliefs (Heerma van Voss and Stokmans 2008).

Second, even though it may seem obvious that religious arguments will not convince in secular Dutch politics, it is not that straightforward at closer look. The Netherlands features a separation between church and state, but this is only *institutional*: churches cannot exercise formal political power, and the state does not have a formal say in church affairs. Nevertheless, religiously inspired positions are quite commonly voiced in politics. In fact, Christian Democrats have been in Parliament from 1918 until the present, with a pause only between 1994 and 2002. Under normal circumstances, no offence is taken from politicians being explicitly Christian or otherwise inspired by religion. The fact that in this case a justification by appeal to Christian morality was so offensive to some therefore needs further explanation.

Rather than keeping to Rawls's normative ideal of public reason, we see here that a form of public reason is itself constructed in political practice, characterized by a pluralism of ideas of the good. Even though all may agree *roughly* on the nature of politics and public reason, we should expect people to give different answers when asked for the *exact* demarcations of public reason. Moreover, despite rough agreement, people have in the backs of their mind *their own specific ideal* of a pure debate. While debating, people actively construct their own pure ideal. I call this *purification* (Valkenburg 2009, ch. 2).

I borrowed this term from Bruno Latour (1993), who uses it to denote the typically modern pursuit of a strict separation between the social and the natural. The former contains all human faculties of morality, politics, and agency. The latter contains the nonhuman parts of the world: nature and, by extension, science and technology. For Latour, purification is a continuous struggle that will never be completely successful, the main reason being that the purification itself is a profoundly human activity. When purifying, we contaminate every seemingly nonhuman thing with a trace of humanness (and vice-versa), thus creating entities that are both technoscientific and political at the same time.

Similarly, elements from the debate is an intrinsically political activity. By disqualifying something for politics, either regarding its scope (e.g., PGD should be left to private decision) or regarding its justification (e.g., biblical arguments are not properly political arguments), the thing becomes contaminated with a trace of politics, even if it was so far not an explicit concern for politics. Moreover, whenever something is removed from politics, it is at least likely if not plausible that this will be a loss *for somebody*—who will therefore protest, again a political action. In the following section, I will expose a second reason why the purification of the public sphere of supposedly private arguments is awkward: because technology contaminates the categories of public and private arguments.

TECHNOLOGY AS A POLITICAL DOUBLE BIND

So far, I have paid little attention to the role of technologies. Indeed, the controversy is chiefly a political mishap and its extensive aftermath. Yet, that is only the front stage, and taking a different perspective will reveal more clearly the roles played by technologies. Indeed, the whole controversy was sparked by the new technology of PGD. However, that was not the only way technology was relevant.

The turmoil over the introduction of a technology such as PGD could be naively explained as follows. Once the technology becomes available and perhaps even proposes itself as an imperative, society faces the choice of whether or not to adopt it. In case the use of the technology conflicts with moral norms, the technology should be dropped. If not, it should be allowed or even promoted. The turmoil itself, as opposed to a peaceful decision, is then explained as a disagreement over the extent to which use of the technology is in conflict with moral norms, or over the question of how conflicting moral norms should be balanced in this specific situation.

There is at least some truth in this naive explanation. Indeed, in the present controversy, PGD appears as a technology that has both attractive sides and downsides that urge moral discussion. However, this explanation leaves two important questions unanswered. First, it leaves unexplained why this controversy was so intense. Indeed, technological developments unfold around us every day, and their call for moral assessment should at least not come as a surprise. Second, the explanation does not account for the fact that arguments of Christian origin were disproportionately discredited, even though Christian comprehensive doctrines have always played an important role throughout Dutch Parliamentary history. I will address these issues by articulating the mechanisms through which technologies not only call for moral debate, but additionally call into question the very rules we abide with while conducting such debate.

Technologies are involved in our relations with the world, and the roles they play can be divided into two dimensions. On the one hand, technologies disclose and conceal options for action. What we *can* do depends to a large extent on the technological options surrounding us. The availability of a hammer allows us to construct things with nails, which would otherwise be very hard. On the other hand, technologies disclose, conceal, accentuate, and amplify ways of looking at the world and interpreting it. We see and interpret the world with the things around us. The existence of a thermometer is a condition for us being able to think of temperature as a quantitative phenomenon (Ihde 1990). These two dimensions, labeled the *existential* and *hermeneutic* respectively (Verbeek 2005), provide nuanced alternatives to the idea of the imperative of the naive explanation. In addition, they each specify one side of technology as a political double-bind.

It is by the existential dimension that the particular technology of PGD urges moral discussion. It enables the selection of embryos based on their genetic predisposition for heritable breast cancer. This use is morally problematic for the Christian's Union, and as their concerns can only be voiced with reference to their comprehensive doctrine, PGD thus directly lures the good life into the public sphere. This is the mechanism that was explained in the first section.

However, through the hermeneutic dimension, technology also takes on momentum in the opposite direction: the privatization of ideas of the good. One's appreciation of PGD depends on the technological culture in which one lives. For example, IVF is largely morally accepted as a technique for overcoming infertility. Like PGD, it involves interfering in early human life. Moreover, it is compatible with thinking of the early embryo as something that we need not unconditionally protect, particularly if it is threatened with serious diseases or disabilities. This is not to deny that most people also recognize the embryo as something human and entitled to some level of respect, nor is it to say that *all* people think this way. Yet by their existence, all these technologies have co-shaped the meanings we give to such things as the embryo and reproductive freedom: first as a technical possibility, and by consequence as a decision space.

Even though these technologies do not coerce us to think in particular ways, they at least provide a background against which we think. They are part of what could be called a *technical code*: a cultural preference for particular technological arrangements in line with particular social goals (Feenberg 2005). In this case, all technologies together point into one direction: PGD should be regulated in a decision-wise privatized way, just as are IVF, abortion, and other issues concerning reproduction.

Arguing against values that are deeply rooted in a culture is politically courageous. It is even more so if those values are consolidated materially. This explains part of the insult that so many people took from the

Christian's Union position. First, the attempt to 'de-privatize' a reproductive issue and bring it into the *scope* of politics conflicted with both the hegemonic idea of privatization of reproductive issues and with the existing technologies consolidating this hegemony. Second, it nuances the observation that Christian arguments were put in disrepute more than other private arguments—which here refers to the residual category complementing dominant visions of politically appropriate arguments. At closer look, it was not the Christian orientation that made the arguments offensive, but the fact that they were so strongly in discord with dominant moral norms, with dominant specifications of scope and justification of politics, and with the dominant technical code supporting all of these. This offers an answer to the two gaps in the naive explanation I started with.

Technology thus offers a double-bind to political debate. Through its existential dimension, PGD urged moral discussion and provoked the use of private arguments. At the same time, existing technologies silence those arguments through their hermeneutic dimension and the hegemony of reproductive freedom they reproduce. The situation carries a great irony: The Christian's Union did not take the initiative to conduct this debate but was forced into it by developments external to its own political program. Yet, once the spark had flown, the Christian's Union had no option but to use arguments that turned out offensive to so many.

It could be questioned here whether technology could not be omitted from the analysis so far. True enough, the whole story could largely be told as a politically dominant ideology of individualism and a consequent privatization of decisions concerning procreation. However, such an analysis would at least have to concede that it is a technology that sparks the controversy. Moreover, it would have to concede that existing technologies show a remarkable resemblance with dominant conceptions of public reason. Finally, it would ignore technology as a particular way to defuse the situation, to which I now turn.

CULTIVATING THE ROLE OF TECHNOLOGY IN DEBATE

In the previous sections, I explained how liberal politics can be interpreted as a privatization of ideas of the good, leading to (partly implicit) rules on what subjects and what arguments can be introduced into politics. Moreover, I explained how those rules are on the one hand challenged by technological developments, whereas on the other hand being contingent on the specific technological background against which the debate takes place. The final question is how such collisions as described can be mitigated.

Technological developments are obviously able to urge ethical consideration. Yet, the thorniest part of this controversy was the dispute over whether particular arguments are appropriate or not in public debate, and over whether particular decisions need political settling. In addition, this acceptance of contributions to the debate is sometimes not the consequence of proper consideration, but instead of drifting conceptual orders and technological changes that go by unnoticed. As technology is part of the cause of such changes, it is logical to find some solutions there as well.

Ethical assessment of technological change needs to extend its focus to include one additional potentiality: technology's capacity to render classes of arguments *non grata* to public debate. Although Christian arguments were not completely silenced, they were at least regarded as highly unwelcome. At the core of any truly democratic system is the belief that no minority, however small, may ever be silenced (Habermas 1996; Mill 1859). Yet this is exactly what technologies are capable of, if not directly, then at least by small steps that each make a small difference to the construction of what can be argued and what cannot.

At issue are not just the specific contributions that were unwelcome to this debate, but the broader concern with the general principle of reasonableness and the unconditional inclusion of points of view. In the end, we may conclude that the Christian's Union's arguments are or are not apt for discussing an issue such as PGD in the 21st century. Yet reasonableness gains if we explicitly account for the fact that the urge to disqualify their arguments stems in part from developments external to the arguments themselves. Parts of those developments are political, social, or cultural. And parts of them are technological decisions that shape the background in which we live our political lives. Discussing the good life in a technological age means also discussing how technologies relate to the very discussion of the good life.

REFERENCES

Bos, Wouter J. (2008, June 2). *Embryoselectie voor Borstkankergen?* [Television interview]. Hilversum, Netherlands: *Eén Vandaag.*

Bussemaker, M. (2008, May 26). "1e Brief van de Staatssecretaris aan de Tweede Kamer," *Kamerstuk (Proceedings of the Lower House)* 31 200 xvi (147).

Bussemaker, M. (2008, May 30). "2e Brief van de Staatssecretaris aan de Tweede Kamer," *Kamerstuk (Proceedings of the Lower House)* 29 323 (41).

Bussemaker, M. (2008, June 27). "3e Brief van de Staatssecretaris aan de Tweede Kamer," *Kamerstuk (Proceedings of the Lower House)* 29 323 (46).

ChristenUnie. (2010). *Kernprogramma* [Political Program].

Etty, Elsbeth. (2008, June 3). "Krijg dan maar Borstkanker," *NRC Handelsblad.*

Evers, J. L. H. (2008, June 3). "Rouvoet haalt Gelijk over de Rug van Zieken," *NRC Handelsblad.*

Feenberg, Andrew. (2005). "Critical Theory of Technology: An Overview," *Tailoring Biotechnologies*, vol. 1, no. 1, pp. 47–64.

Habermas, Jürgen. (1996). *Between Facts and Norms.* Cambridge, MA: MIT Press.

Heerma van Voss, Sandra, and Derk Stokmans. (2008, June 7). "Wij Fanaten en Tirannen? Ongelofelijk," *NRC Handelsblad*, pp. 4–5.

Herderschee, Gijs. (2008, May 28). "Gezondheid niet Overwaarderen," *De Volkskrant.*

Ihde, Don. (1990). *Technology and the Lifeworld: From Garden to Earth.* Bloomington: Indiana University Press.

Latour, Bruno. (1993). *We Have Never Been Modern* (H. Wheatsheaf. trans.). Cambridge, MA: Harvard University Press. Originally published *Nous n'avons jamais été modernes: Essais d'anthropologie symmétrique* in 1991)

Lower House. (2008, June 5). "Handelingen 2e Kamer," *Kamerstuk (Proceedings of the Lower House)* vol. 93, pp. 6569–6591.

Mill, John Stuart. (1991). *On Liberty*. Oxford: Oxford University Press. (Originally published 1859)

Niermeijer, M. F., C. E. M. De Die-Smulders, G. C. M. L. Page-Christiaens, and G. M. W. R. De Wert. (2008). "Erfelijke Kankersyndromen en Kinderwens: Ouders en Politiek in Dialoog over Pre-implantatiegenetische Diagnostiek," *Nederlands Tijdschrift voor Geneeskunde*, vol. 152, no. 27, pp. 1503–1506.

Rawls, John. (1993). *Political Liberalism*, vol. 4: The John Dewey Essays in Philosophy. NewYork: Columbia University Press.

Valkenburg, Govert. (2009). *Politics by All Means: An Enquiry into Technological Liberalism*. Simon Stevin Series in the Ethics of Technology. Deflt, the Netherlands: 3TU.Center for Ethics and Technology.

Verbeek, P. P. C. C. (2005). *What Things Do: Philosophical Reflections on Technology, Agency, and Design*. University Park: Pennsylvania State University Press.

Contributors

Kelli Barr: Graduate Student, Department of Philosophy and Religion Studies, University of North Texas, U.S.

Robert Biswas-Diener: Instructor, Department of Psychology, Portland State University, U.S.

Philip Brey: Professor, Department of Philosophy, University of Twente, the Netherlands

Adam Briggle: Assistant Professor, Department of Philosophy and Religion Studies, University of North Texas, U.S.

Bengt Brülde: Associate Professor, Department of Philosophy, Linguistics, and Theory of Science, University of Gothenburg, Sweden

Luigino Bruni: Associate Professor, Department of Economics, University of Milan-Bicocca, Italy

Mark Coeckelbergh: Assistant Professor, Department of Philosophy, University of Twente, the Netherlands

Robert Frodeman: Professor, Department of Philosophy and Religion Studies, University of North Texas, U.S.

J. Britt Holbrook: Research Assistant Professor, Department of Philosophy and Religion Studies, University of North Texas, U.S.

Justine Johnstone: Lecturer, Science and Technology Policy Research, University of Sussex, U.K.

David M. Kaplan: Associate Professor, Department of Philosophy and Religion Studies, University of North Texas, U.S.

Diane P. Michelfelder: Professor, Department of Philosophy, Macalester College, U.S.

Carl Mitcham: Professor, Division of Liberal Arts and International Studies, Colorado School of Mines, U.S.

Lindsey Patterson: Graduate Student, Department of Psychology, Portland State University, U.S.

Omar Rosas: Postdoctoral Fellow, Department of Political, Social and Communication Sciences, University of Namur, Belgium

Joo Yeon Shin: PhD Student, Department of Psychology, Colorado State University, U.S.

Johnny Hartz Søraker: Assistant Professor, Department of Philosophy, University of Twente, the Netherlands

Edward Spence: Senior Lecturer, School of Communication and Creative Industries, Charles Sturt University, Australia

Michael Steger: Assistant Professor, Department of Psychology, Colorado State University, U.S.

Tsjalling Swierstra: Professor, Department of Philosophy, University of Maastricht, the Netherlands

Valerie Tiberius: Professor, Department of Philosophy, University of Minnesota, U.S.

Anton Tupa: Visiting Assistant Professor, Department of Philosophy, Auburn University, U.S.

Govert Valkenburg: Postdoctoral Fellow, Faculty of Arts and Social Sciences, University of Maastricht, the Netherlands

Shannon Vallor: Assistant Professor, Department of Philosophy, Santa Clara University, U.S.

Jeroen van den Hoven: Professor, Department of Philosophy, Delft University, the Netherlands

Ibo van de Poel: Associate Professor, Department of Philosophy, Delft University, the Netherlands

Ruut Veenhoven: Professor, Faculty of Social Sciences, Erasmus University Rotterdam, the Netherlands

Peter-Paul Verbeek: Professor, Department of Philosophy, University of Twente, the Netherlands

Pak Hang Wong: PhD Student, Department of Philosophy, University of Twente, the Netherlands

Index

Made in the USA
Monee, IL
13 September 2026

62978959R00208